# RECREATING TITANIC AND HER SISTERS

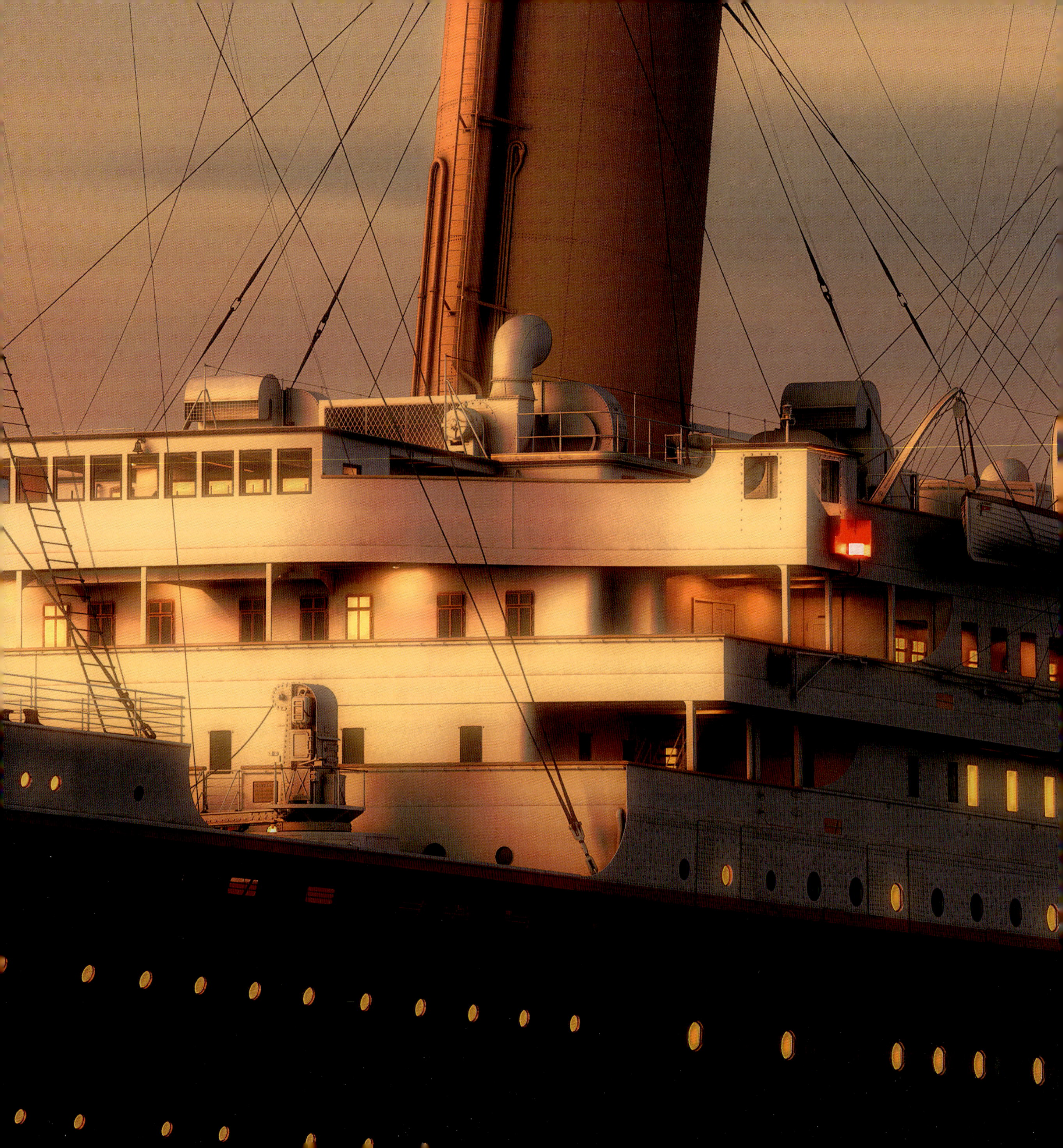

TAD FITCH, J. KENT LAYTON, AND BILL WORMSTEDT

# RECREATING TITANIC AND HER SISTERS

## A VISUAL HISTORY

WITH A FOREWORD BY KEN MARSCHALL

The History Press

First published 2022
Reprinted 2022, 2023 (twice)

The History Press
97 St George's Place
Cheltenham
GL50 3QB

www.thehistorypress.co.uk



British Library Cataloguing in Publication Data.
A catalogue record for this book is available from the British Library.

ISBN 978 0 7509 9868 0

Typesetting and origination by The History Press
Printed by Thomson Press, India

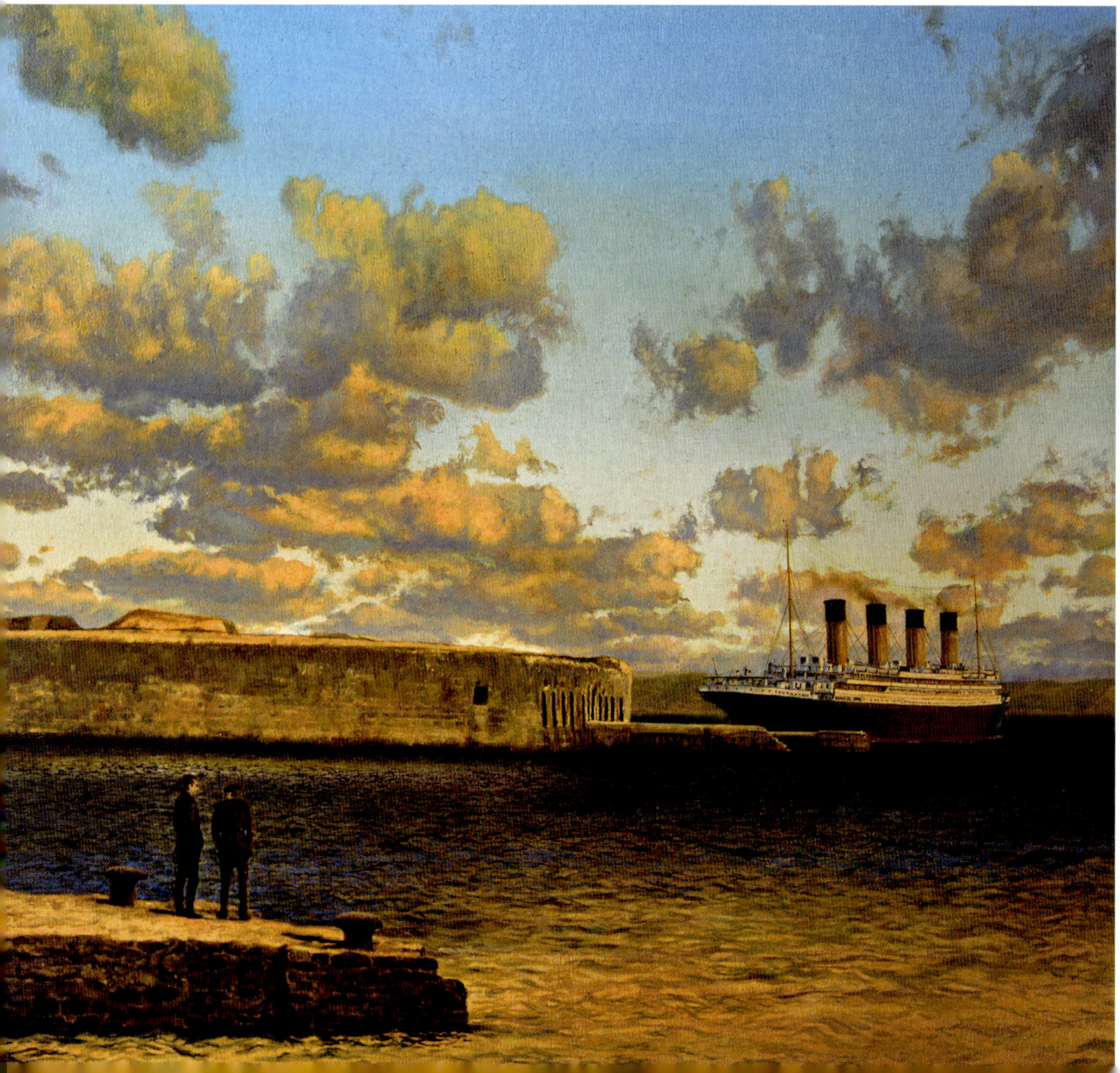

*Frontispiece:* The setting sun bathes *Titanic*'s forward decks in a warm glow on the evening of Sunday, 14 April 1912. Lights come to life throughout the ship as passengers prepare for dinner. (Vasilije Ristović)

*Left: Olympic* in Cherbourg Harbour at dusk during 1911 or early 1912. (David Olivera)

# CONTENTS

# FOREWORD
## BY KEN MARSCHALL

The aura of the spectacular, dramatic, almost unbelievable *Titanic* story captivates nearly all who hear of it. The epic tale of invincibility versus vulnerability, hubris versus the indifferent power of nature, seems too poetic to be true, like an almost laughably unrealistic screenplay that turns out not to be imaginative fiction but actually happened. It's irresistible.

But *Titanic* was only one of three nearly identical sister ships, each with its own fascinating and important history, if not quite so gripping. While most are satisfied with reading, watching movies or documentaries, a few creative souls are so inspired that they are motivated to do more, producing drawings, paintings or architectural renderings, building computer-generated models and realistic scenes, crafting remarkable physical replicas and even full-sized reconstructions.

Beautifully supplementing the highly respected research of the authors of this book, a concise and fascinating summary of the history of the three *Olympic*-class sisters, is the work of a number of talented people who, through various media, are keeping the memory of these liners alive. Several of them were first enthralled by James Cameron's 1997 film. Others had their interest sparked when the wreck was discovered in 1985 or by reading a book. And some even flatter me by saying that my artwork was one of the catalysts.

I did my first painting of *Titanic* after watching the 1953 movie on television in the mid-1960s, then reading the book *A Night to Remember* by Walter Lord. And I was spellbound by artist Willy Stöwer's highly dramatic 1912 painting showing the ship sinking. I was hooked.

In trying to bring *Titanic* back to life with my paintbrush, the artwork soon attracted attention, and requests began arriving to paint the ship for others. My desire for detail and accuracy resulted in amassing a vast archive of research and reference material. This, in turn, led to my illustrating numerous books, enabled my supervising of many model projects from plastic kits to motion pictures, created opportunities to work with film directors and explorers and, ultimately, to dive several times to *Titanic* and *Britannic* over four expeditions. Amazingly, doing a small painting of *Titanic* in high school literally spawned a lifelong career.

Originally, *Olympic* was of little interest to me, and *Britannic* even less so. I soon understood, however, that in most respects the three ships were almost identical, built side by side at nearly the same time, their panelling and woodwork carved from the same trees and by the very same hands. I began collecting fittings from *Olympic*, and my interest in her grew, as it did for *Britannic*, which I used to consider the ugly stepsister. But I now fully appreciate what a wondrous achievement she was. Had *Britannic* been completed as intended she'd have been an even more lavish and mature *Titanic*.

The legacy of these three historic liners is being nurtured through the resourceful talents of artists and craftspeople like those featured in the following pages, and it's humbling to know that some of them were inspired by my work.

Ken Marschall, August 2021

(Ken Marschall's *Titanic* artwork can be viewed at KenMarschall.com, and prints can be ordered from TransAtlanticDesigns.com.)

# INTRODUCTION

There is something truly magical about *Titanic*. It is difficult to define, perhaps impossible; yet its perplexing lack of definability can by no means be taken as evidence that its magic doesn't exist. It captures the imagination of people from around the world, of all age groups, means, cultures and races. And once *Titanic*'s spell has been cast over you, it is thoroughly impossible to break.

Admittedly, for us historians, there have been moments when our love for *Titanic* has been tested. In recent years, we have been forced to work ever harder at preserving her history inviolate, preventing it from becoming distorted into some sort of 'fairy tale' through myth and innuendo. Savvy media outlets and journalists have also learned an axiom: *Titanic* sells. The very name '*Titanic*' in a headline sells copy; uttering it keeps viewers tuned in past the next advert break. A juicy-sounding new *Titanic* conspiracy theory, or a sensationalised programme that claims to be – but could only loosely be referred to as – a 'documentary' is simply cash in the pockets for those behind them. Better yet, there always seems to be someone or other who is thoroughly *convinced* that they know better – usually because they saw it in a badly researched show, read it in a headline, or because they make a hobby of peddling nonsense on an unsuspecting public eager for anything new from *Titanic*.

Disproving the carefully crafted yet utterly tangled webs of conspiracy theories, distortions, half-truths and outright lies is so complex that it can truly test one's love of the ship, of fact, of truth, and of the memory of the people who lived and died on that liner. As historians, we soldier on, but even for us there are times when it starts to feel like nothing more than a mental exercise, words in fading ink sitting on a yellowing page, spoken or written by people who have long since passed away, and having little bearing on the grim world we live in today.

Every now and then, however, one feels the thrill of why we first fell in love with *Titanic*. In those moments, the very soul of the ship feels as if it is stirring anew. This book began with such a moment: one afternoon, one of our team stumbled across a set of photographs taken in a recreated stateroom of *Titanic*, populated by a woman in period garb. At that moment, time seemed to stop … you could almost hear the refrain of 'Barcarolle', from *The Tales of Hoffman*, echoing faintly from down the corridor just outside the stateroom as the ship's band played their last song in the Reception Room that Sunday evening. You could almost hear the steady throb of *Titanic*'s great engines pulsing through the deck plates. It was as if the woman was about to spring to life and would head back up to play bridge with her friends in the Lounge. In that moment, *Titanic* was alive again.

It didn't stop there. Today there is an entire crop of new, enthusiastic artists working in various media to recreate *Titanic* and her sisters. Next, we found a set of photographs from a talented model-maker who made authentic-looking images of *Titanic* being fitted out in Belfast; then there was a series of paintings by another artist; then a series of images from a digital model of *Titanic*. One after another, we kept finding more and more pieces that simply thrilled and fired the imagination.

Just recently, we worked with a team of digital and graphic artists to create an all-new computer model of the *Titanic* and sink it in real time for an audience of thousands, while we and our team members narrated the events. Available on YouTube, it was entitled *On a Sea of Glass – LIVE!* after the title of our 2012 volume.

Making this film (and the *Real-Time Sinking 'Historian Edition'* that we released afterward with improved audio and visual effects) was like nothing that any of us had ever experienced before. As we worked and watched, we were able to see all of the different elements of the disaster play out in real time and all together in one place for the first time since 1912. It was an exciting and simultaneously deeply moving experience for us.

Then it hit us: it was time to bring *Titanic* back to life again. To accomplish this feat, we would use the artwork and combine it with well-chosen words and photographs in order to recreate *Titanic* and her sisters using the most up-to-date and accurate research available.

As a note, we are featuring the artwork of numerous up-and-coming artists, whom we will introduce to you at the end of the book. You may notice that each artist has a slightly different take on the colour or design of this feature or that of the ship; quite simply, there is some room for interpretation in colours or certain, little-known features that were on *Titanic*. We are not going to jump in and say that it had to be 'this way or the wrong way'. Instead, we are going to have the artists present their own vision, but we want you to enjoy the attention to detail and skill that went into each piece. These are remarkable artists, and we are very impressed with each one of them.

As you read through this book, try to put the modern world from your mind for a little while. Turn off the mobile phone. Perhaps play some period music. Try to regain the feel of that entire era once more. Perhaps, just perhaps, you will feel the stirrings of what first brought you to *Titanic*. We can't guarantee that you will actually slip away from this century and find yourself in 1912, as Jack Finney's fictional character Si Morley did in the novel *Time and Again*, but who knows? At the very least, we hope that you will come away with the same childlike feeling of wonder and awe that we felt while putting this book together. We hope that *Titanic* will come back to life anew for you as you look through these pages.

Some twenty-five years ago, James Cameron asked the world, 'Are you ready to go back to *Titanic*?' The question is as valid today as it was then, but it is a personal question that only you can answer for yourself. We hope that your answer is 'yes'. *Bon voyage.*

Tad Fitch, J. Kent Layton & Bill Wormstedt, January 2022

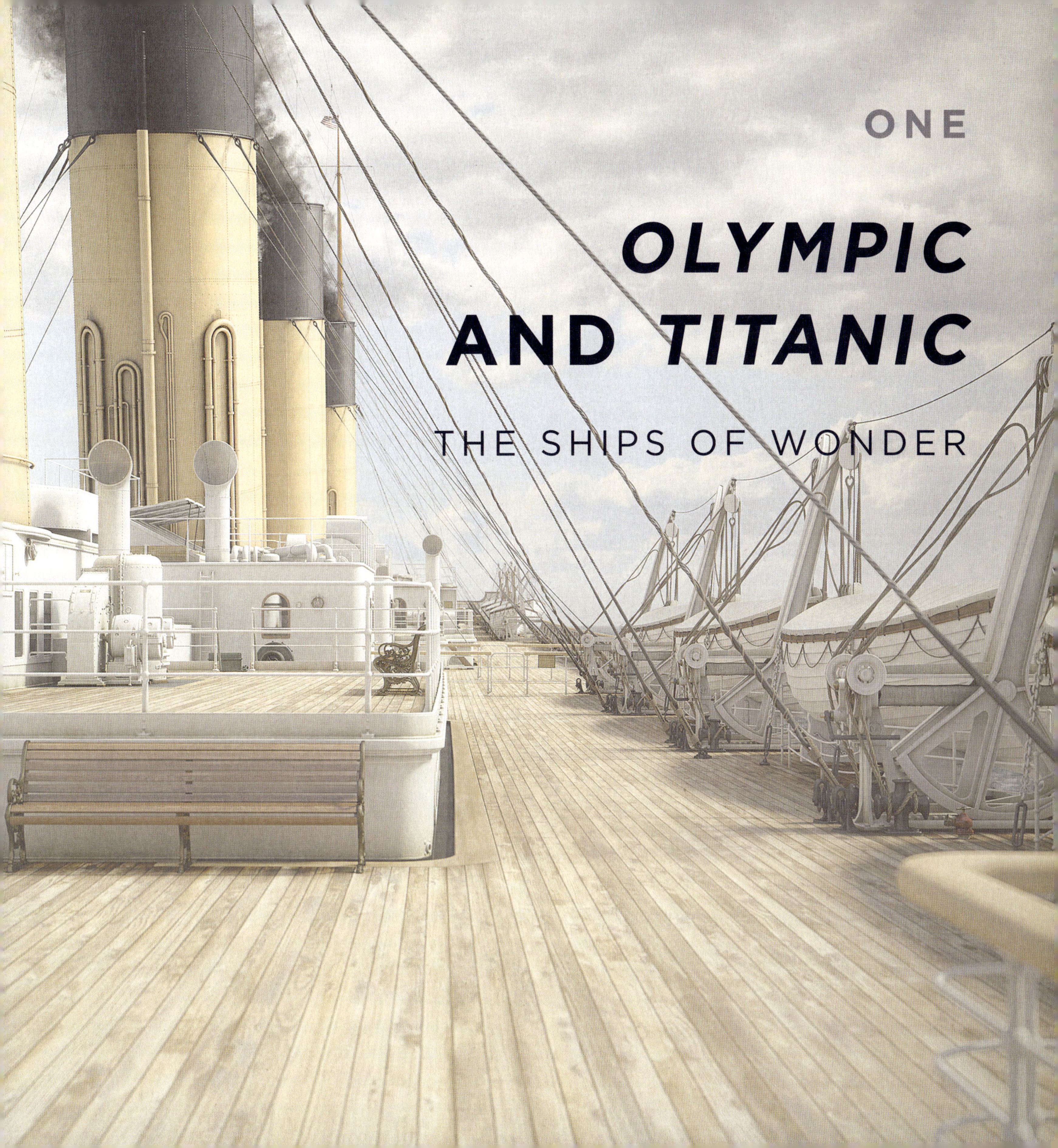

ONE

# *OLYMPIC* AND *TITANIC*

## THE SHIPS OF WONDER

Storm clouds were brewing on the horizon, and no one knew that better than J. Bruce Ismay, Chairman of the prestigious White Star Line. Ismay was the son of Thomas Ismay, the driving force behind the creation of the White Star Line, which had begun operations on the North Atlantic in early 1871. After that, White Star had clawed its way to a very prestigious position in the crowded North Atlantic shipping trade – yet it was still a tenuous one.

White Star was one of just two premiere steamship lines with a long British history; the other was the legendary Cunard Steamship Company, which had thirty years' more experience in the trade than White Star. Without a doubt, Great Britain was the dominant world power as the nineteenth century was drawing to a conclusion and her empire was truly global. Her Royal Navy was the strongest and greatest in the world and, being an island nation, trade across maritime routes was vital to her interests. It was, thus, not surprising to find that Britain had long enjoyed unquestioned dominance in merchant shipping, which included crack passenger liners such as those in the Cunard and White Star fleets.

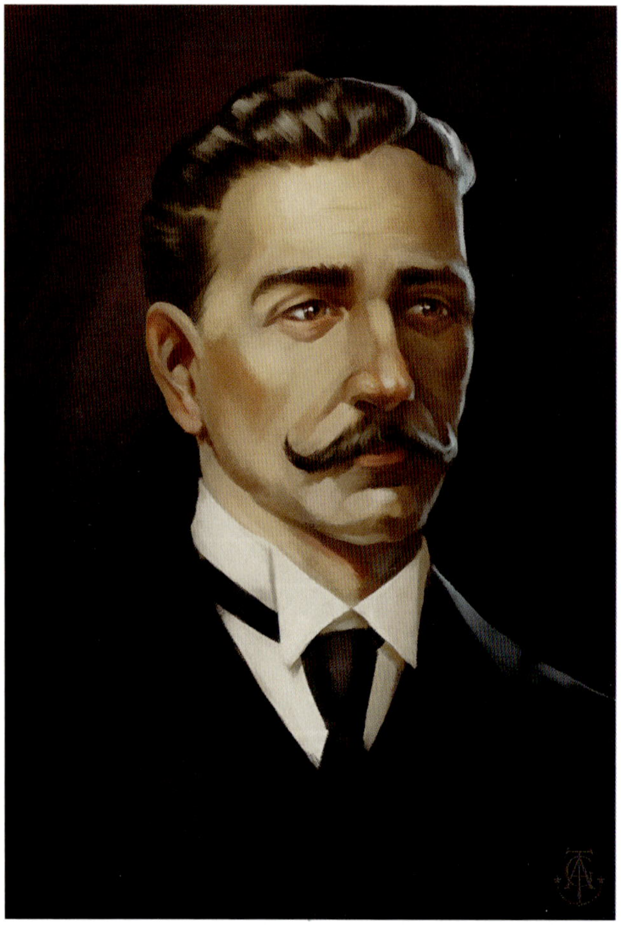

Joseph Bruce Ismay, Chairman of the White Star Line and son of the company's founder. (Tatiana Yamshanova)

However, less than a decade earlier, English dominance at sea was challenged by Germany, which had constructed the all-new High Seas Fleet. Though smaller than the Royal Navy, it was more modern. Also, with the backing of the German Government, the Norddeutscher Lloyd and Hamburg-Amerika Lines had built a fleet of super-fast, super-luxurious ocean liners that usurped the prestige of Britain's liners.

At the time, the White Star Line was wholly British owned, but it did not remain so. American financier J.P. Morgan had begun building an international conglomerate of merchant shipping lines under the American banner of the International Mercantile Marine. In 1902, Morgan made a very generous offer to the White Star Line, which was accepted, making White Star ships American owned. Bruce Ismay retained the title of White Star's Managing Director, and even became President of the International Mercantile Marine, helping to bring his company's prestige and experience to Morgan's fledgling combine. Even though the agreement stipulated that White Star ships were to remain British-flagged, be available to the British Government for charter in time of war, and their crew and officer corps must remain British, the change in ownership was another badly timed sting to British maritime prestige.

Cunard had been in a relatively poor financial position at the time. However, they worked out a deal with the British Government for a massive loan and ongoing subsidy to build two enormous new superliners, the biggest and fastest in the world. Their names, even before they entered service in the final third of 1907, were legendary: *Lusitania* and *Mauretania*.

Meanwhile, White Star's five greatest liners were relatively new vessels and the last of these five had not yet entered service as 1907 dawned. First had come *Oceanic* of 1899, a 704ft-long ocean liner of 17,272 gross registered tons (GRT). Bearing a service speed of 21 knots, *Oceanic* was a splendid vessel that stood up well to the new German liners. Next came a quartet of sister ships who earned the nickname 'The Big Four'. First was the *Celtic* of 1901, at 20,904 GRT. Then came the *Cedric* of 1903, slightly larger at 21,073 GRT. The *Baltic* followed in 1904, coming in at 23,876 GRT. Finally, even as the *Lusitania* and *Mauretania* were nearing completion in their respective shipyards, the fourth big White Star liner, *Adriatic*, entered service in May of 1907, the largest of the class at 24,541 GRT.

### WHEN A TON IS NOT A WEIGHT

Atlantic liners were measured by two methods. One was Gross Registered Tonnage (GRT), and the other was displacement. Both of these measurements used the term 'tons', but confusingly, only one of them actually referred to weight.

A ship's displacement is actually a measurement of its weight because it measures how much seawater the hull displaces when it is afloat. *Olympic* and *Titanic*'s displacements were measured in imperial tons of 2,240lb each, not the 2,000lb ton that many are familiar with today.

While the displacement, or weight, of the two sisters was identical, in reality, the figure constantly changed throughout a voyage. The deeper the ship sat in the water, such as at the start of a crossing with a heavier load of coal and provisions, the more water she displaced. At the end of a crossing, she was in a 'light' condition, with fuel and provisions having been expended throughout the trip, and would ride higher in the water. Their draught was constantly changing throughout every trip.

The figures most frequently referred to as *Olympic* and *Titanic*'s 'tonnage' actually cite their GRT, which was a measurement of enclosed volume, not weight. In this instance, a 'ton' of enclosed space was a measurement of 100 cubic feet of enclosed space.

There were many complex rules for measuring a ship's GRT, however, and various spaces on ships were either counted or not counted based on their intended purpose; the final result of the measurements, after all deductible spaces were removed from the gross figure, was referred to as net tonnage.

As we shall soon see, misunderstandings on what tonnage referred to has led to over a century of confusion on whether *Titanic* really was the largest moving object in the world when she was launched, or at the time of her maiden voyage.

These four liners were praised for their size and comfort, and they proved quite popular with fare-paying passengers. However, they were not even remotely in the same league as the new Cunard sisters. Not only were the Cunarders longer – four-funneled beauties of 787.2ft and 790ft overall length, as opposed to the 709.2ft of the *Adriatic*, for example – but they were also much bigger. Each of the Cunarders came in at over 30,000 GRT, and their speed was contracted to be an astounding 25 knots. The 'Big Four', comfortable as they were, ran at speeds of only 16–17 knots, even slower than the *Oceanic*.

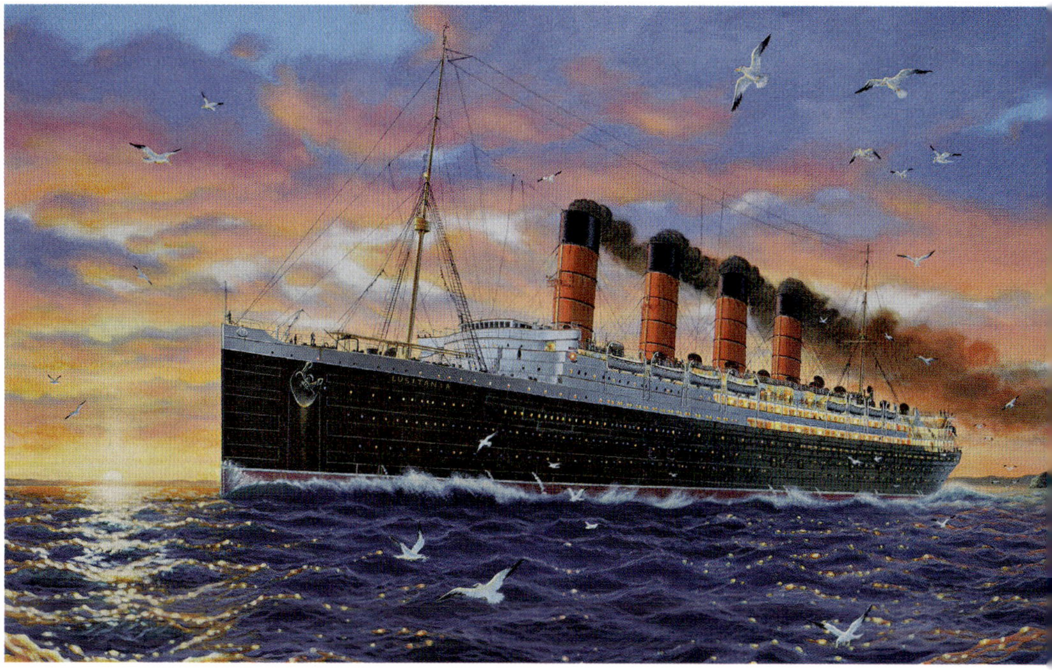

The mighty *Lusitania* entered service with Cunard in September 1907. She and her sister *Mauretania* were the largest, fastest and most luxurious in the world – but they were not always the most comfortable at sea. (Elang Erlangga)

White Star and Bruce Ismay knew they could not rest on their laurels. Ismay had to have known this for years before the two new Cunarders actually entered service. Each of those vessels had been laid down in August 1904, and the general details of their speed, size and amenities were well known to the public when the loan and subsidy were agreed to back in September 1902. Regarding the competition they posed, Ismay must have consulted very closely not only with the members of his own company's upper ranks, but also with Lord William Pirrie, Chairman of the prestigious Harland & Wolff shipyard in Belfast, Ireland, and Pirrie's top assistants.

The histories of Harland & Wolff and White Star were inextricably intertwined, going back to the line's first transatlantic ships in the early 1870s. Their working relationship was so close that they always worked through an honourable 'cost-plus' arrangement: Harland & Wolff built all of White Star's liners, and they were assured of receiving the amount of money that they had outlaid in so doing, plus a percentage of profit over and above that. This level of trust did not always work, as Cunard painfully discovered when *Lusitania* and *Mauretania*'s builders ran severely over budget in 1907.

Ismay and Pirrie also had a long personal history. Pirrie was some fifteen years Ismay's senior and had been a partner at Harland & Wolff since 1874. He had worked closely alongside Ismay's father, Thomas; that relationship must have helped ease Bruce's mind as he took the reins of his father's company, after Thomas passed away in November 1899.

Oddly enough, we do not know exactly when Ismay and Pirrie first decided how they would respond to the threat posed by the *Lusitania* and *Mauretania*. Popular history has it that this occurred over after-dinner drinks at Pirrie's home in the summer of 1907, but the facts do not support that legend. In fact, on 30 April 1907 – months before this alleged dinner – White Star had formally requested Harland & Wolff to design a pair of new liners that would combat Cunard.

Yet much had already gone into preparing for the project even before that. As early as late 1906, Pirrie had begun modernising his shipyards in preparation for this undertaking. This costly upgrade included the demolition of three major slipways and the construction of two new ones in their place, and the construction of an enormous 'gantry' structure that would facilitate work on two massive new ships. Work on building an enormous new graving dock at the shipyard, eventually named the Thompson Graving Dock, had begun even earlier, back in 1904. Clearly, White Star and Harland & Wolff were planning a response to the Cunard threat as far back as when construction on the *Lusitania* and *Mauretania* had commenced, and long before is popularly alleged.

With the stakes high and the competition fierce, the plans for White Star's two new liners were carefully conceived. It would not be a question of beating Cunard in terms of speed. Speed was expensive – making a jump from 21 knots to 25 took enormous quantities of coal and had necessitated the development of a whole new system of propelling machinery, the marine steam turbine.

Speed had other drawbacks. *Lusitania* and *Mauretania* were designed with very fine hull forms to cut through the water like arrows. This did not allow them to be as large in terms of enclosed space as slower ships with more traditional hull forms. It also ensured something of a rough ride for their passengers as they sliced through waves rather than climbing over them, and they frequently reached port battered.

Another major issue was vibration. When *Lusitania* and *Mauretania* entered service, they were also plagued with vibration at various operational speeds. To combat this, *Lusitania*'s Second Class accommodation was built up with enormous strengthening braces; on both ships, panelling creaked, domes rattled and lift doors clattered in their tracks as the ships shook. During a high-speed test on *Mauretania*'s sea trials, her Captain was said to have rung down for 'Full Stop' on the engines; when asked why, he tartly replied that he was being shaken off his Bridge. A running series of propellers of various designs were installed on both Cunarders during the first few years that they were in service in an attempt to make them both more efficient and to combat the vibration.

As it turned out, there were also problems with reliability in the turbine engines of the *Lusitania* and *Mauretania*. Both vessels suffered some significant teething issues with them. Although they retook the top speed honours, it was not uncommon in those early years for one or other of them to have turbines out of commission, requiring costly repairs and lay-ups.

White Star astutely learned from all of these painful lessons. Although they had at one time competed for the prize for the fastest Atlantic crossing, known as the Blue Riband, the company had since honed its focus on passenger comfort, amenities and luxury.

In the end, it was decided that Harland & Wolff would build a pair of sister liners simultaneously. Construction would be slightly staggered between the two to sidestep certain logistical issues. If the two ships proved satisfactory, White Star retained the option to build a third ship of the class at a later date to ensure a balanced weekly service by three ships of a single class.

The designs for the two ships underwent many refinements. On 29 July 1908 came a remarkable step toward the commencement of construction for the pair. White Star representatives met with Lord Pirrie and staff from Harland & Wolff in Belfast to consider a set of plans that the builders had drawn up. Termed 'Design D', this plan gave a good idea of what the new liners would look like.

Although many features would evolve through the design and construction phase, these plans gave an early glimpse into what Harland & Wolff was proposing to build for White Star. Excitement must have been incredibly high: no one had ever tried to build a ship on this scale – half again as big as the new Cunard speedsters – let alone building two of them at the same time in the same shipyard.

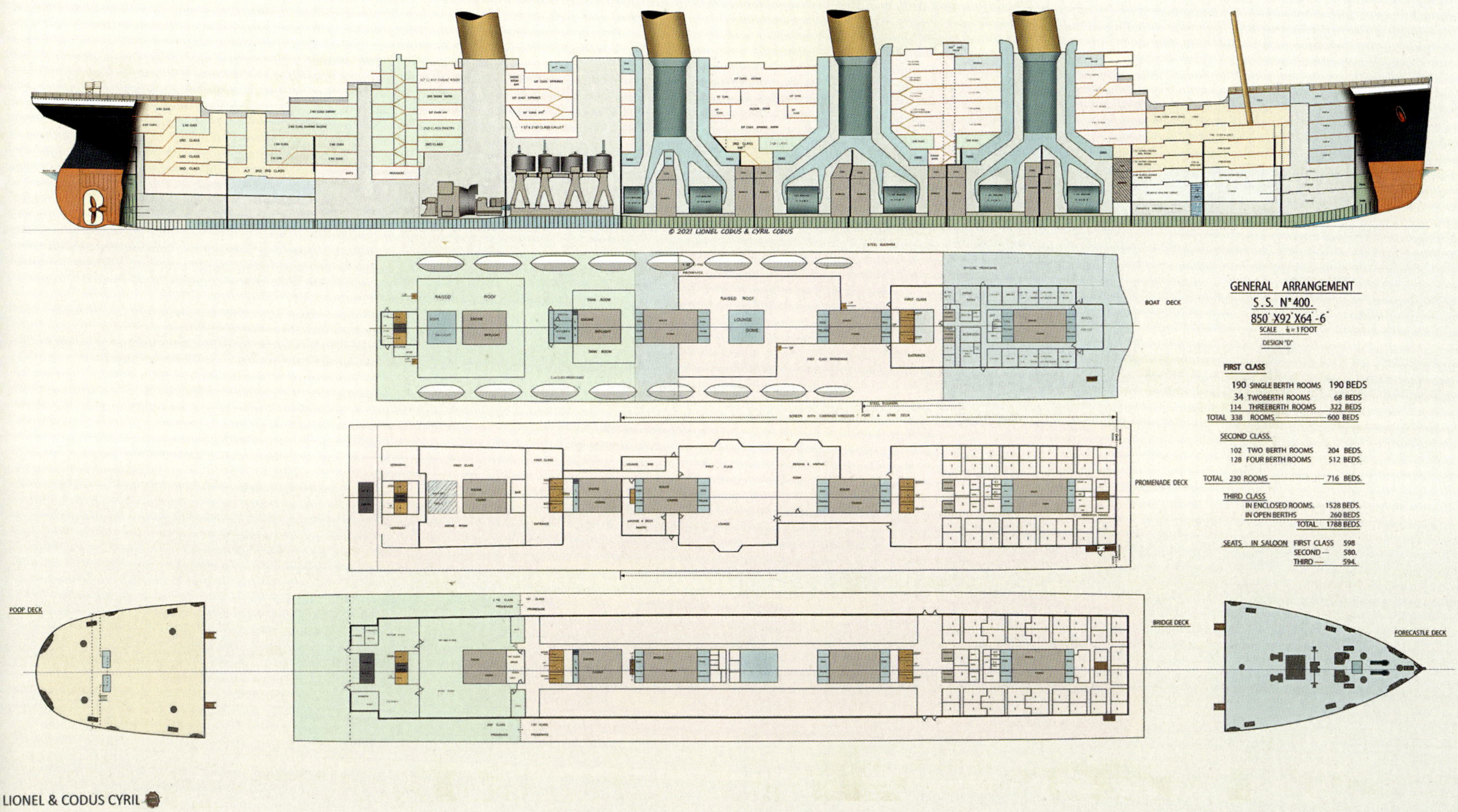


## *OLYMPIC*-CLASS 'DESIGN D'

These plans closely represent the drawings that were inspected by White Star personnel in their meeting with Harland & Wolff representatives on 29 July 1908. Named 'Design D', it was a close approximation of the two ships' final appearance. However, the 1908 plan differed from the final designs of the ships in some crucial areas. For example, the sixteen lifeboats shown in this plan are all placed aft on the Boat Deck, whereas the final design had the boats in the fore and aft corners of the deck on both sides of the ship.

As depicted here, the First Class Smoking Room had a dome behind the fourth funnel, and the Lounge had one as well. The main staircases in First Class were not nearly as 'grand' as they were in finished reality; on the other hand, the finished lifts did not run all the way up to the Boat Deck, meaning that passengers had to climb the stairs to arrive on that level. In the end, the Gymnasium was brought up to the Boat Deck and placed abaft the Entrance, on the starboard side of the No. 2 funnel casing, giving it no end of natural light.

Each of these changes, though small in themselves, showed the refinements being made to the liners' design as the project was moving forward. (Plans by Cyril and Lionel Codus)

Thomas Andrews, nephew of Harland & Wolff Chairman Lord Pirrie and a key man in the team that designed *Titanic*. He worked alongside his uncle and his uncle's brother-in-law, Alexander Carlisle, and many other draughtsmen and maritime architects. (Tatiana Yamshanova)

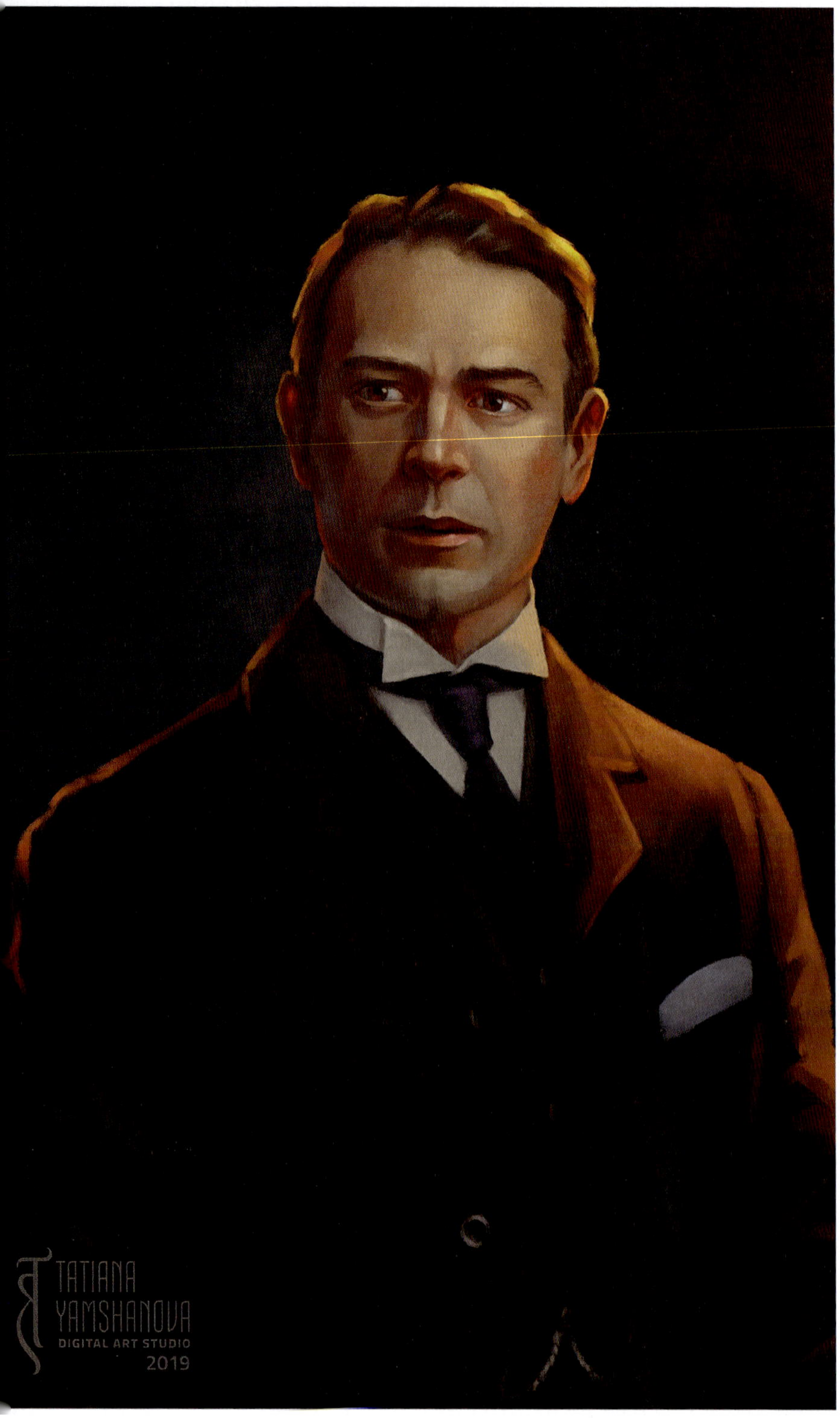


## WHO DESIGNED *OLYMPIC* AND *TITANIC*?

Thomas Andrews has often been called the 'designer' or 'builder' of *Olympic* and *Titanic*. From these titles, one could be forgiven for believing that he drew all the plans and built the ship himself. However, Andrews was really only one member of the design team, and he was not even the project leader.

Lord Pirrie closely oversaw the details of all the ships his company built. His brother-in-law, Alexander Carlisle, worked at Pirrie's right hand for decades as the firm's Chief Naval Architect. Carlisle later bragged about having been 'responsible for the designing and building of our greatest ships, up to and including the *Titanic*'. Yet, he also said that the two superliners 'were entirely designed practically by Lord Pirrie', while the 'details, the decorations, the equipments and general arrangements' fell under Carlisle's oversight.

Carlisle oversaw a team of the shipyard's managing directors and draughtsmen as they worked. Among these were Thomas Andrews, J.W. Kempster, George Cumming, Draughtsman Roderick Chisholm, Assistant Chief Draughtsman W.J. Smith and, of course, Andrews' right-hand man, Edward Wilding. White Star's Bruce Ismay also oversaw even the smallest details of the new liners, which had to represent the finest his company could offer potential clients.

Carlisle retired on 30 June 1910; it has been claimed that this was because of an argument over how many lifeboats the new liners were to have, but such notions have no basis in fact. Instead, it seems that the retirement was planned.

Any clashes in personality between Pirrie and Carlisle had plenty of fodder beyond lifeboats: the two men had differing politics, and Carlisle was known to be outspoken, while Pirrie liked to retain firm control of matters himself. When Carlisle departed, *Olympic* was nearing launch and *Titanic* was well under way. His position was filled by Thomas Andrews, Pirrie's nephew, and the man who, it is said, was being groomed to take Pirrie's place as the head of Harland & Wolff when Pirrie retired.

Because Andrews was in a key role at the shipyard by the time of *Titanic*'s maiden voyage and headed the builder's delegation that went on the trip, his capacity is often blown out of proportion: while it would be fair to say he was a critical and high-level figure at the shipyard, the reality is that he was but one part of a larger team.

The 29 July presentation was a success, and on 31 July, White Star signed the formal letter of agreement to commence construction. The keel for Harland & Wolff Hull No. 400, which would eventually be named *Olympic*, would be laid down on Wednesday, 16 December 1908. The keel for the second ship, Hull No. 401, was laid on Wednesday, 31 March 1909, thus offsetting each stage of their construction by just over three months.

## WHO FUNDED CONSTRUCTION OF *OLYMPIC* AND *TITANIC*?

It has often been said that American millionaire J.P. Morgan and his International Mercantile Marine (IMM) 'bankrolled' the construction of *Olympic* and *Titanic*. Although wholly untrue, this has led to conspiratorial portrayals of the role Morgan played in the design and construction of the two ships.

White Star was accustomed to funding the construction of expensive ships. Their *Oceanic* (1899) was said to have cost over £1 million – an extraordinary sum at the time. The combined final cost of the 'Big Four' ships was reportedly over £2.3 million, a veritable bargain by comparison, yet still an enormous expenditure. When completed, *Olympic* cost about £1.5 million, and *Titanic* cost slightly more at £1,564,606 because of the modifications and improvements she underwent. So, how did White Star come up with over £3 million to fund construction of the two new giants?

In 1908, White Star authorised £2.5 million in bonds for public subscription at a rate of 4.5 per cent interest – the value of their existing fleet of twenty-three ships, totalling just over £4.8 million, was the security for the bonds. White Star also enjoyed tremendous profits between 1907 and 1911.

Clearly, White Star's financial position was much better than Cunard's had been when they took the £2.6 million loan from the British Government for the construction of *Lusitania* and *Mauretania*. In the end, those ships were highly unusual, with special features intended to prepare them to serve as armed auxiliary cruisers in time of war. They suffered terrible cost overruns, and their final cost was a shocking £3,437,715 – £837,715 above the original £2.6 million government loan.

By comparison, despite their unprecedented size, *Olympic* and *Titanic* were rather ordinary ships. Their hulls did not need reinforcing to carry enormous deck guns, nor did they need to pioneer exotic engines for high speed. They were, instead, essentially scaled-up versions of what Harland & Wolff had built before. The yard used tried-and-tested methods and materials, despite their size. This prevented the astounding cost overruns that Cunard suffered with *Lusitania* and *Mauretania* and ensured that White Star could finance the construction of *Olympic* and *Titanic* independently of J.P. Morgan and the IMM.

The names for the two liners had been publicly announced as *Olympic* and *Titanic* as far back as April 1908, even before the paperwork to start construction was signed. White Star had intended to use the name *Olympic* for a sister ship to *Oceanic* of 1899 that never came to pass. The name *Titanic* was entirely new for the company, but fitting. In Greek mythology, the Olympians and the Titans had been two successive generations of gods who vied for control of the universe; in an odd twist, however, the *Olympic* would precede *Titanic*, while in Greek mythology the Olympians had come to power only after overthrowing the Titans.

Thousands of men were employed in every facet of work on the ships, from design to riveting, to joiners and carpenters, right through to those building the ships' engines. There was no question that the new White Star ships were the pride of Belfast.

In the draughting offices, workers finalised the plans for the new ships. In the moulding lofts, skilled workmen upscaled the drawings to full size. Then the steel that would be used in the two ships' hulls was ordered and delivered to the shipyard. Soon, the keel of each ship was laid, and construction was progressing out and up.

## BADLY DESIGNED, CHEAPLY MADE?

It has often been alleged that in order to cut costs, *Olympic* and *Titanic* used substandard materials in their construction and that this caused the sinking of the *Titanic*. However, nothing could be further from the truth. Harland & Wolff needed to create products that satisfied their clients; deliberately building ships of questionable quality would alienate customers and destroy the firm's own reputation in the highly competitive shipbuilding trade.

Nor would White Star have wanted substandard ships in their fleet, as they were intended to last for over twenty years. It was less expensive to build them correctly from the start rather than to try to fix them as they went along. The shipbuilder and shipowner also had a long and trustworthy working relationship, with no evidence that the shipowner ever pressured the shipbuilder to cut costs at the expense of quality or safety.

The two new liners were actually built to industry standards of the day. Comparisons to *Lusitania* and *Mauretania* are not fair, because the Cunarders were built for a dual purpose to serve as warships, and some of the high-quality materials used in their hulls were actually used to combat worrying issues with their stability – issues *Olympic* and *Titanic* never faced. Nor did White Star or Harland & Wolff have final say in cutting corners and building the ships 'on the cheap'. Although neither sister was Lloyd's of London certified, they were built to Lloyd's standards, and Lloyd's did certify the quality of steel used in both ships. Furthermore, the British Board of Trade had on-site surveyors, headed by Francis Carruthers, who pored over every element of the liners' design and construction.

The members of this team were not 'yes men'. Carruthers himself could often be found climbing through ships under construction in the shipyard, inspecting and examining the quality of materials and workmanship. On more than one occasion, Carruthers stepped in to question certain details and the written record shows that the Board gave him ultimate authority if he was not convinced of the safety of any feature of the two new liners.

In short, while *Olympic* and *Titanic* were not perfect ships, they were of a high calibre of design and materials for their time. Harland & Wolff was an experienced shipbuilder; they used quality materials and methods, and they never allowed the 'shortcuts' that were common among other shipyards. Although *Titanic*'s shattered wreck suggests deficiencies to some today, the truth is that in the sinking and descent to the sea floor, she was subjected to strains and stresses that no ship was designed to sustain.

Indeed, the long service record of *Olympic*, built from the same stock of steel hull plates and rivets, shows that *Olympic* had far fewer issues with her hull over the years than many contemporaries, and that includes Cunard's beloved and long-lived *Aquitania*. In fact, the German liners then under construction, *Imperator*, *Vaterland* and *Bismarck*, would prove to have far greater issues over their careers. *Olympic* never had such concerns and was noted to be superior. The same type of steel used in *Olympic* and *Titanic* was used over twenty years later on *Queen Mary*, which is still afloat after eighty-five years, in spite of years of missed maintenance on her hull.

Much has been made of the damage patterns seen in *Olympic*'s hull plates after her collision with the *Hawke*. However, it must be remembered that the *Hawke* was a naval cruiser with a reinforced concrete bow that was designed to sink ships by ramming them. Yet when she rammed *Olympic*, the liner's overall safety was never in doubt; she was able to offload all of her passengers and return to Belfast for repairs under her own power.

Meanwhile, *Olympic* and *Titanic*'s watertight subdivision was actually superior to that of *Lusitania* and *Mauretania*; the latter ships had wing coal bunkers and longitudinal bulkheads that caused loss of stability in any but the most minor flooding scenarios. Both of those Cunarders were, in essence, ticking time bombs of bad design. In the end, this ensured that when *Lusitania* was struck by a torpedo, she sank violently in a mere eighteen minutes, with the loss of nearly 1,200 people; *Mauretania* herself was nearly lost at least twice during her career due to bad watertight subdivision. Either Cunarder would almost certainly have sunk more dramatically and more quickly had they struck an iceberg the way *Titanic* did.

Interestingly, when Harland & Wolff later built *Justicia*, they more or less designed and built it as a smaller *Olympic*-class ship for another line. While of comparable size to *Lusitania*, she was of similar materials and overall design to the larger White Star vessels. In July 1918, *Justicia* was attacked by German submarines and stayed afloat for over twenty-two hours despite a gruelling assault in which it was repeatedly torpedoed. In the end, it took six torpedo strikes to sink the vessel. Clearly it was a well-built ship.

Ultimately, claims of bad design, poor materials or cost-cutting on *Titanic* are latter-day efforts to scapegoat Harland & Wolff or the White Star Line. No real evidence – save for what has been taken out of context or misconstrued and splashed in headlines around the world – has ever come to light to indicate that *Olympic* and *Titanic* were anything but well-made vessels.

Work on the liners progressed with surprising speed, considering their enormous size. Framing on *Olympic* – the process of setting each of her over 300 vertical ribs in place and connecting these to her keel and double bottom – was complete by 20 November 1909. Next, shell plates were placed over these frames and to form the watertight 'skin' of the ship. Eventually, *Olympic* stood proudly above her slip, No. 2, nearly filling her half of the 'Great Gantry'.

At *Olympic*'s starboard side, on Slip No. 3, sat the identically sized shape of *Titanic*. Her framing was complete on 6 April 1910, a little more than a year after the keel was laid, and she was fully framed by 19 October, just a day before *Olympic* was to be launched into the River Lagan.

Thursday, 20 October 1910 was a proud day for everyone connected with the construction of *Olympic* and *Titanic*. When *Olympic* began to slide down the ways toward the water, she became the largest moving object in the world, wresting the title from *Mauretania*. Crowds gathered to watch, and the launch was even filmed for newsreels, which was something of a novelty at the time.

Through the summer, autumn and winter of 1910, work continued on both ships. Eventually, by May 1911, *Olympic* was nearing completion.

This painting of *Titanic* in the new graving dock, prior to its forward promenade being enclosed, gives an idea of what *Olympic* woud have looked like as it neared completion in early May 1911. The latter would be ready for her maiden voyage in mid-June. (David Olivera)

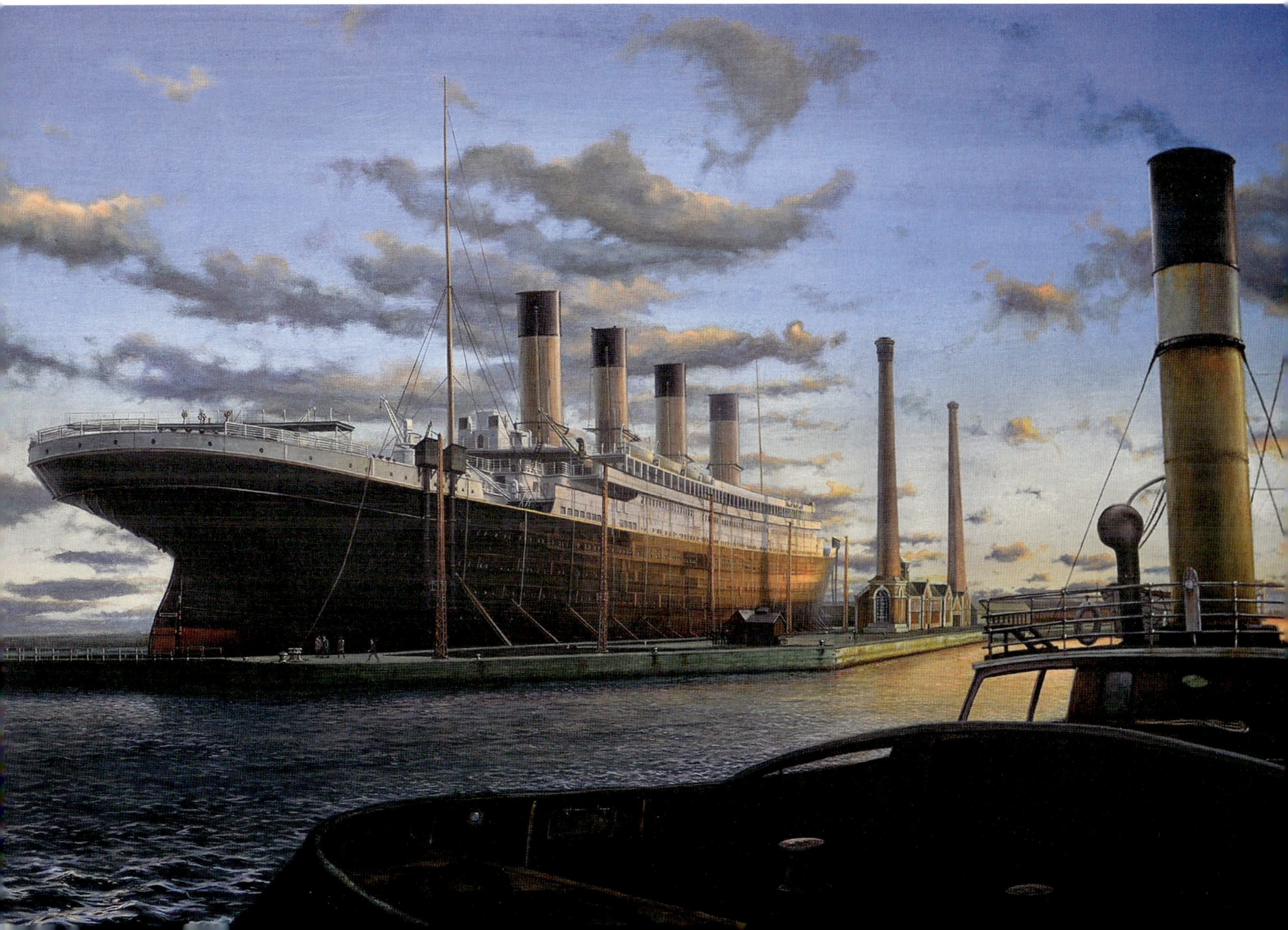

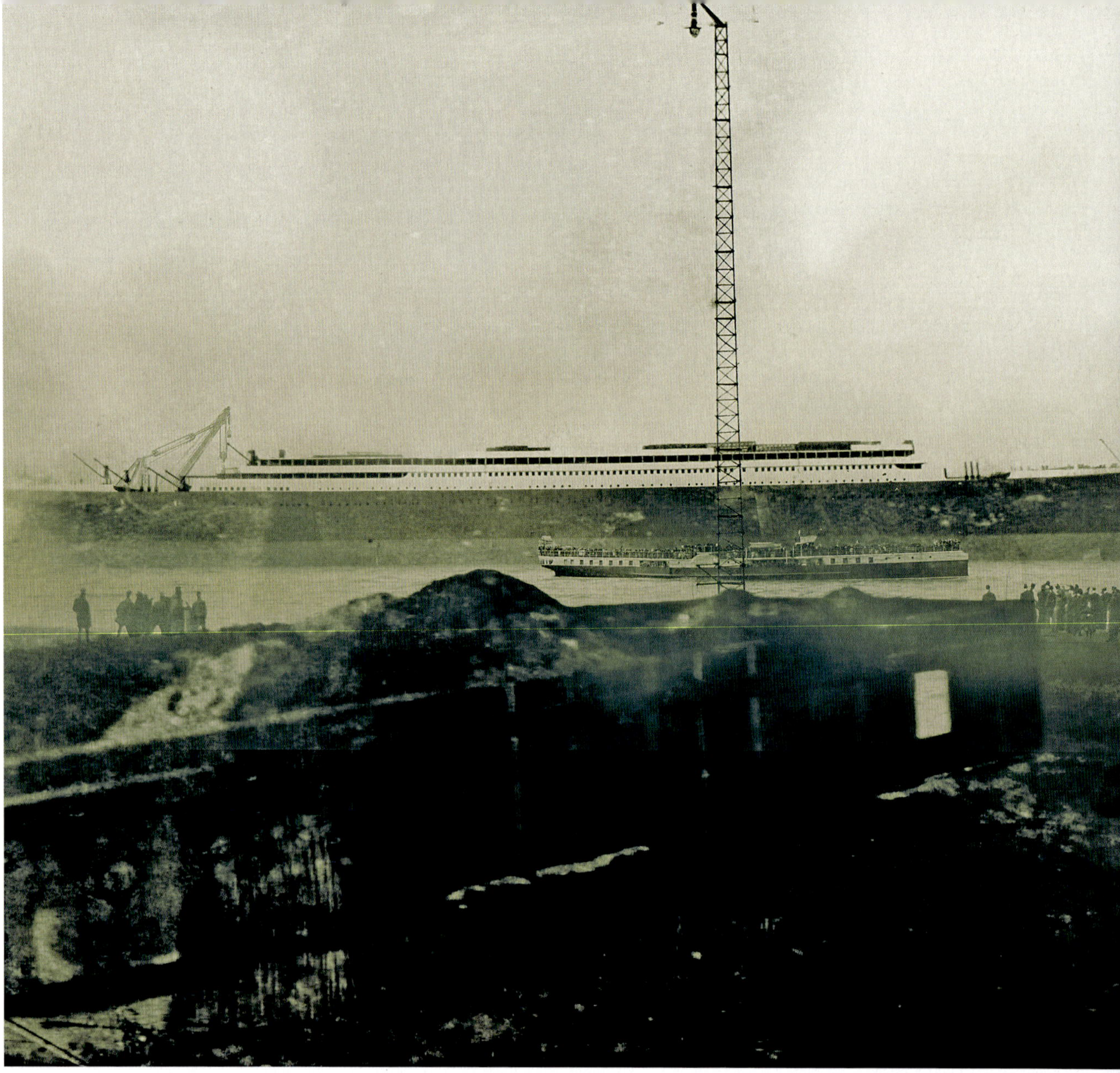

*Titanic* is afloat for the very first time. Spectators line the banks and watch as the vessel's hull is towed from the main channel toward the new fitting-out wharf in this artistic recreation of the event. (Ala Kermiche)

On 29 May, crowds of spectators lined the banks of the Belfast Lough to cheer *Olympic* on as she took to open waters for the first time on her builder's trials. Over the course of two days, *Olympic* was put through her paces. Her engines, machinery and manoeuvrability were all thoroughly tested, setting a benchmark for what could be expected from her younger sister. Although designed for a speed of 21 knots, she achieved 21¾ knots at some points during these tests, a high mark that was expected to be surpassed as the engines were broken in. Her certificate of seaworthiness from the British Board of Trade was signed at the end of these trials, and she returned to Belfast late on 30 May.

The following day, Wednesday, 31 May, came the launch of *Titanic*. The event was scheduled for 12.15 p.m., and up to 100,000 spectators gathered to observe. American financier J.P. Morgan was on hand, along with Bruce Ismay and any number of other VIPs. As with *Olympic*'s launch, a newsreel crew was on hand to record the event, although the footage has since disappeared.

At 12.13 p.m., *Titanic* began to move for the first time. The men on the ship began to cheer wildly once the launch was complete and a cacophony erupted as the crowds on shore cheered. The steam tugs that were standing by to take her to the fitting-out wharf let loose with their whistles. The bedlam went on for several minutes, and soon workmen began to detach the hull from the drag chains and take her to the fitting-out wharf as the crowds began to disperse.

## *TITANIC*: THE WORLD'S LARGEST SHIP?

*Titanic* is often referred to as 'the world's largest ship' at the time of her maiden voyage. This was technically true, but not by a wide margin. *Olympic* and *Titanic* were significantly longer, larger and heavier than *Lusitania* and *Mauretania*. However, the comparison between *Olympic* and *Titanic* is more complex.

Both were 882ft 9in in overall length, and 92ft 6in wide. Their dimensions were identical, as was their rated displacement of 52,310 imperial tons. Because of this, at the time *Titanic* was launched, *Olympic* had been completed and so *Titanic* did not take the title of 'the world's largest ship' on 31 May 1911.

When completed in early 1912, however, *Titanic* did technically become the largest ship in the world. However, this was in terms of GRT only; she was registered at 46,328.54 (rounded up to 46,329) GRT, while *Olympic* was 45,323.82 (rounded up to 45,324). This 1,004.72-ton advantage led to *Titanic* justifiably being called the largest, and White Star played up this publicity angle. However, the difference was merely a matter of remeasuring internal volume (GRT) based on how certain spaces were utilised, not any real size variation.

Yet had *Titanic* not sunk, she would not have held the title of 'world's largest' for very long. Indeed, when *Olympic* emerged from her 1912–13 winter refit, she was remeasured at 46,358.70 GRT, some 29.88 tons larger than *Titanic*. Not to be forgotten, Cunard's *Aquitania* had begun construction in June 1911. She would be nearly 20ft longer than *Olympic* and *Titanic*, at 901ft 6in in overall length, yet smaller than *Titanic*'s 1912 tonnage measurements. Even bigger ships were then under construction, such as the 909ft, 52,117-ton German liner *Imperator*, which was launched on 23 May 1912, followed in short order by two even longer and larger sister ships.

So, while *Titanic* could technically be referred to as the 'world's largest ship' in April 1912, had she survived, her own sister would briefly have regained the title when she emerged from her 1912–13 overhaul. Meanwhile, other liners were then under construction or in the works that would soon challenge or easily exceed her brief claim to supremacy on this front.

(Vasilije Ristović)

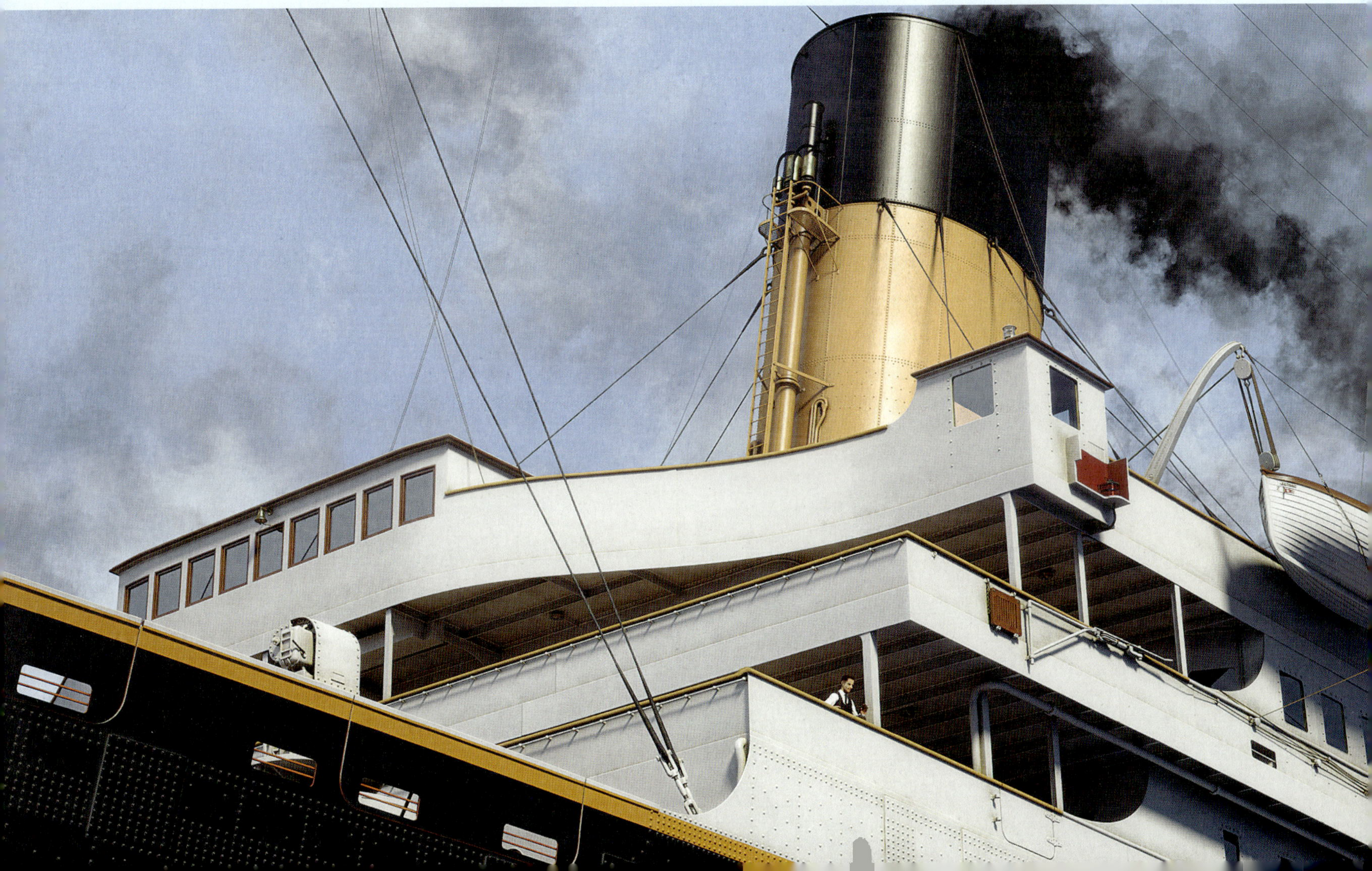

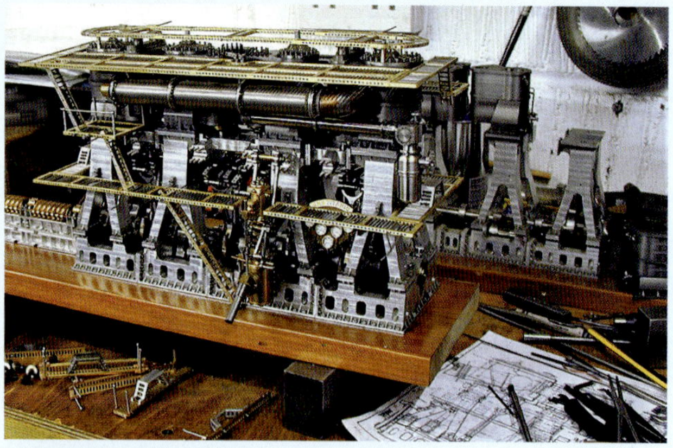

Three views of the working reconstruction of *Titanic*'s port-side reciprocating engine, as built by Derek Staines.

This photo shows the entire engine, including the double bottom platform. (Derek Staines)

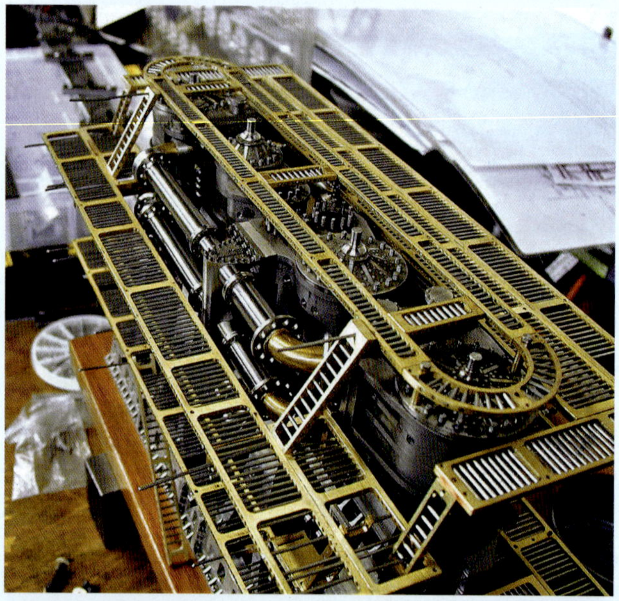

An overhead view of the rebuilt engine, showing the catwalks and ladders surrounding the structure. (Derek Staines)

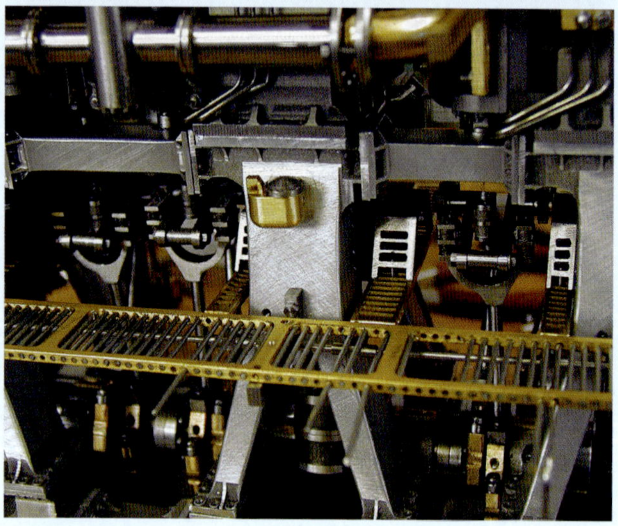

A close-up view of the rebuilt port-side reciprocating engine. (Derek Staines)

*Olympic* left Belfast before the day was out, and her maiden voyage began on 14 June. She arrived in New York on 21 June after making the crossing at an average speed of 21.43 knots – faster than the 21.17 knots her souvenir log card for the trip stated, apparently due to a miscalculation in the run time.

*Olympic*'s maiden voyage was such a success that on the day before she arrived in New York, 20 June, Bruce Ismay – who was aboard the ship – decided to exercise the option to order the third ship of the class, which would eventually be called *Britannic*. After the maiden voyage, *Olympic* began to settle into her career, and proved enormously popular with passengers.

Back in Belfast, *Titanic*'s fitting out was proceeding. The first steps involved the installation of her boilers and engines; only after that could her upper works and deckhouses be finished.

## *TITANIC*'S MIGHTY ENGINES

The *Olympic*-class liners were driven by a truly remarkable hybrid powerplant design. Up until *Lusitania* and *Mauretania*, all previous ocean liners had channelled the steam from their boilers into reciprocating engines. These massive structures had enormous cylinders, pistons and crankshafts; they took up a lot of space within the ships' hulls, and they had many moving parts to maintain. Yet Harland & Wolff had grown very good at building them, and they were both relatively inexpensive and efficient. Cunard had boldly eschewed reciprocating engines in favour of the new marine steam turbine in their sister speedsters, and the venture had succeeded, albeit with drawbacks. The Cunarders made record speeds but suffered teething issues and were not very fuel efficient.

Harland & Wolff and White Star bestowed *Olympic* and *Titanic* with a hybrid-style powerplant. Both ships were given two massive four-cylinder triple-expansion reciprocating engines, each of which was rated for 15,000 nominal horsepower. Then they added a low-pressure turbine to the design, fed by what would otherwise have been waste steam after the reciprocating engines were finished with it. That 'waste' steam thus provided an additional 16,000 nominal horsepower, an increase of over 50 per cent beyond what the reciprocating engines could do alone, and with no increase in fuel consumption.

The reciprocating engines installed in each of the *Olympic*-class ships were of identical size, and were the largest ever installed in ocean liners. They were, without a doubt, the beating 'heart' of *Titanic*. Indeed, to many, the stopping of the engines is what caused the most alarm on the night of 14 April 1912.

Derek Staines, from the United Kingdom, undertook to build operating scale models of *Titanic*'s port-side reciprocating engine in exquisite detail. The engine stands roughly 8in tall and is 11.4in long at the engine base, and 16.75in long including the thrust block, at about 1/65 scale. It weighs 15.4lb and will eventually run on steam power. Even now, running on compressed air, watching it spin up and begin to move evokes childlike sensations of wonder. When filmed operating at extremely high speeds and then slowed down to watch and listen to, these model engines actually begin to recreate the rhythm of the *Olympic*-class liners' beating hearts once again.

Before June 1911 was out, significant modifications to *Titanic* had been approved, based on operational experience with *Olympic*, which would mean *Titanic* would be an improvement over her sister's already high standard. *Olympic*'s enclosed B Deck Promenade was not as popular as had been hoped, but it was found that the *à la carte* Restaurant was a smash hit, as was the Reception Room down on D Deck, which grew too crowded during performances by the ship's band, leaving some patrons sitting on the stairs of the Grand Staircase. The layout of the Turkish Bath complex could see improvement, and one of the Electric Baths could be removed because it wasn't as popular as hoped.

*Titanic*'s B Deck layout was completely redesigned. The enclosed promenade there was removed and replaced with fine new staterooms, including two Parlour Suites boasting Private Promenades. The expansion of the Restaurant aft on the same deck, the inclusion of a new Restaurant Reception Room, and of a new French sidewalk café, all added to the list of upgrades.

This spectacular photo was taken in September 1911 and shows the extent of the project to rebuild B Deck. In areas, one can see right through the ship. Even the Second Class Promenade is affected, as it is being shortened. (Günter Bäbler Collection)

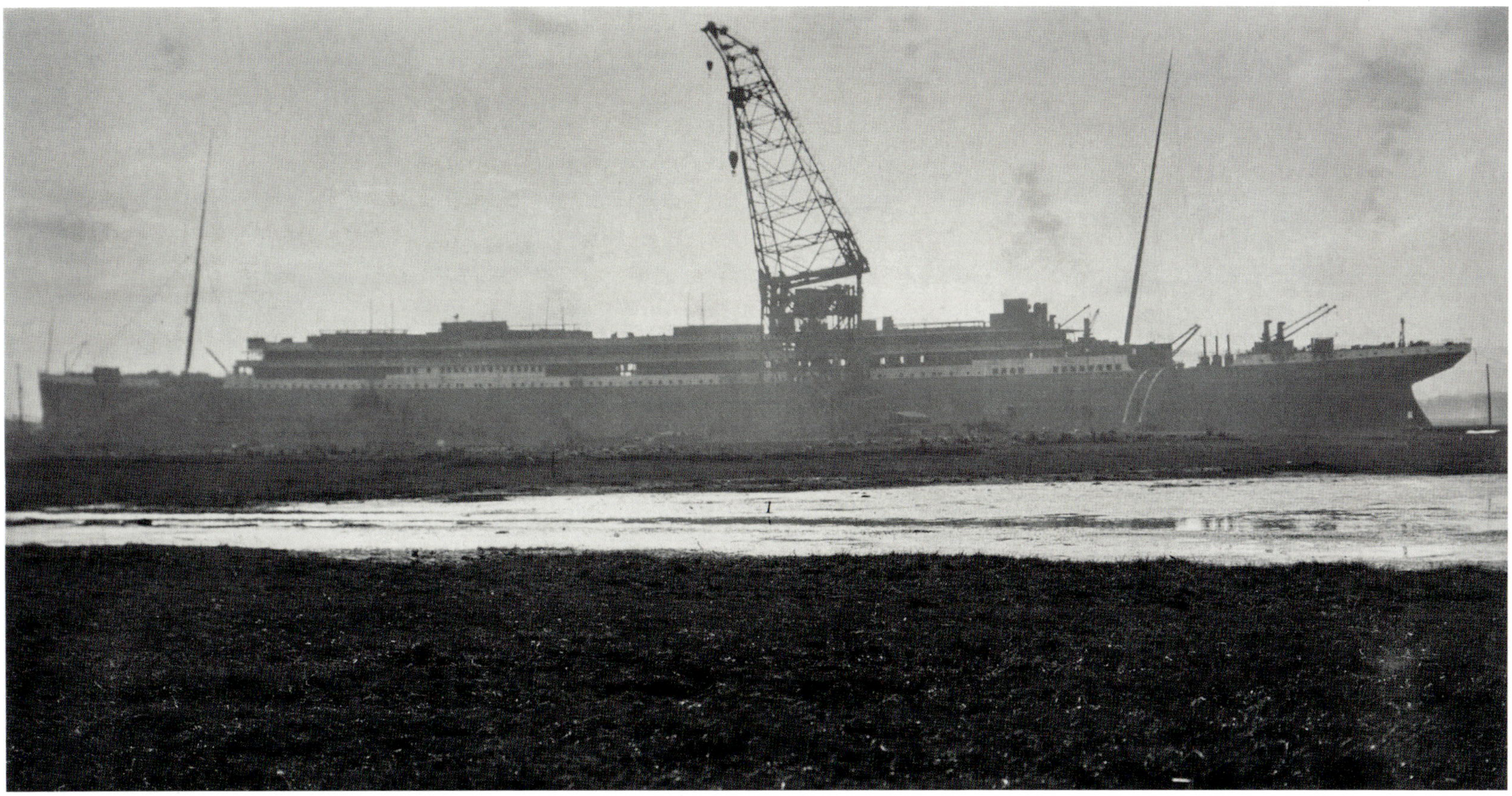

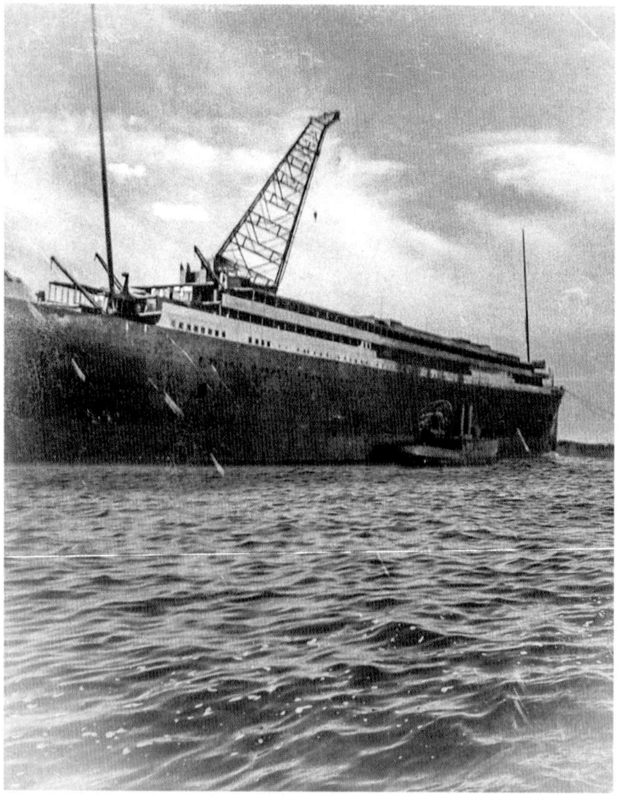

*Right:* This artistic depiction of the work on the starboard side of B Deck shows the Harland & Wolff tug *Jackal* alongside *Titanic*, then serving as a floating auxiliary power station. Workers ran cables aboard *Titanic* from the shore and from the tug to supply power to aid their work. (Ala Kermiche)

*Below:* This artistic rendering, set around late summer 1911, shows the gaping opening in the Boat Deck where the Grand Staircase and its dome would eventually be. The stairs were in place by mid-October 1911. (Chris Walker)

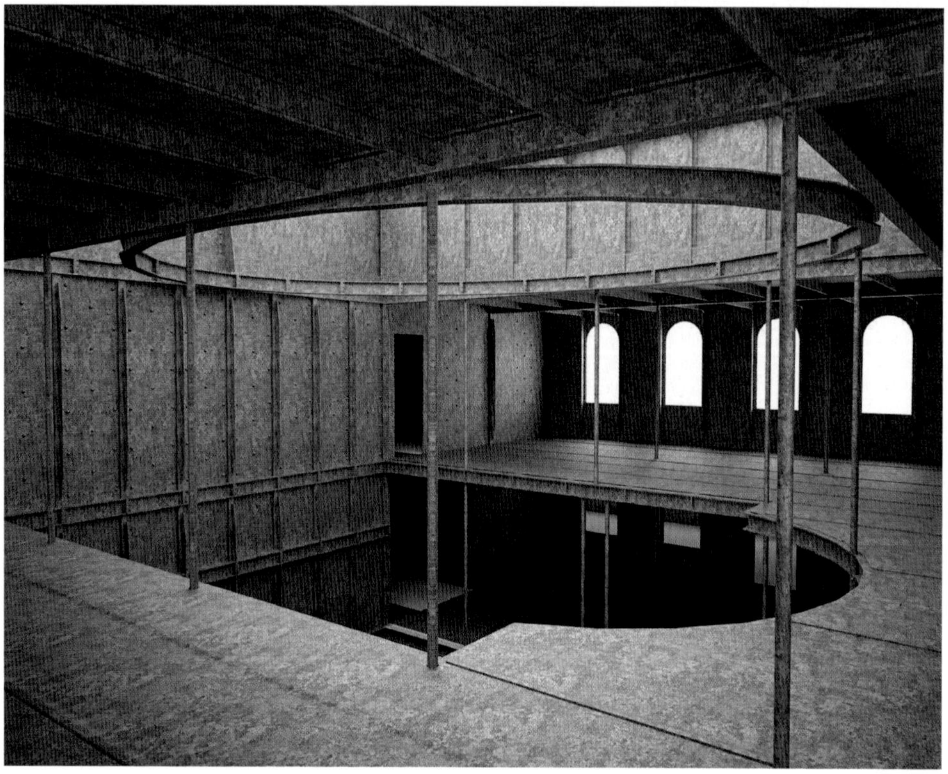

These extensive modifications would take six weeks. The result was a postponement of *Titanic*'s intended maiden voyage from January 1912 to 20 March 1912.

By September 1911, the side shell plating along much of B Deck was being torn off to make way for a completely new pattern of screen for the new staterooms and expanded Restaurant. Next came the funnels: first to be installed was the No. 2, which was in place by early October. By the time it was installed, however, *Olympic* was shortly to arrive back in Belfast, for she had undergone an astounding mishap on 20 September.

The incident occurred while *Olympic* had been departing Southampton on her fifth westbound crossing. Before clearing Southampton waters, near the Isle of Wight, she was struck by the cruiser HMS *Hawke*. The warship's prow scythed its way deep into the liner's starboard-stern quarter. There was significant damage, but fortunately no one was killed and the ship's watertight bulkheads prevented any threat to her overall safety. *Olympic* underwent temporary repairs in Southampton, then limped back to Belfast for involved and expensive repairs. She arrived there on 6 October.

Bearing only her No. 2 funnel, *Titanic* was moved from the deep-water fitting-out jetty, where she had lain since 31 May, down past the graving dock and into the Alexandra Wharf. *Olympic* tied up at the fitting-out jetty *Titanic* had just vacated, where she was significantly lightened. On 10 October, preparations were complete and *Olympic* was dry-docked.

After examination, it was found that it would take longer to repair than hoped. After conferring with Harland & Wolff, White Star announced on 11 October that *Titanic*'s maiden voyage would be delayed by another three weeks, from 20 March to 10 April 1912, the date originally scheduled for her second departure from Southampton. The tug *Egerton* was subsequently brought in from Liverpool to help move *Titanic* back to the fitting-out wharf, so that work on her could proceed with a minimum of interruption.

By 19 October a progress report on *Titanic*'s fitting out ran in the press. Although only one funnel was in place, a lot of other work was proceeding: aft on G Deck, much of the Second and Third Class accommodations were complete; on E and F Decks, most of the wooden stateroom bulkheads were in place; on D Deck, the stateroom bulkheads were being installed, while the joinery frames of the First and Second Class Dining Saloons were all up, and things were ready for framing, panelling and ceiling work. On C Deck, the framing for the First Class staterooms was 'well in hand', and the Third Class General Room and Smoking Room joinery frames were finished. The Grand Staircase was in place, although work was still being done on the lifts.

Work continued both on repairing *Olympic* and fitting out *Titanic* side by side. By 10 November, the carefully choreographed ballet had been reversed and *Olympic* lay at the fitting-out wharf while *Titanic* was back at the Alexandra Dock. By this point, the main damage to *Olympic*'s hull and machinery had been addressed and the finishing touches were being put on. She was successfully turned around on Sunday, 19 November, preparatory to leaving, and departed the next day. *Titanic* was to be returned to her fitting-out wharf the day after, on 21 November.

On 3 February, *Titanic* was moved into the graving dock for the first time. The event was so momentous that a newsreel photographer was brought in to film it, even though it was a Saturday. Here, *Titanic* would have her propellers fitted, her hull thoroughly cleaned and painted, and have numerous other details seen to. She was removed from the dry dock on 17 February.

*Titanic*'s No. 2 funnel was placed first, in early October 1911. By early November, the No. 1 funnel had also been fitted. This artistic view shows the No. 1 funnel being lifted aboard by the giant crane at the fitting-out wharf. (Ala Kermiche)

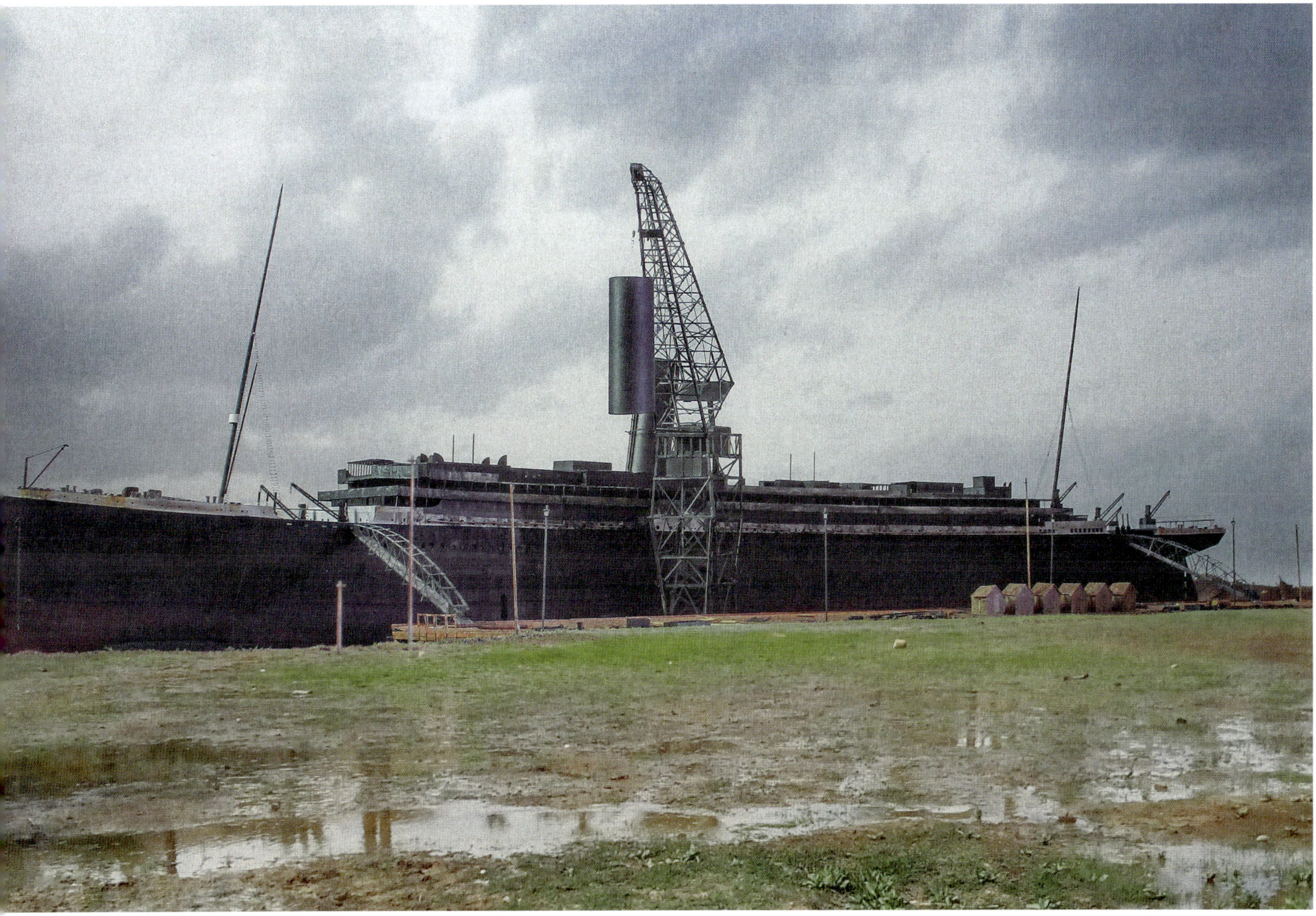

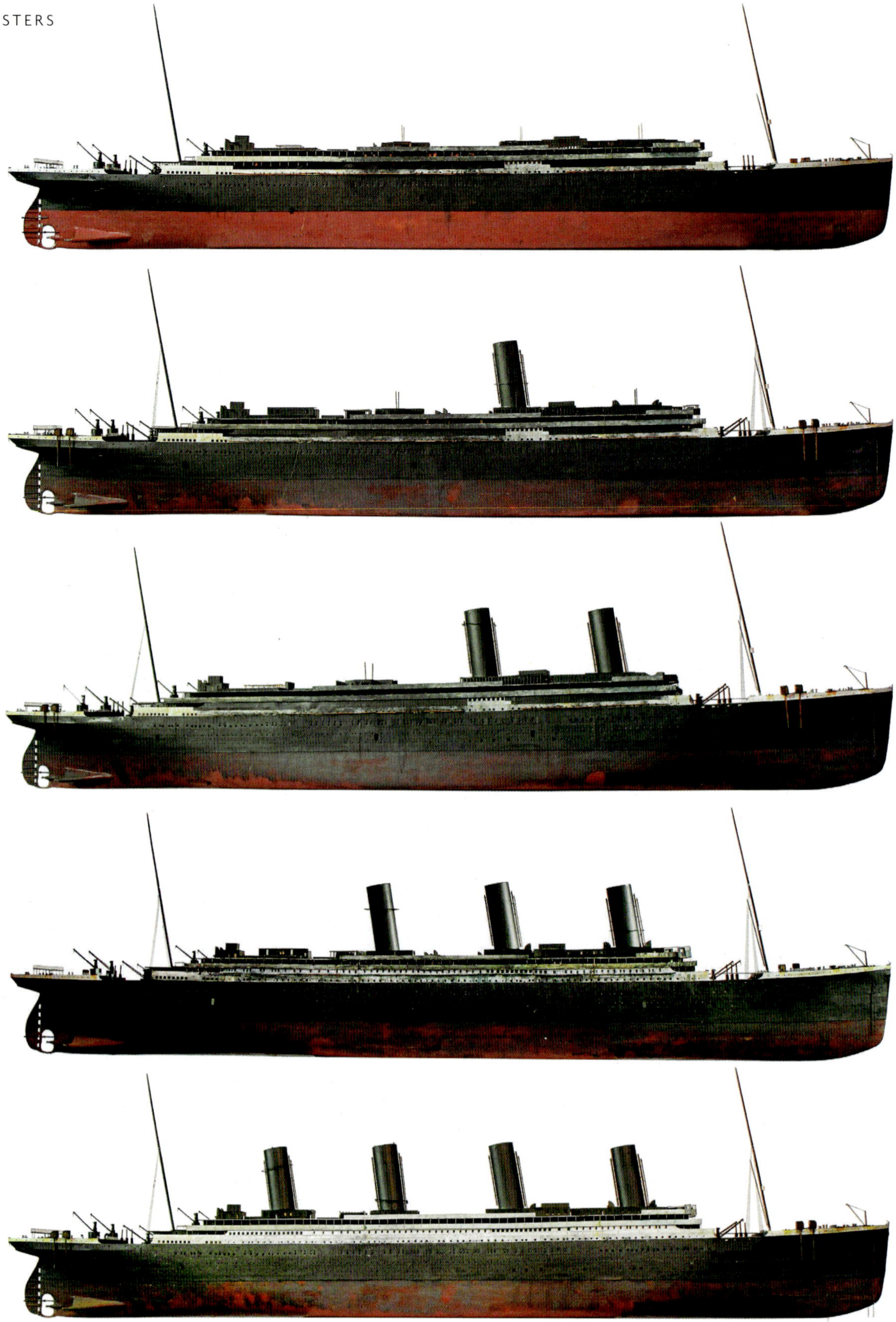

Metamorphosis: how *Titanic* became a liner. On 8 December 1911, it was reported that the No. 3 funnel had been fitted, with only the No. 4 funnel yet to be placed. When it was installed in early January 1912, *Titanic*'s profile at last resembled its final form. (Ala Kermiche)

Before *Titanic* had left the graving dock, on 14 February 1912, another last-minute change was made. A notation on her plans records a decision to add a bulkhead with a door and window to the forward end of the First Class Promenade on A Deck; this timing likely corresponds to the decision to enclose the forward half of that deck on either side of the ship with new 37in x 31in screens. Photos taken on 2 March showed that work on this had not yet been started in earnest.

It has often been said that this change was because passengers on *Olympic* complained that they were too exposed to the elements in this area, but no documentation of such complaints survives. Instead, it seems that the causative factor behind this modification was the conversion of *Titanic*'s B Deck Promenade to stateroom space; previously, that enclosed Promenade would have offered strolling passengers a weatherproof alternative in bad weather, as it did on *Olympic*. None of the screens were in place by 8 March, but somehow, Harland & Wolff managed to get the job completed in time – despite other scheduling issues that arose.

This recreation shows *Titanic* illuminated by electric lights during one of the nights of her February 1912 stay in the graving dock. (Ala Kermiche)

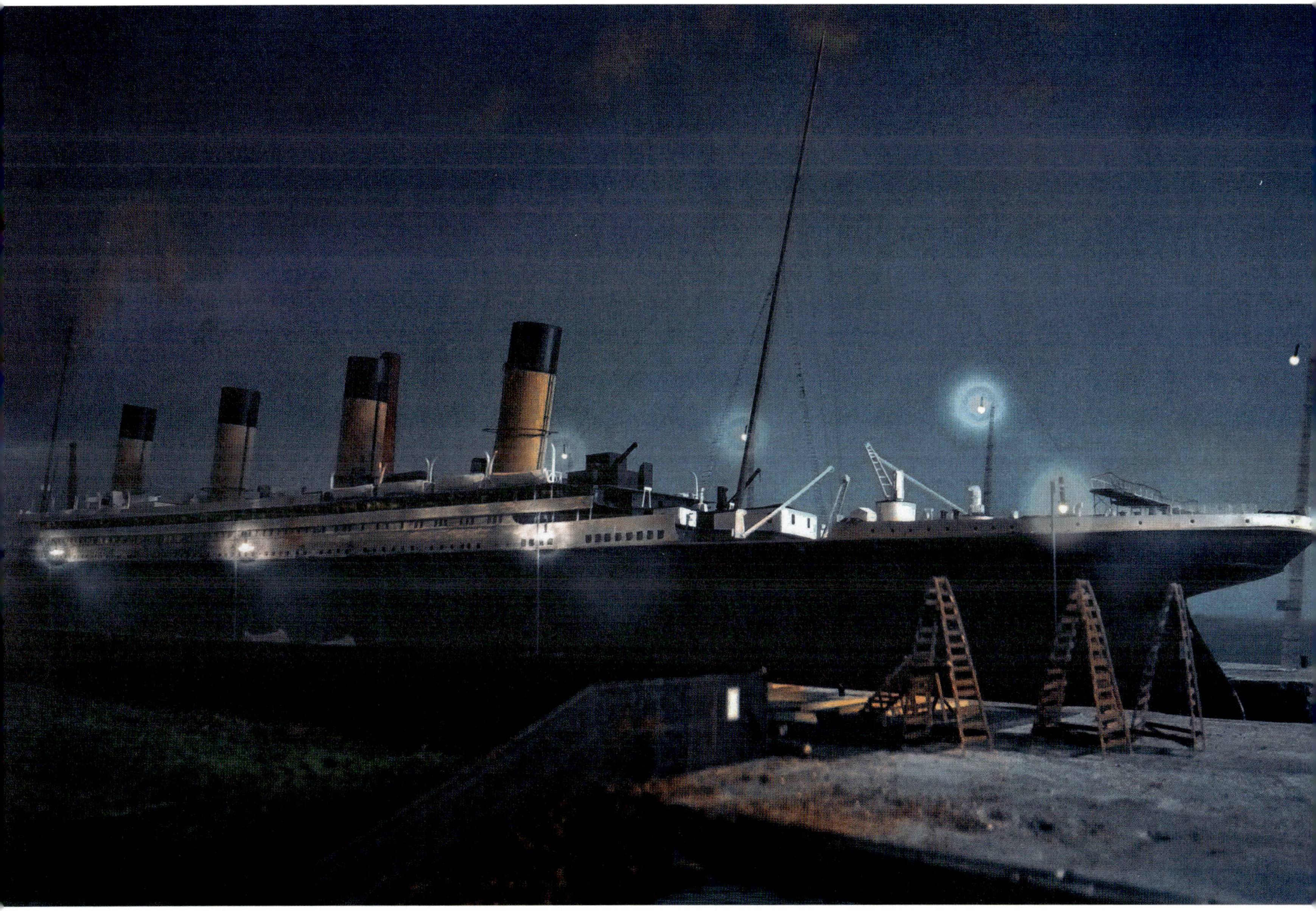

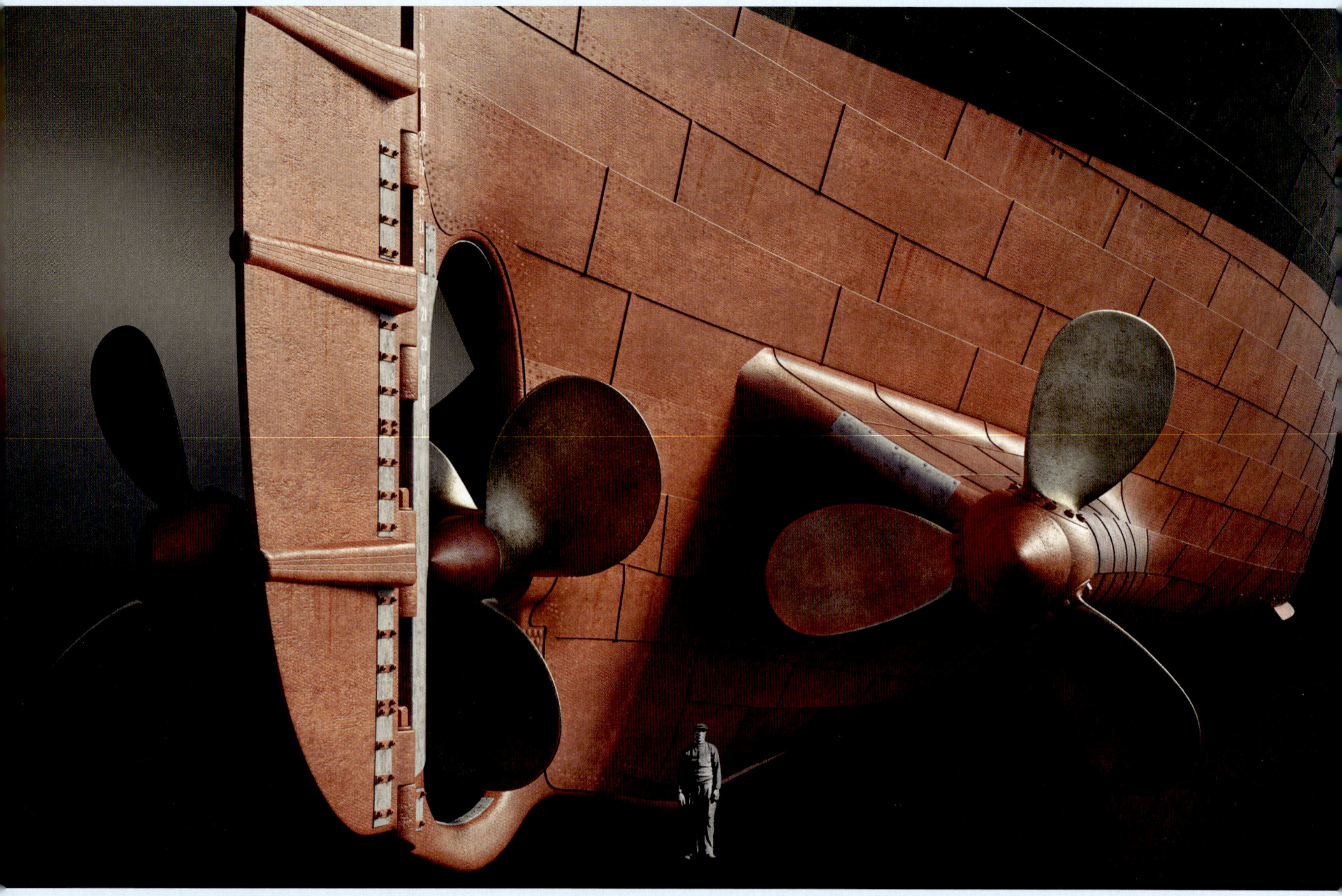

This rendering shows a Harland & Wolff worker standing beside *Titanic* with her three-bladed centre prop in place. Although no photographs exist of *Titanic*'s centre prop installed aboard ship, all existing documentation proves that she was to receive one that had three blades. (Vasilije Ristović)

## THE MYSTERY OF *TITANIC*'S CENTRE PROPELLER

It has long been assumed that each of the *Olympic*-class ships' centre propellers were made from a single four-bladed casting. This assumption was repeated in every *Titanic* book and shown on every *Titanic* model, physical or digital, including those shown in films and documentaries. Photographs of *Olympic* and *Britannic*'s propellers showed this configuration and matched the documentation available on their design. Furthermore, photos of *Olympic* and *Britannic*'s propellers are frequently miscaptioned as being of *Titanic*. Why would *Titanic* have been any different?

No one ever really seems to have asked: do we know for certain what *Titanic*'s propeller configuration was? Up to now, no photographs have surfaced of *Titanic*'s propellers installed on the ship – all photographs claimed to be such are actually of her sisters.

In 2007, maritime historian Mark Chirnside became aware of a notebook from Harland & Wolff which recorded the engineering particulars of each ship that they completed, including their propelling machinery and specifications. It is the only contemporary primary source that recorded *Titanic*'s complete propeller specifications alongside those for *Olympic*. Astoundingly, it recorded that *Titanic* was fitted with a three-bladed centre propeller, rather than the consistently assumed four-bladed design. The document showed:

YARD NO. 400, *OLYMPIC*:

- Wing propellers: 23ft 6in diameter, 33ft pitch (pitch increased to 34ft 6in in November 1911), 160ft 2in area, 3 blades.
- Centre propeller: 16ft 6in diameter, 14ft 6in pitch, 120ft area, 4 blades.

YARD NO. 401, *TITANIC*:

- Wing propellers: 23ft 6in diameter, 34ft 6in pitch (crossed out and amended to 35ft), 160ft area, 3 blades.
- Centre propeller: 17ft diameter, 14ft 6in pitch, 120ft area, 3 blades.

The recorded modifications to *Olympic*'s propellers in late 1911 should not be surprising, since there was then a lot of study going into the best designs for large, high-speed liner props. *Lusitania* and *Mauretania* went through numerous different propeller designs in their early years, so frequently that they are difficult to keep track of.

For example, buried deep in a 30 July 1908 Cunard memo, reference was made to the fact that when *Mauretania* had fractured her stern bracket, 'one of the two spare three-bladed propellers' had been 'fitted to the starboard wing', after which the ship had successfully made three voyages running on just three props. Meanwhile, *Lusitania* had also already received new propellers. This sort of experimentation within months of both Cunarders' entry into service shows that such alterations were made frequently and documentation on the matter can be hard to find.

A good comparison might be to car tyres: the vehicle and engine itself never really changes, but performance, road grip, comfort and ride noise are all greatly affected by the quality and design of the tyres. Ships' propellers of the era were changed nearly as frequently – and in some cases, even more often than vehicle tyres are today.

Indeed, we really should have been looking for such alterations earlier. *Olympic* sported a three-bladed centre prop after her 1912–13 refit. Although she later reverted to a four-bladed prop, there can be no question that she bore *Titanic*'s configuration for a time. *Britannic* was also given a four-bladed centre prop. There were further clues. Bruce Ismay had testified at the American inquiry that *Titanic* was supposed to be slightly faster than *Olympic*. *Finance, Commerce and Shipping* had also noted on 11 April 1912 that there was 'no doubt' that improvements had been made to *Titanic*'s propellers over *Olympic*'s original design.

In January 2020, another primary source document came to light. First referenced by João Gonçalves, it was a notebook written by Stephen Pigott, who worked for several months in the turbine department of John Brown & Company, who constructed the centre turbines for *Olympic* and *Titanic*. His notebook offered two proposed propeller designs for *Olympic* and *Titanic*'s centre shafts – the ones connected to the turbines he was helping to create.

One design, for a four-bladed centre propeller, was the same diameter and surface area as that *Olympic* was given in 1911; only its pitch varied slightly from the final design. The second proposed design was for a three-bladed propeller of the same diameter, surface area and number of blades – namely three – as that recorded in Harland & Wolff's own notebook for *Titanic*. Again, the only variation came in the pitch of the blades, something that we know that Harland & Wolff was experimenting with at the time. So instead of finding primary source evidence to the contrary, more documentation is coming to light proving that assumptions about *Titanic*'s centre prop having four blades were incorrect.

The discovery that *Titanic* had a three-bladed centre prop touched off a firestorm of controversy. Emotional and often very nasty arguments based on now disproven assumptions were levelled at Chirnside's unassailable research. Since *Titanic*'s centre propeller is buried deep into the seabed, it cannot be examined. No doubt, some of the more emotional on the matter will not believe it until the actual prop is recovered from the wreck and placed at their front doorstep. However, actual primary documentation on the subject cannot be clearer: *Titanic* sported a three-bladed centre propeller.

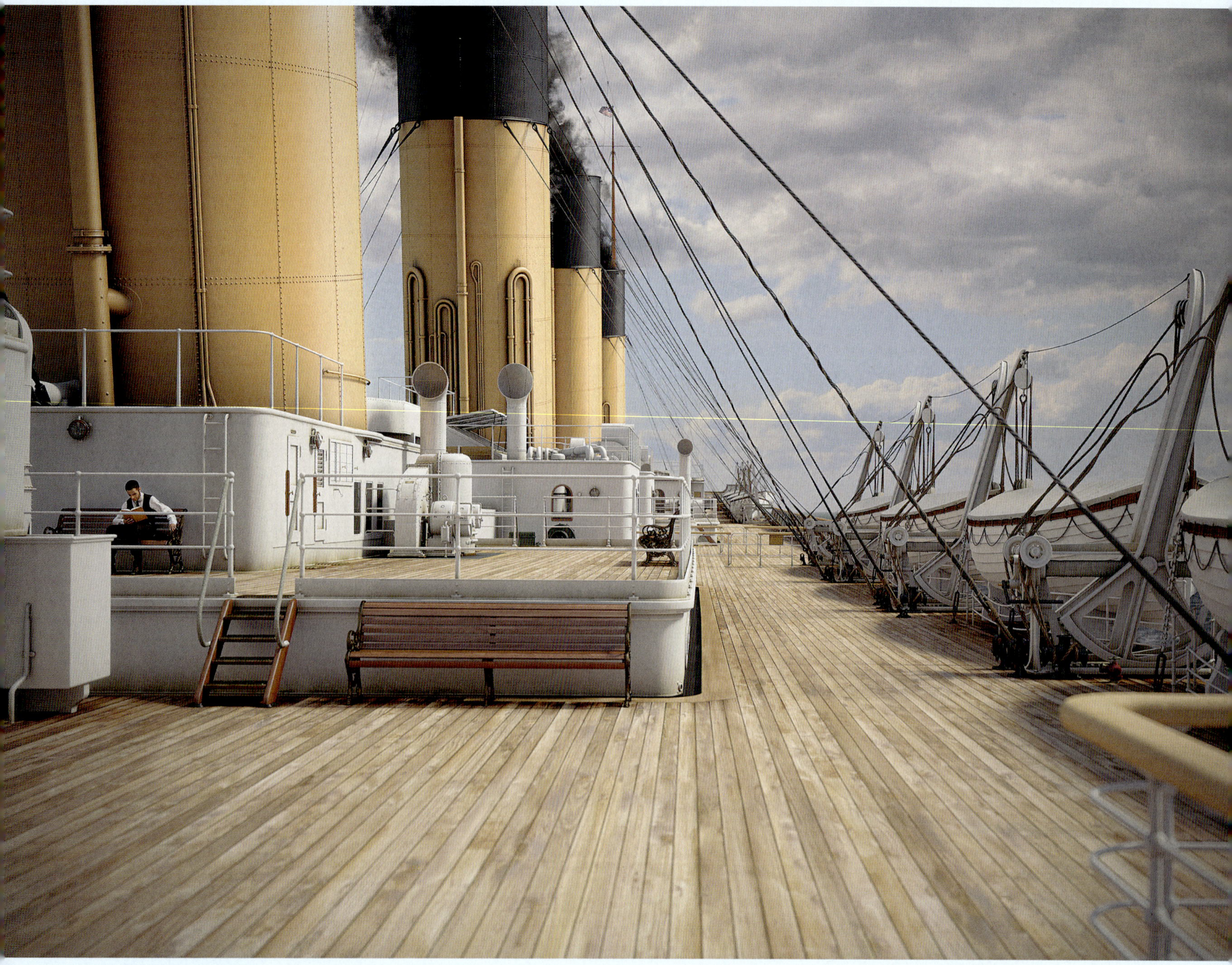

The Second Class Boat Deck, looking forward along the starboard side, showing a line of lifeboats. (Vasilije Ristović)

## TOO FEW LIFEBOATS – A SYSTEMIC ISSUE IN THE TRADE

During her stay in the Belfast Graving Dock, *Titanic*'s lifeboats were fitted. One of the most popular condemnations of *Titanic* is that she and *Olympic* were allowed to set sail with lifeboats for just half those aboard. Most would be shocked to learn, however, that not only were *Olympic* and *Titanic* not alone in this, but that it was a systemic issue in the trade at the time.

The regulations then in force had come into effect in 1897 and said that any liner of over 10,000 GRT needed to carry a specific cubic capacity of space in their lifeboats. Yet *Olympic* and *Titanic* were four and a half times that large. *Olympic* was certified to carry 3,511 passengers and crew, and *Titanic* 3,547. However, each ship had lifeboat capacity for only about 1,178 persons, some 2,369 fewer than *Titanic* could legally carry.

Other ships then in service were just as bad: *Lusitania* and *Mauretania* carried only sixteen lifeboats each. Their boats could hold only 986, but each ship could carry 2,900 passengers and crew. The German liner *George Washington* could carry nearly 3,500 persons but carried boats for fewer than half that capacity; the *Auguste Victoria* would similarly have left over 1,300 of her passengers and crew stranded after the boats were launched.

This industry-wide issue was well known long before *Titanic* sank, and there were concerns about it. During the construction of *Olympic* and *Titanic*, Harland & Wolff and White Star personnel were all expecting an increase in regulations by the time the ships entered service; Alexander Carlisle even developed the new Welin quadrant davits and commissioned them specially for installation aboard the two new superliners.

Ismay and Pirrie, meanwhile, discussed the subject in meetings and agreed to the expense of installing the new davits so that if the regulations were changed, they would easily be able to meet them. Contrary to widespread popular depictions on the matter, the historical record gives no indication that Ismay or anyone else at White Star pressured Pirrie or others at Harland & Wolff to stick to a 'bare minimum' of boats, nor does it give any indication of an internal struggle on the matter at Harland & Wolff between Pirrie and Carlisle, or others. Quite the opposite: everything was being made ready to implement the changes if and when they came.

Using the new davits, up to four boats could be accommodated per each of the sixteen lifeboat stations, giving a maximum capacity of sixty-four boats. If the revised regulations weren't that stringent, then two or even three boats per station could be carried, for a total of thirty-two or forty-eight lifeboats. Plans were drawn up showing how these boats could be placed, ready for quick implementation if the regulations were changed.

Alexander Carlisle retired from Harland & Wolff in June 1910 and went to work on the Merchant Shipping Advisory Committee, which had been set up by the Board of Trade to arrive at a recommendation on the matter. In May 1911, just days before *Olympic*'s maiden voyage, none had yet been made. Carlisle advised the committee he thought the regulations should be raised, and later claimed that he told them he did not believe *Olympic* and *Titanic* carried enough boats. But when the committee decided not to recommend an upward revision in regulations, Carlisle signed off on it. Although he later claimed it was against his better judgement and that he had been very 'soft' when he did so, it was easy to make such claims after the disaster. At the time, it was simply felt that new ships were being built so safely that, if they ever did sink, it would be so slowly that there would be plenty of time for help to arrive.

*Olympic* and *Titanic* actually carried more lifeboats than the sixteen originally planned for in the 'Design D' general plan dating to July 1908, for on 5 May 1911, a quartet of Engelhardt collapsible lifeboats had been ordered for each of the new ships. Simultaneously, the new superliners' carrying capacity was actually reduced from that original 'Design D' vision by roughly 20 per cent. This means that instead of plans for high numbers of boats being reduced during construction, modifications reducing passenger-carrying capacity and adding extra lifeboats led to a far better lifeboat capacity as a proportion of overall carrying capacity than had been initially conceived.

Statements to the effect that the ships were originally intended to carry thirty-two or sixty-four boats, and that the numbers were cut back to save deck space or costs, are grossly oversimplified. The capacity estimates for *Olympic* and *Titanic*'s lifeboats are recorded below, and by any measurement they voluntarily exceeded the standing regulations:

| BOAT TYPE | QTY | CAPACITY (BOT) | CAPACITY (H&W) |
|---|---|---|---|
| Standard open boats | 14 | 65 each (910) | 64 each (896) |
| Emergency cutters | 2 | 40 each (80) | 33 each (66) |
| Engelhardt collapsibles | 4 | 47 each (188) | 40 each (160) |
| **TOTAL CAPACITY** | **20** | **1,178** | **1,122** |

Lastly is this vital detail: extra lifeboats would have done little or no good on the night *Titanic* sank. There was neither the manpower nor the time to launch more boats; in fact, the last two boats had to be floated from the deck and proved dubious havens for those who reached them. Many of those who initially boarded the crafts died in the course of the frigid night.

Those at White Star and Harland & Wolff were not purposely endangering lives to save a few pounds; they were merely caught up in a way of thinking that was bound to end in disaster sooner or later.

On 24 February 1912, *Olympic* threw a blade from her port propeller while returning from New York, necessitating yet another trip back to Belfast. She arrived on 1 March and was placed directly in the dry dock the next morning. The prop was replaced quickly, and on the morning of 4 March she was removed from the dock and attempted to leave. However, due to a strong storm that was then blowing, the tugs had difficulty manoeuvring her, and she was brought back into the graving dock and moored there overnight.

The next morning, the weather had not improved, so on the following day, 6 March, on a single high tide, *Olympic* and *Titanic* swapped positions. Moving *Titanic* gave *Olympic* more room to turn around to depart. *Titanic* was placed in the graving dock and *Olympic* tied up at the fitting-out wharf. The following day, *Olympic* successfully left and, on 8 March, *Titanic* was returned to the fitting-out wharf.

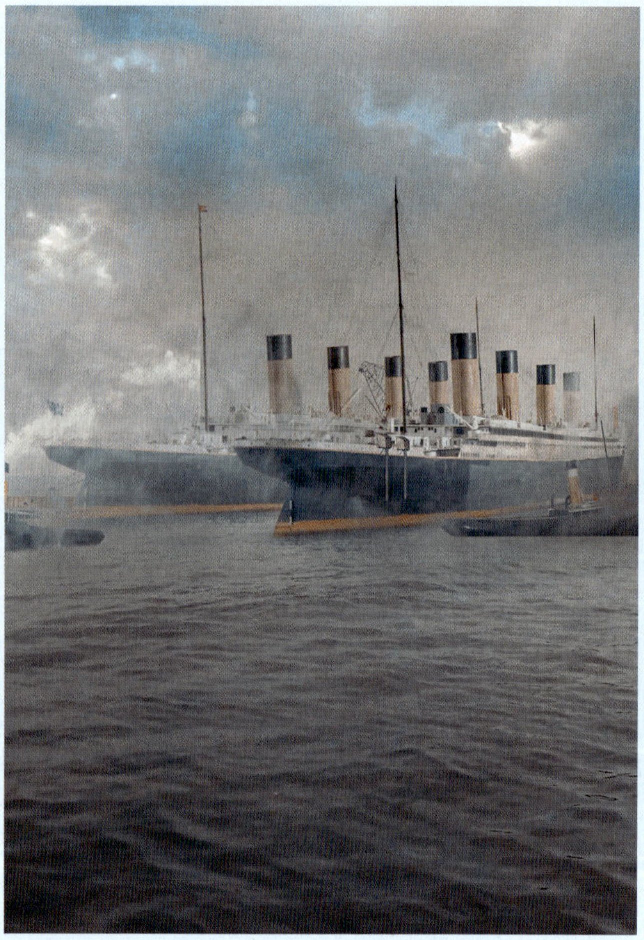

This art shows a bold move in progress: on 6 March 1912, *Titanic* was placed in the graving dock and *Olympic* was moved to the fitting-out wharf, all on a single tide and in tight quarters. *Olympic* would leave the following day: the last time that the sisters were together. (Ala Kermiche)

## THE INFAMOUS *TITANIC / OLYMPIC* SWITCH

One of the most popular conspiracy theories about *Titanic* is that she was switched with *Olympic* as part of an insurance scheme by White Star. As the theory goes, when HMS *Hawke* rammed *Olympic* in September 1911, either the repair costs were prohibitive or the damage was too extensive to be fixed, to the point where *Olympic* could not remain profitable. By switching the two ships, it is said, they could send the almost completed *Titanic* out to sea, now substituted as *Olympic*. They could then take a disguised *Olympic*, now called *Titanic*, and deliberately sink her as part of an insurance scam, reaping enough money to build a replacement liner.

In the end, the scenario is absurd. First, just to get *Olympic* out to the sinking location, extensive changes and repairs would need to be made to the severely damaged ship. Why throw good money after bad? Worse yet for this theory, while it had cost over £1.5 million to build *Titanic*, when she sank, she was insured for only £1 million. The remaining difference was carried by the company's insurance fund. The insurance fraud concept thus sinks, because it would have been a money-losing idea, not a gainful one.

Furthermore, there were only two opportunities for a switch to be made: first between 6 October and 20 November 1911, and second between 2 and 7 March 1912. *Titanic* was not ready in time to take *Olympic*'s place during the first window and the second window was too narrow to carry out the extensive modifications that would have been required.

Every major piece of wooden panelling was marked with the ship's hull number and, when *Olympic* was scrapped, her fittings were sold, and even today bear the correct number for *Olympic* – 400. It is impossible that every piece of panelling was swapped between the two ships.

The extensive modifications to *Titanic*'s B Deck took six weeks to carry out in the summer of 1911. There is no way that these could be made on *Olympic* in just a few days. These new staterooms were no 'mock-ups' – they were occupied by wealthy passengers and they have been thoroughly explored by ROVs since the wreck's discovery, documenting the lavish suites, which *Olympic* never had. Even the simple enclosure of *Titanic*'s A Deck Promenade took some weeks to carry out – longer than either opportunity to make the switch allowed.

Furthermore, not only were there some 15,000 people employed at the yards, but the public could actually see the two ships when they visited. Yet not one reliable account has come to light detailing the switch. Not a scrap of official correspondence on any of this work has been found from White Star or Harland & Wolff. Many of the items recovered from the wreck of *Titanic* also bear her correct hull number – 401 – and *Titanic*'s name can be seen, etched into her bow plates.

In spite of the theory's popularity, it was literally impossible to have switched the ships. The wreck on the bottom of the ocean is definitely *Titanic*.

In the end, it was a close thing. In fact, certain aspects of *Titanic*'s accommodations were not yet finished by the time of her formal trials, which were scheduled to begin at 10.00 a.m. on Monday, 1 April 1912. Only one thing could put a dampener on the enthusiasm for events that day: poor weather.

In the end, it did. The smoke pouring from *Titanic*'s funnels was flattened by strong winds and it was simply too dangerous to try to turn *Titanic* around in confined waters in that sort of weather. Harland & Wolff had taken no risks with *Olympic* the previous month, and they took no risks with *Titanic*: the weather would surely be better the next day, and it was.

At around 6.00 a.m. on 2 April, *Titanic* was being moved. By 9.30 a.m., she was heading up and out of the Victoria Channel. Once she reached open water, the tugs cast off and the trials commenced; they took the remainder of the morning and afternoon. These were not by any means as extensive as the tests carried out with *Olympic* the previous year, nor did they need to be.

*Titanic*'s performance was expected to closely match the results of the more extensive tests made with *Olympic*. However, the ship had to meet certain parameters in order to be accepted by the White Star Line, and so that Board of Trade Surveyor Francis Carruthers could sign off on her passenger certificate. A variety of manoeuvring tests were made at various speeds. Two full turning circles were carried out at high speed, one with both engines at 'Full Ahead', and another with one of the main engines reversed to aid the turn.

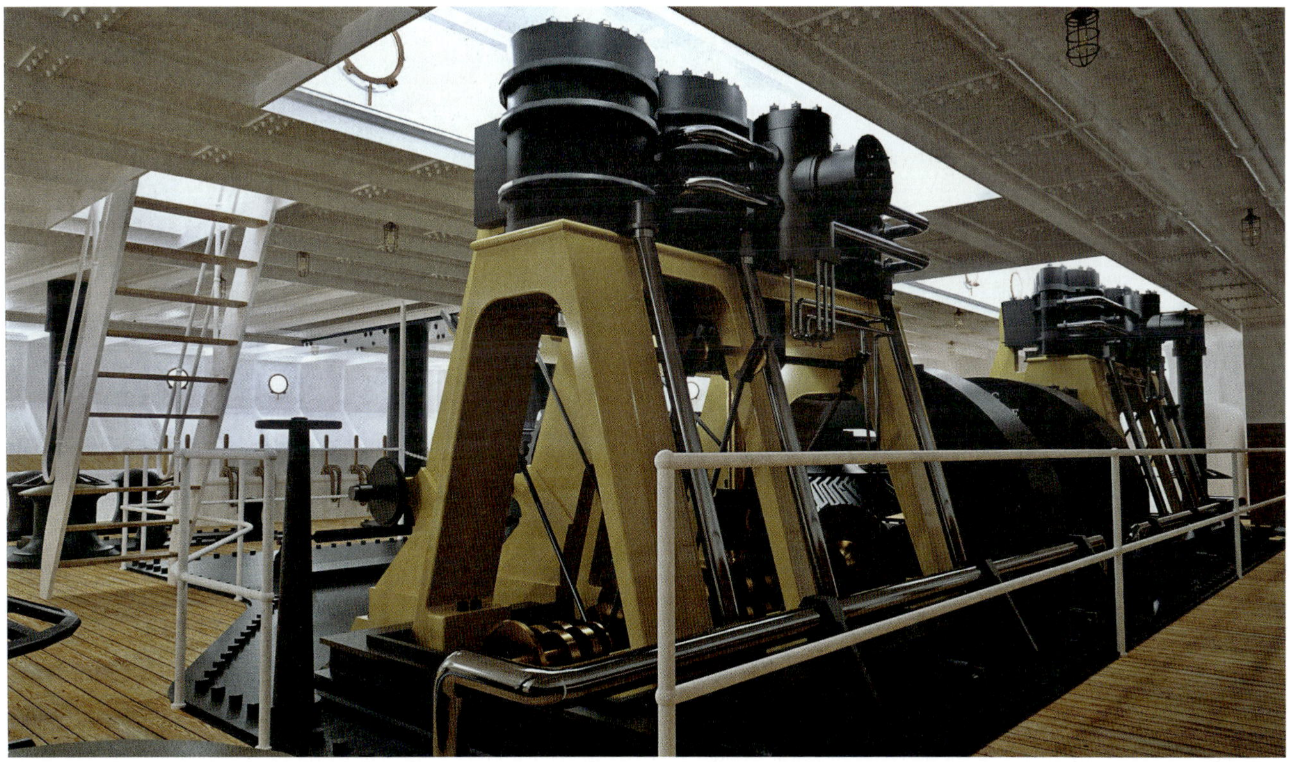

As *Titanic*'s manoeuvrability was tested on 2 April, these steering engines housed aft under the Poop Deck at the very stern took input from the helm and moved the ship's massive rudder. The steps on the left lead up beneath the Aft Docking Bridge. (Chris Walker)

A full-stop test was also conducted. From 18 knots with the engines engaged ahead, she was brought to a dead stop with the engines going full astern in just over 3,000ft, or just three and a half times her own length.

## *TITANIC*'S RUDDER: NOT TOO SMALL

It has been repeatedly claimed that *Titanic*'s rudder was too small and she was handicapped in her turning ability when the iceberg was spotted. However, a mathematical analysis of the size of *Titanic*'s rudder turning surface, when compared with the overall size of her underwater surfaces, shows that her rudder was only marginally smaller than would be acceptable on even the most modern ships, and this is so minimal that it would have made little difference in her ability to turn.

We know the turning characteristics of *Olympic* and *Titanic* very well. The two sisters had an identical turning circle of between 3,800 and 3,900ft, or about four and one-third times their overall length. By comparison, battleships of the era sported turning circles of three to four times their length; older cruisers of the period had turning circles between five and a half and six and a half times their length, while even newer cruisers of the time sported circles of between four and a half and five times their length. The turning circle of *Lusitania*, by comparison, was a smaller 2,850ft, roughly three and two-thirds her overall length. However, these extraordinary turning abilities were only accomplished by 'cutting away' a large area of their underwater surface ahead of the rudder, and did not typify passenger liners of the time.

Importantly, the rudders of both *Olympic* and *Britannic* were never altered from their original design. This is strong evidence that the manoeuvrability of the *Olympic*-class ships was never in any real doubt throughout their careers. Indeed, Board of Trade Surveyor Francis Carruthers, who was aboard for *Titanic*'s trials, later recalled that she was put through 'a number of circling trials', and that the ship 'made a very small circle', a fact he 'observed' and which was 'noticed' by others.

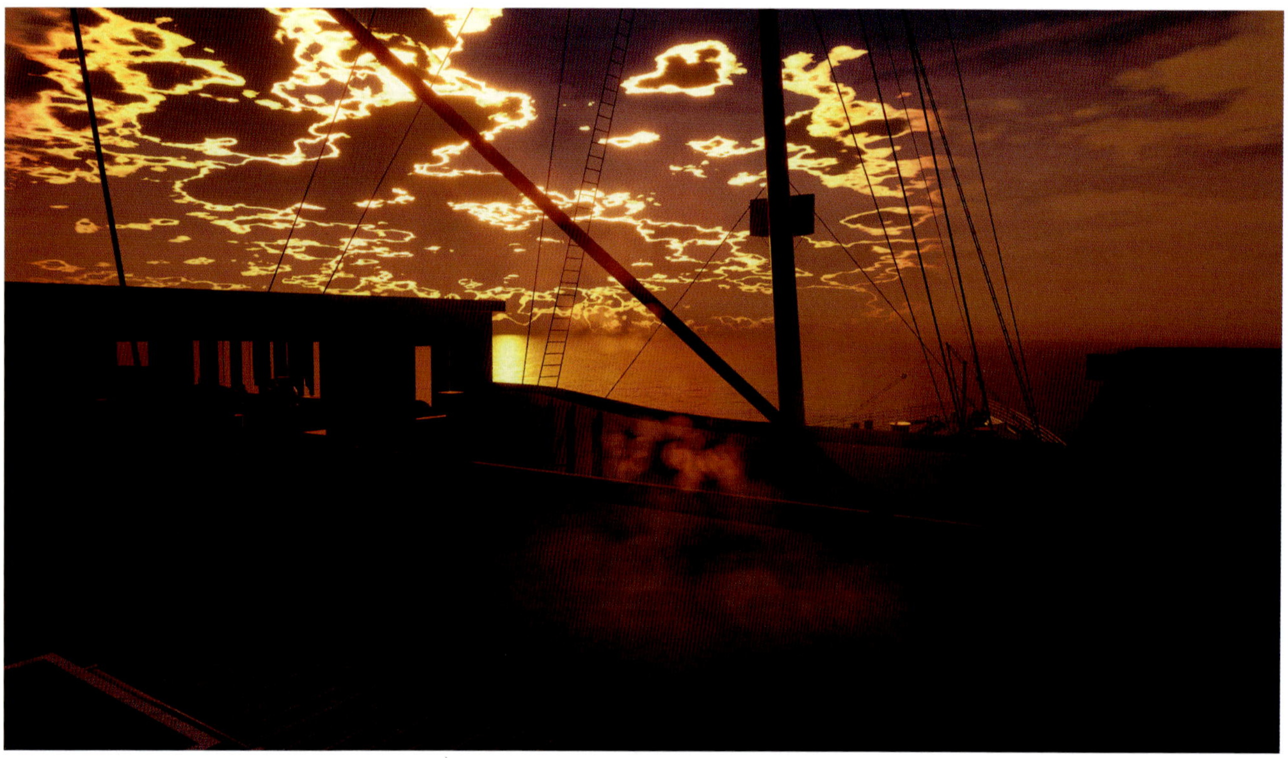

Sunrise, 3 April 1912. From two to six o'clock that morning the ship encountered fog, but when it broke, the day became 'fine and bright', with a fairly smooth sea. (HFX Studios)

*Titanic* passed her trials with flying colours. She stopped back in Belfast at 7.00 p.m., her passenger certificate was signed, those who were not going to stay with the ship down to Southampton disembarked, some last-minute equipment and supplies were brought aboard, and by about 8.00 p.m. *Titanic* was once again departing Belfast, this time for good.

Many Harland & Wolff workmen and representatives stayed aboard. Foremost among these was the nine-man 'Builder's Delegation' or 'Guarantee Group'. It was headed by Thomas Andrews and consisted of a number of hand-selected men who stood out as being among the shipyard's most promising. There was Roderick Chisholm, a draughtsman; two electricians, Ennis Watson and William Parr; a carpenter, or joiner, William Campbell; three fitters, Alfred Cunningham, Anthony Frost and Robert Knight; and plumber Francis Parkes. A tenth man, painter Joseph Thompson, would stay with the ship down to Southampton, but would not continue the journey to New York. They ranged in age from 18 to 43, and this team of men would stay with the ship for the transatlantic journey to ensure that everything went smoothly, and that *Titanic* represented the best that Harland & Wolff could offer.

Overnight and through 3 April, *Titanic* steamed south in the Irish Sea, bound for her new home port on a trip that encompassed roughly 570 miles. That evening she approached Southampton, and Trinity House Pilot George Bowyer came aboard to guide her through the darkness. He oversaw as she was swung around and safely backed stern-first into Berth 44, a manoeuvre that was intended to make it easier for the ship to depart on her maiden voyage the following Wednesday.

She was tied up to the berth at 1.15 a.m. on Thursday, 4 April. Ahead lay a six-day stay that would see an almost round-the-clock effort to finish the ship in time for her first trip. Traditionally, new ships were opened to public inspection before their maiden voyage, with the proceeds going to local charities, but so much needed to be done to *Titanic* that this was foregone. Instead, the ship was 'dressed out' in an array of flags as a salute to the city.

It was just as well that the public at large was not brought on board. Carpet was still being laid, paint was still being applied, crates of china and many pieces of furniture were still being brought aboard.

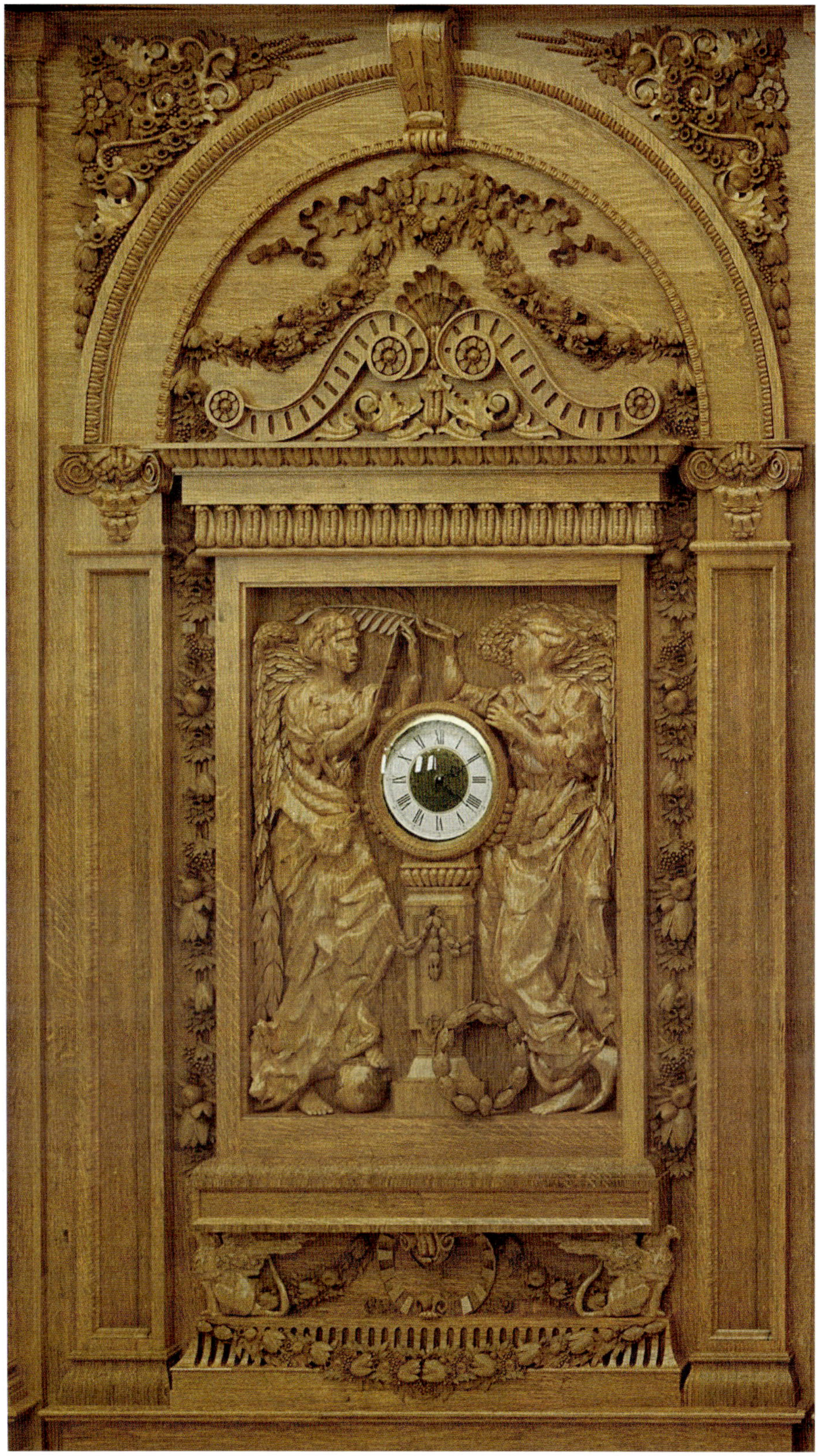

Charles Wilson, who carved the panel representing 'Honor & Glory Crowning Time', recalled that when the ship left Belfast the clock itself had not yet arrived and a mirror had to be put in place temporarily. No further information has emerged on whether the clock was installed before sailing. (Chris Walker)

Thomas Andrews oversaw all of this work, which was in addition to the usual frenzy of activity for any large ship in port preparing for her next crossing. An unusually large order of flowers was placed with local florists and brought aboard the night before sailing, apparently to attempt to mask the smell of fresh paint that lingered throughout the ship. Miraculously, the workers pulled it off. Even as ticket-holding passengers began to converge on Southampton that night and early in the morning of Wednesday, 10 April, the finishing touches were being applied.

However, although some details remained incomplete, she was able to sail as scheduled. What might have seemed impossible some weeks before had been accomplished.

## SHORT OF COAL?

It has been claimed repeatedly that *Titanic* was short of coal on her maiden voyage and this in some way contributed to the tragedy. This claim is wholly false and is often repeated by those with an accusatory or conspiratorial agenda. Excluding her reserve coal bunker, *Titanic* was able to carry some 6,611 tons of coal. She sailed with 5,892 tons of coal, or 89 per cent of her standard fuel-carrying capacity. Considering the coal strike that was then affecting shipping throughout England, this was a remarkable feat.

Using *Olympic*'s coal consumption figures as a model, it has been estimated that *Titanic* would easily have had 1,070 tons of coal remaining in her bunkers when she arrived in New York after a very high-speed trip. This was enough to drive the ship at a more reasonable 21 knots for nearly two days. When Bruce Ismay testified that she had 'sufficient coal to enable her to reach New York, with about two days' of spare consumption', he was not exaggerating.

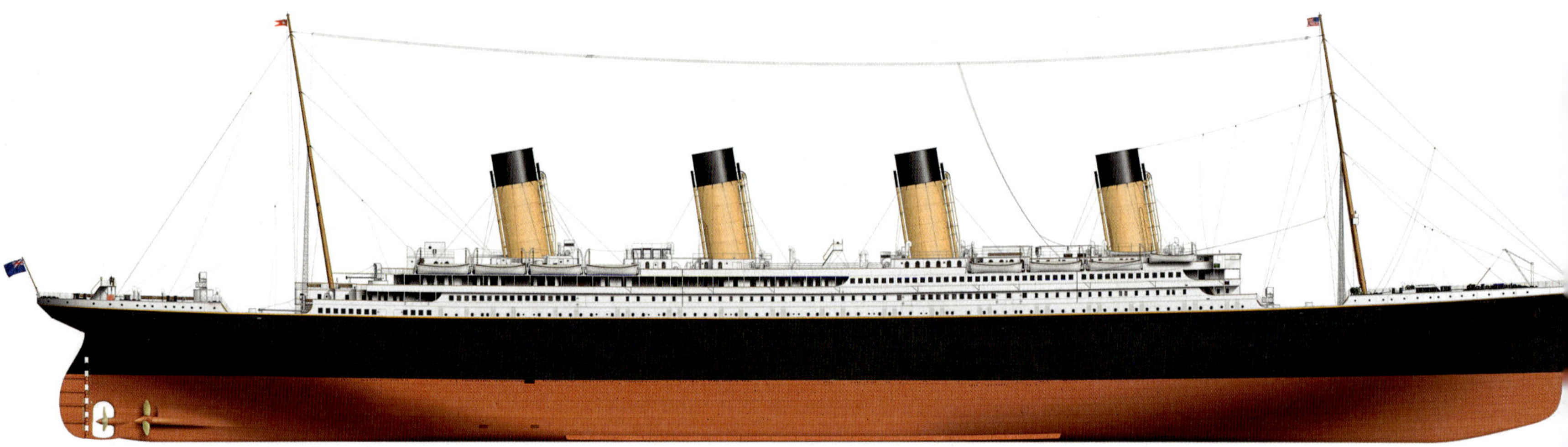

A splendid starboard profile view of the finished *Titanic*. (Michael Brady)

TWO

# *TITANIC'S* MAIDEN VOYAGE

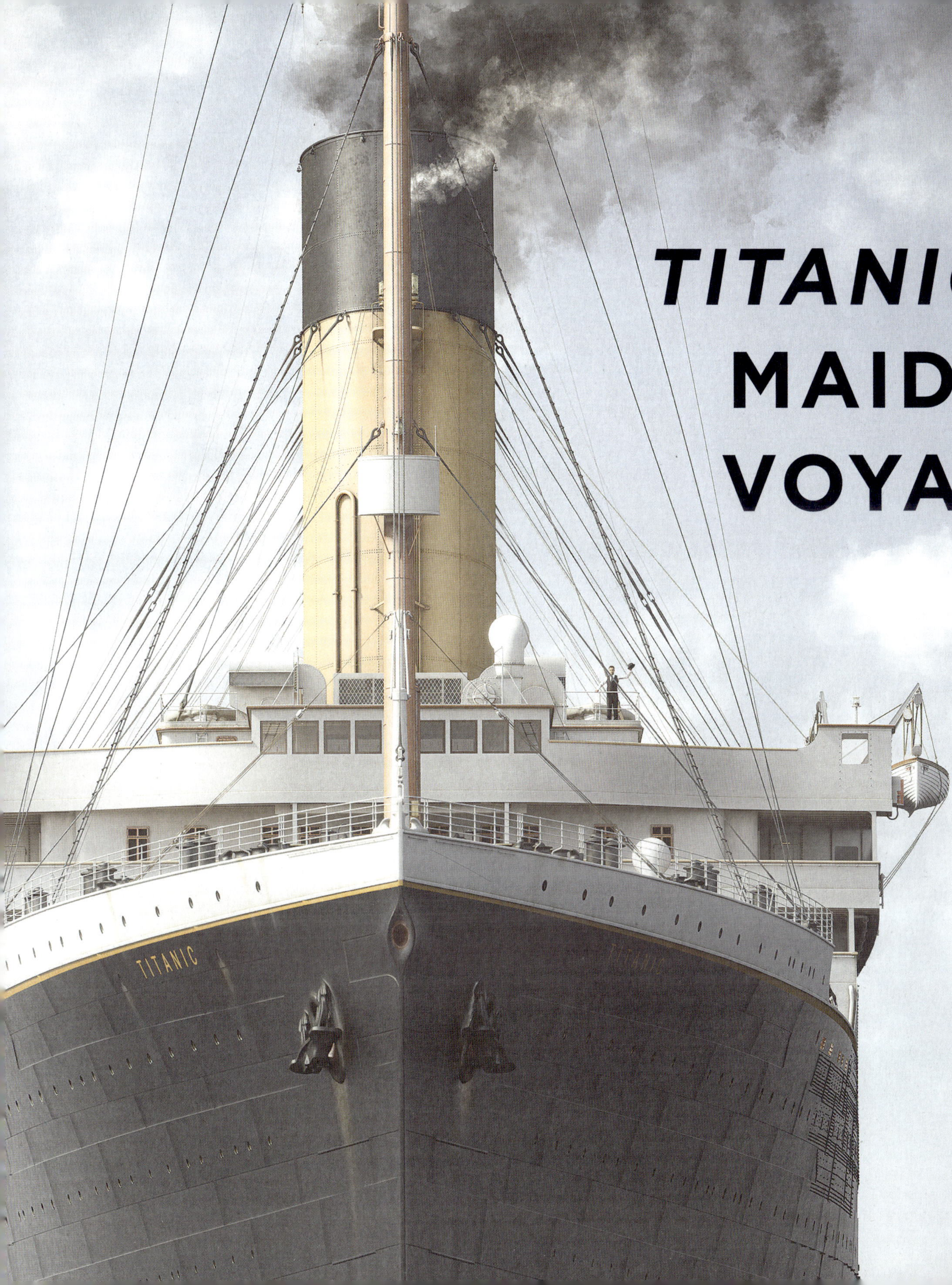

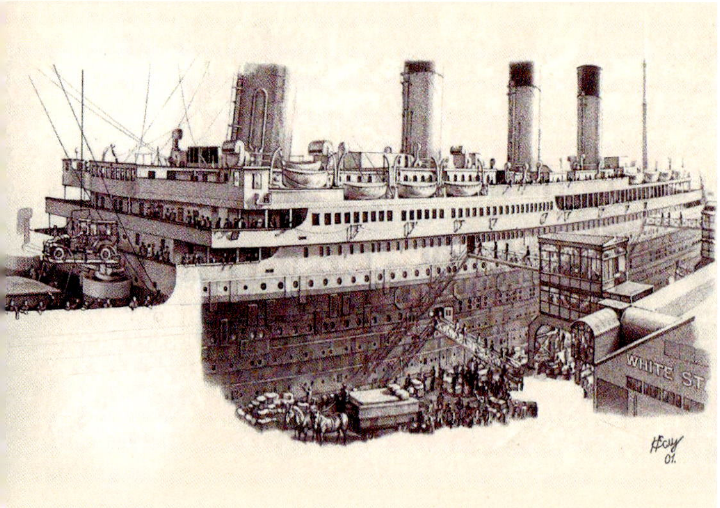

The maiden voyage of *Titanic* was scheduled to begin at noon, and that Wednesday morning was a busy one in Southampton. Dawn came at 5.23 a.m., and by 6.00 a.m. Thomas Andrews had come aboard, as had many of *Titanic*'s crew. Captain Smith arrived at about 7.30, and there was a formal muster at about 8.00. An hour later, there was a boat drill. Boat Nos. 11 and 13 were uncovered and swung out, and eight crewmen wearing lifebelts and an officer boarded each craft. They were then lowered to the water and rowed around the dock without issue. Bruce Ismay boarded *Titanic* at about 9.30 a.m., as it was his custom to take the maiden voyage of each new ship.

For passengers who had not stayed overnight in Southampton, like Ismay and Andrews, many would come by means of the two special boat trains that would arrive right on the dock, direct from London's Waterloo Station. One was for First Class passengers, and the other for Second and Third Class. Each of these trains suffered delays, and eventually arrived late.

Visitors and a number of reporters paid a visit to the great ship that morning, adding to the chaos that began to build as they traipsed over the decks alongside passengers and busy crew members, recording their impressions of the new ship.

*Above:* A sketch of passengers boarding *Titanic* at Southampton on the day of sailing. (Eugene Nesmeyanov)

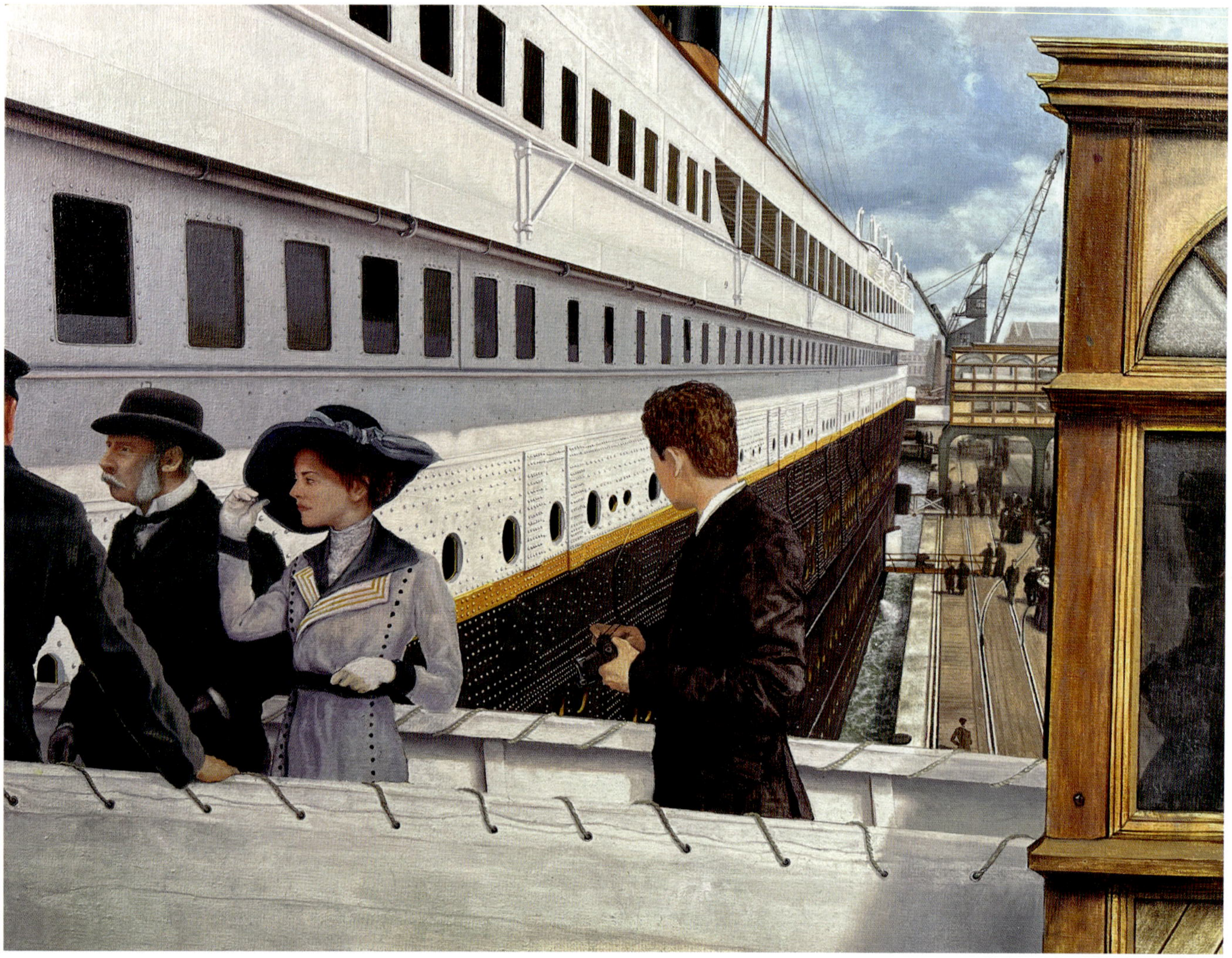

*Right:* Francis Browne pauses on the gangplank to snap a famous photo looking along the towering side of *Titanic*. (David Olivera)

## WELCOME ABOARD *TITANIC*!

As passengers were welcomed aboard *Titanic* by smiling crew members, they were ushered into a splendid world that offered great comforts and luxuries – even in Third Class, whose patrons were often unacquainted with indoor plumbing or water closets. Let's step aboard *Titanic* and experience her wonders, class by class, as her passengers did, and as seen through stunning interior recreations by artist Chris Walker.

## BOARDING *TITANIC* AT SOUTHAMPTON

Southampton's primary deep-water dock, Berth 44, was known as White Star Dock in 1912. It had undergone significant upgrades to prepare for the *Olympic*-class liners, including the construction of enormous pier-side sheds. Boat trains carrying passengers would arrive directly on the pier, and passengers could disembark the trains and enter the sheds to prepare for boarding, all protected from the elements. The sheds had comfortable waiting rooms and lavatories, and a veranda some 18ft above the ground, from which passengers could view the ship they were about to board.

Owing to the unprecedented height of *Olympic* and *Titanic*, the passenger gantries for boarding the liners were double-deckers, so that they could be used no matter the state of the tide. The steel-framed, timber-covered gantries were 40ft high and 20ft wide and were moved into position by hand. On 10 April, one was connected to the aft of the two First Class gangway doors on the port side of B Deck, and the other gave entry to the Second Class Promenade aft on C Deck.

## FIRST CLASS: A TOUR

First Class passengers boarding *Titanic* that Wednesday morning climbed to the upper level of their gantry and crossed over to the ship via a gangplank with canvas-covered rails. As passengers crossed that divide, they left land and entered the B Deck vestibule, stepping onto its yellow pine decking. Chief Steward Andrew Latimer would have been on duty there, along with some of his staff members, to help welcome passengers and direct them either to their cabins or to the Purser's Enquiry Office, one deck below. Before them were two sets of double doors that led into the First Class Entrance, and here embarking passengers caught their first glimpse of the forward Grand Staircase.

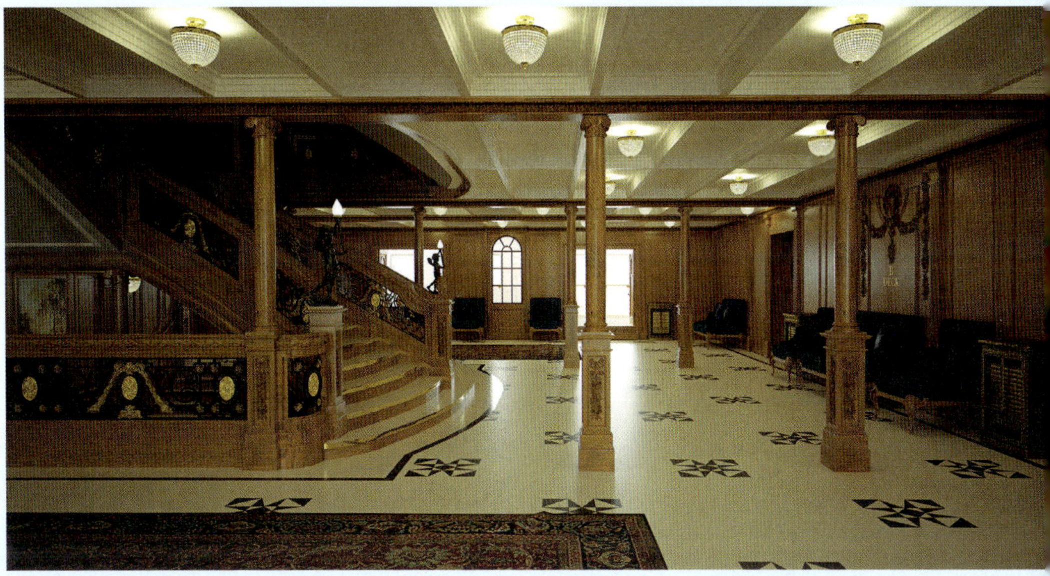

The B Deck Entrance, forward, as *Titanic's* First Class passengers would have seen it upon boarding the ship at Southampton. This view was their introduction to *Titanic*. (Chris Walker)

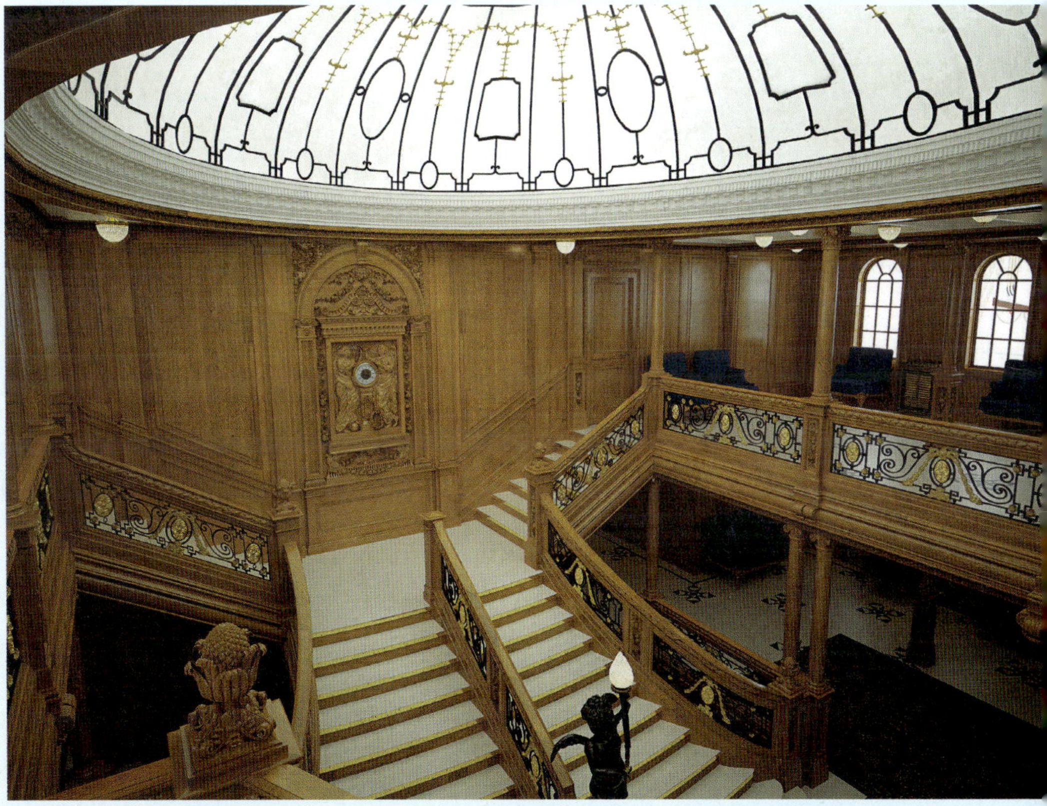

Looking down on the forward Grand Staircase from the Boat Deck to A Deck. The dome was oval in shape, 19ft by 25ft, and contrary to popular legend, was not backlit at night. The ship's band played here from 11.00 a.m. to 12.00 noon each day. (Chris Walker)

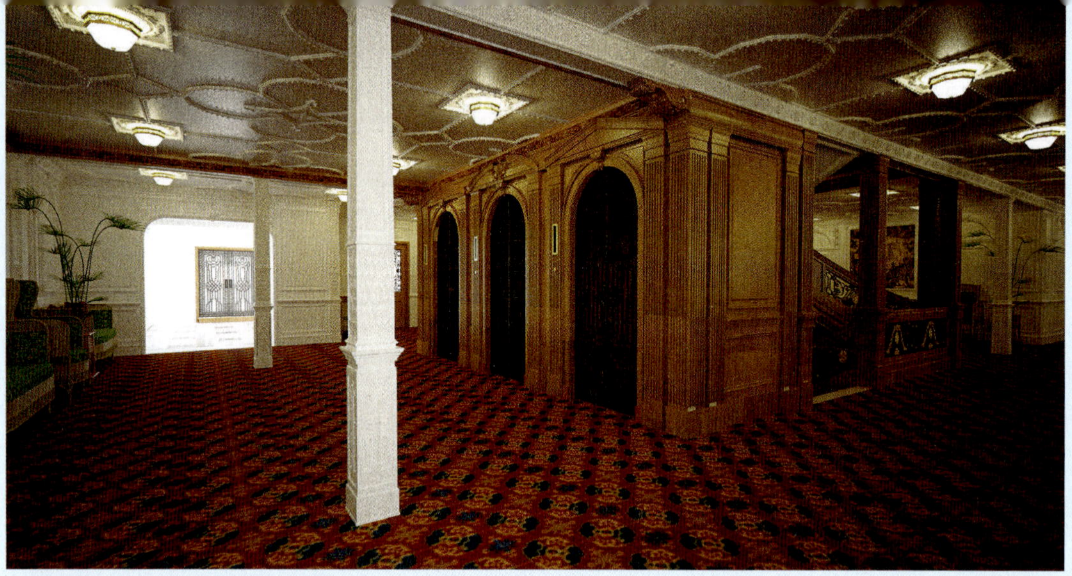

For passengers weary of vertical travel via the stairs, these three lifts ran from A to E Decks. Located just forward of the Grand Staircase, this view shows the lifts on D Deck. (Chris Walker)

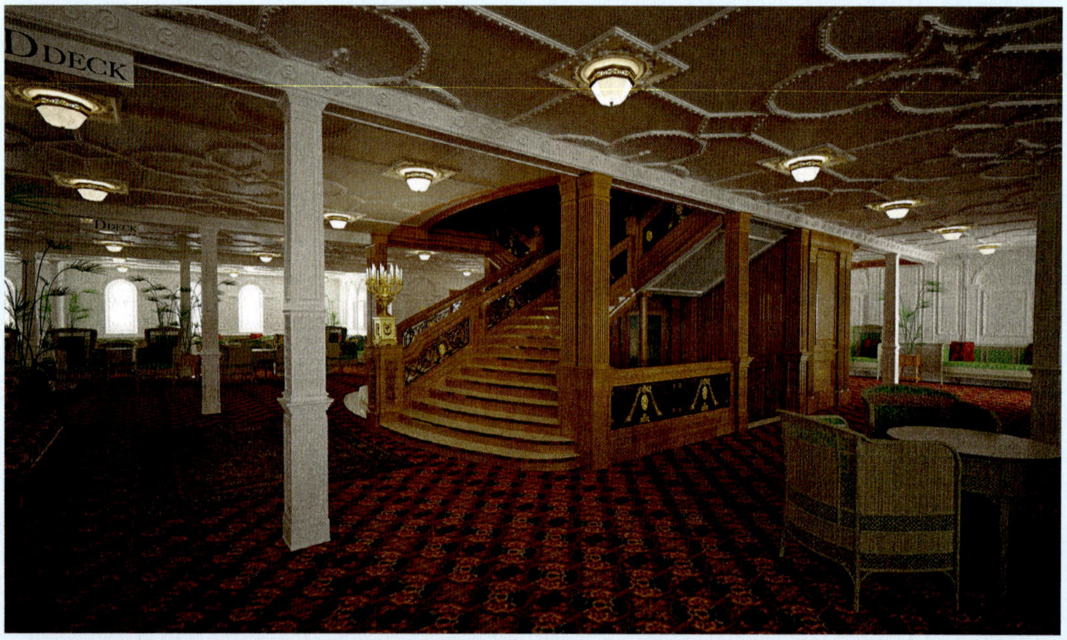


The D Deck Reception Room, 90ft wide and with leaded-glass windows, which were backlit during the daytime by exterior light and at night by electric lighting. The room opened at 8.00 a.m. and remained open until 11.00 p.m., and passengers could obtain tea and light refreshments here throughout the day. (Chris Walker)

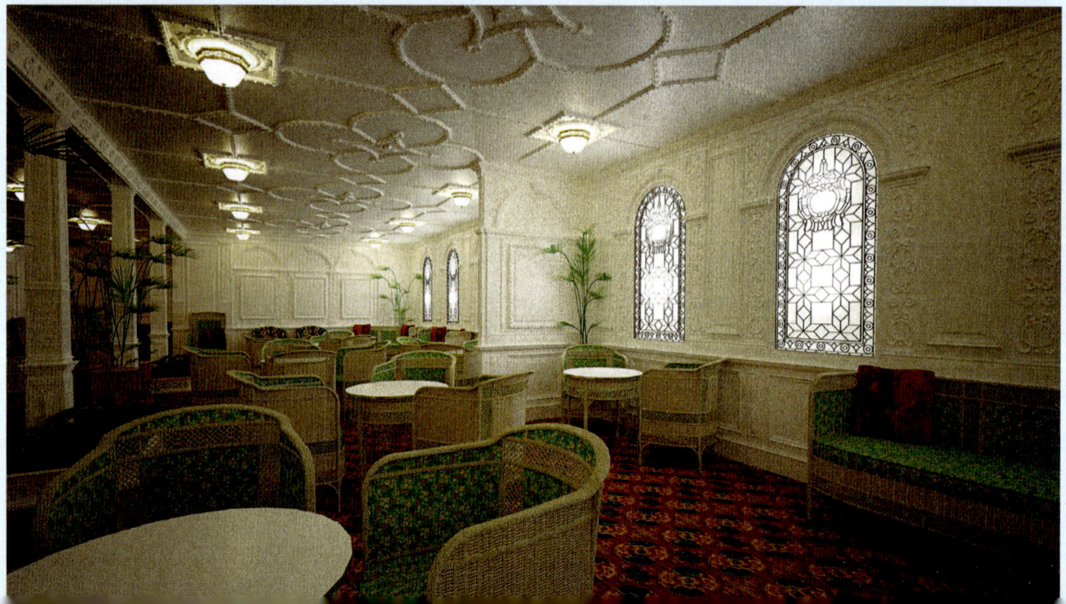

The forward starboard corner of the Reception Room. The ship's band played here daily from 4.00 to 5.00 p.m., and again after dinner from 8.00 to 9.15 p.m, utilising a Steinway Model B grand piano located on the port side. (Chris Walker)

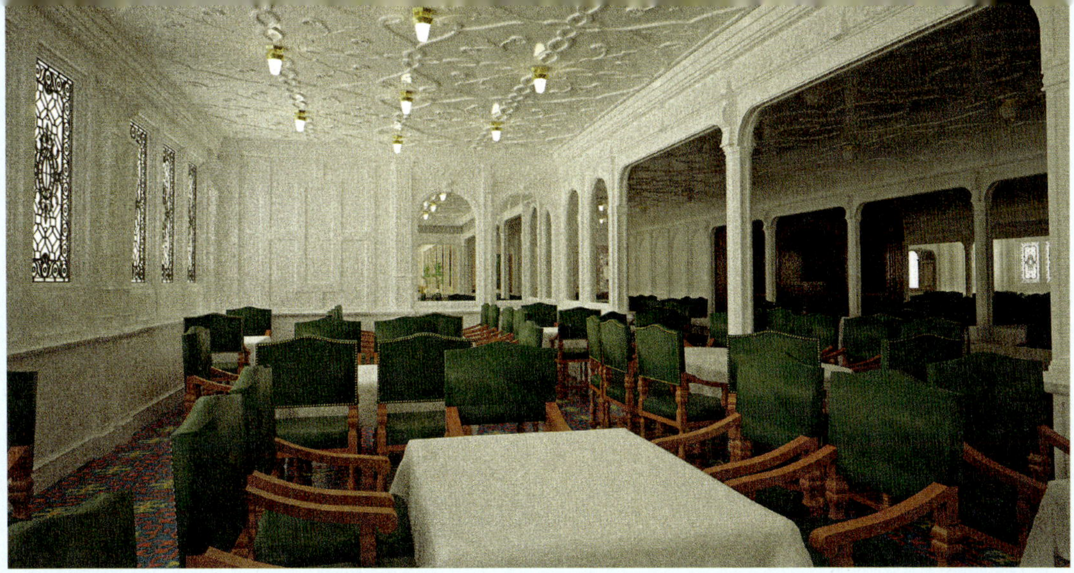

Aft of the Reception Room was the First Class Dining Saloon. Both rooms were rendered in Jacobean style, featuring white enamel. The chairs at each table were not bolted to the deck, as in earlier liners, and separate alcoves created a more intimate dining experience than other liners offered. (Chris Walker)

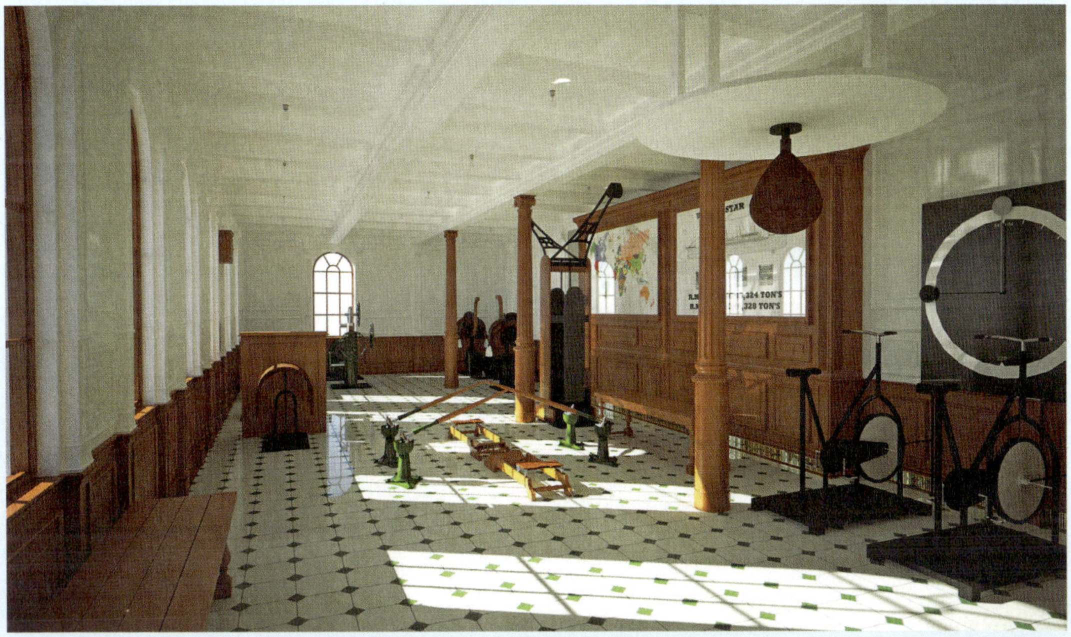

The First Class Gymnasium sat on the starboard side of the Boat Deck, just behind the Entrance. It was 45ft 6in long and 14ft 3in wide. A variety of exercise equipment was available to occupy passengers, while a total of seven windows gave light and, in good weather, fresh air. (Chris Walker)

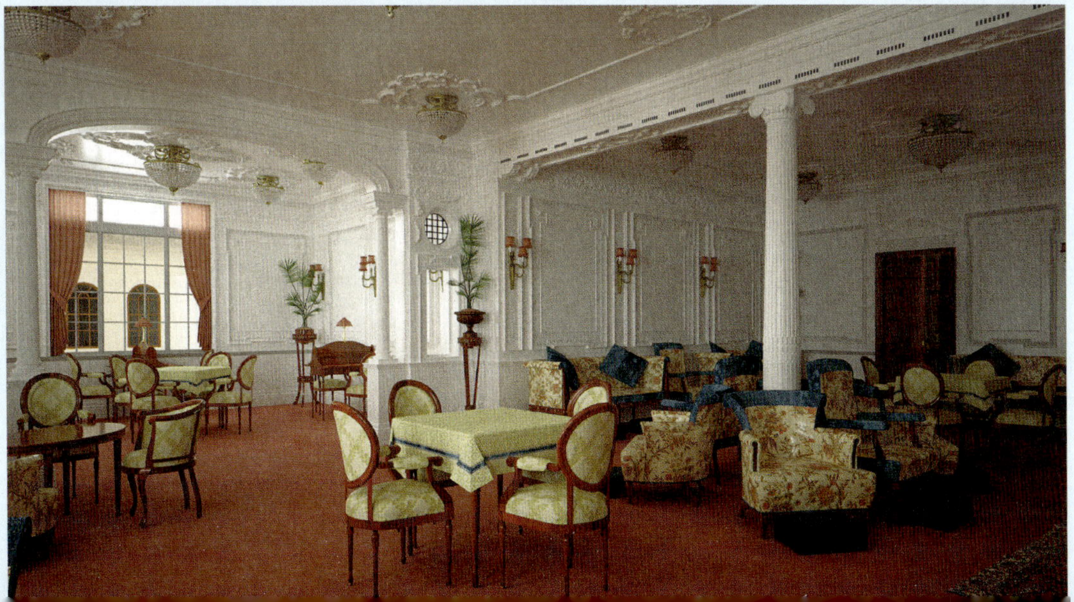

The Reading & Writing Room sat on the port side of A Deck. It was intended primarily for ladies' use. The ceilings here, in the Lounge and in the Smoking Room were raised to 12ft 3in by extending them above the Boat Deck, which brought in more natural light. (Chris Walker)

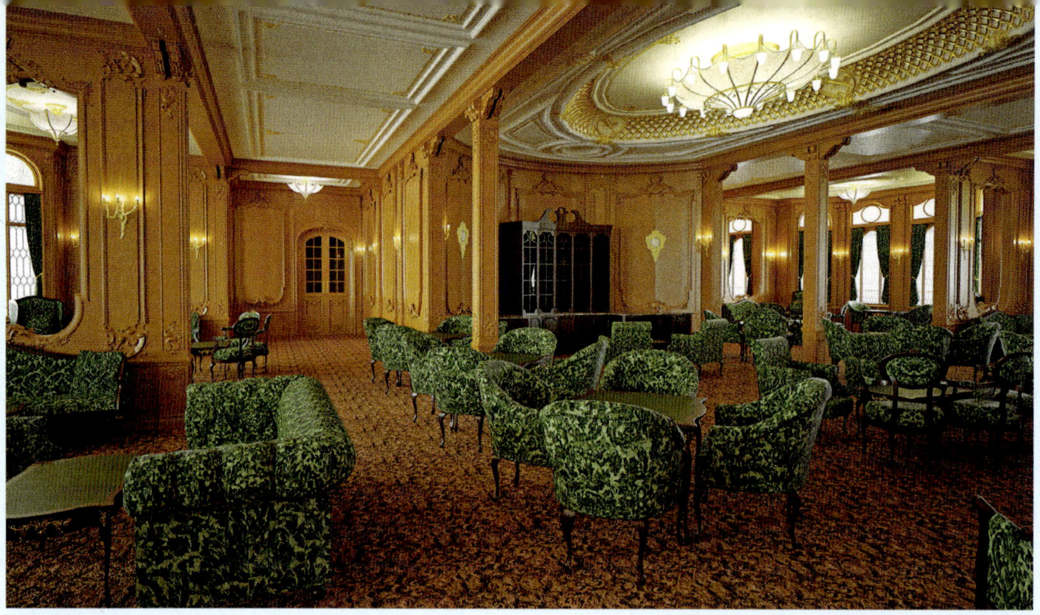

The First Class Lounge was located amidships on A Deck. It was in the Louis XV style, and was intended for the use of both sexes. It was open from 8.00 a.m. to 11.30 p.m., but there was no piano here and the band did not ordinarily play in this room. (Chris Walker)

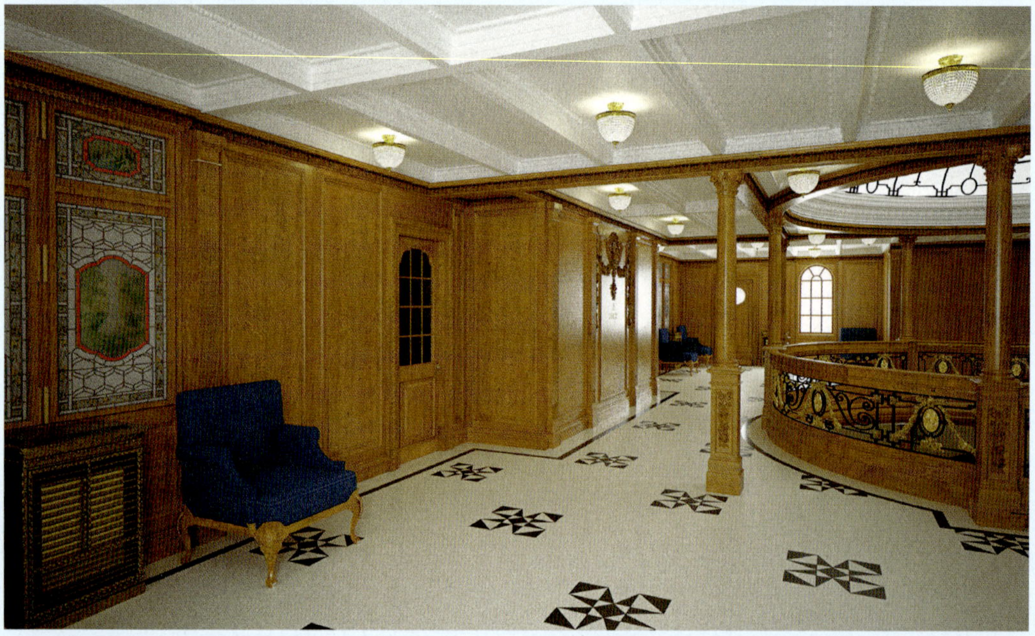

Moving aft from the Lounge was the aft First Class Entrance, with its own Grand Staircase. Two staterooms had been added here, a further change from *Olympic*'s original design. The doors on the aft wall led into the First Class Smoking Room. (Chris Walker)

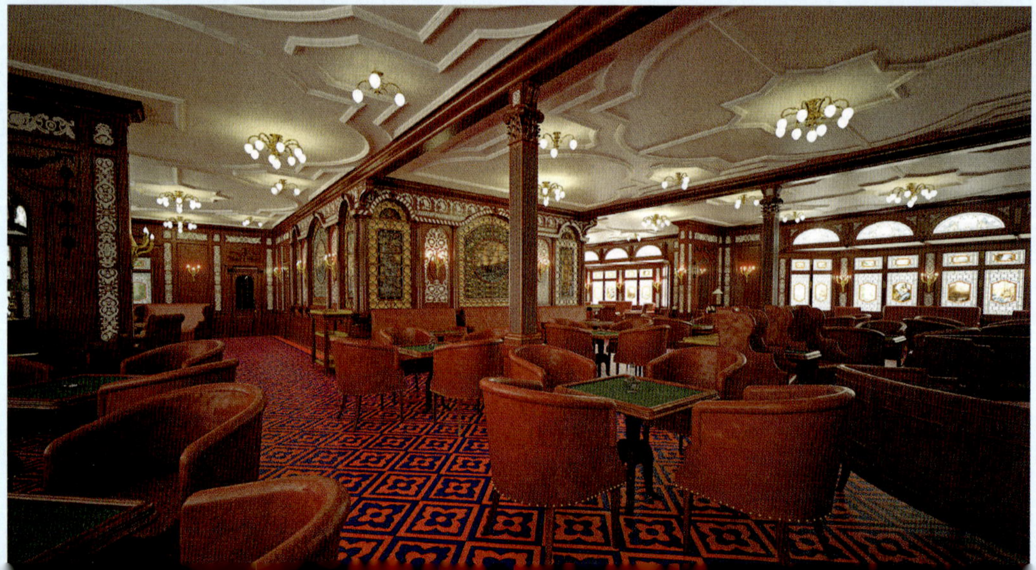

The Smoking Room was reserved for the use of gentlemen. It was adapted from the early Georgian style, but with inlaid mother-of-pearl. On *Titanic*, the floor tiles located here were red and blue, suggesting to some that the chairs might have been a deep burgundy instead of green. (Chris Walker)

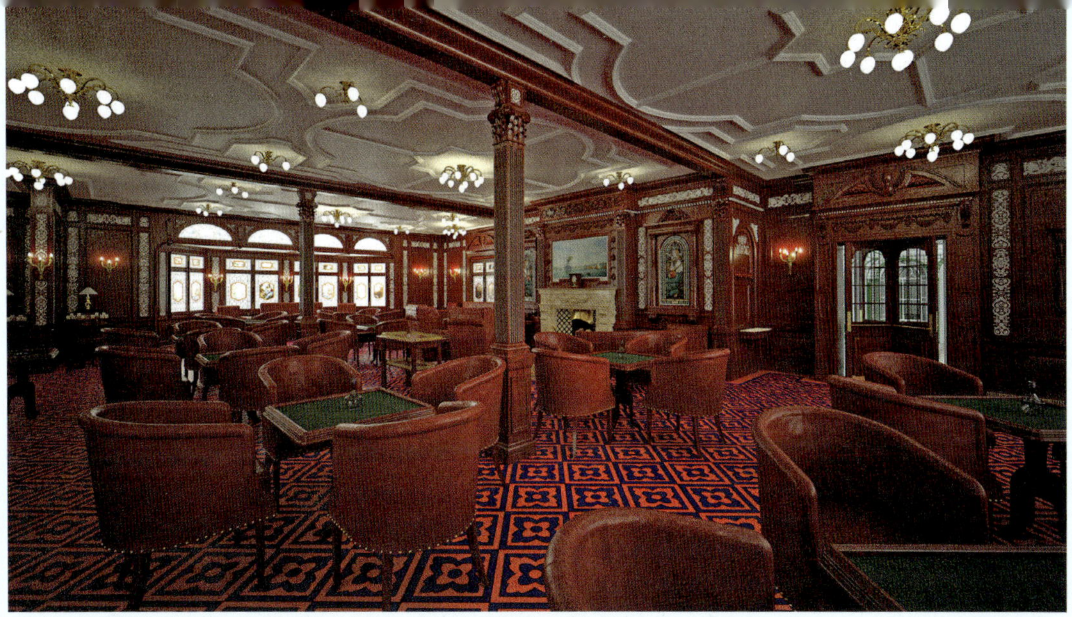

At the aft end of the Smoking Room stood a coal-burning fireplace, the only real one on the ship. On *Titanic*, the painting over the mantle, by marine artist Norman Wilkinson, was entitled 'Plymouth Harbour'. (Chris Walker)

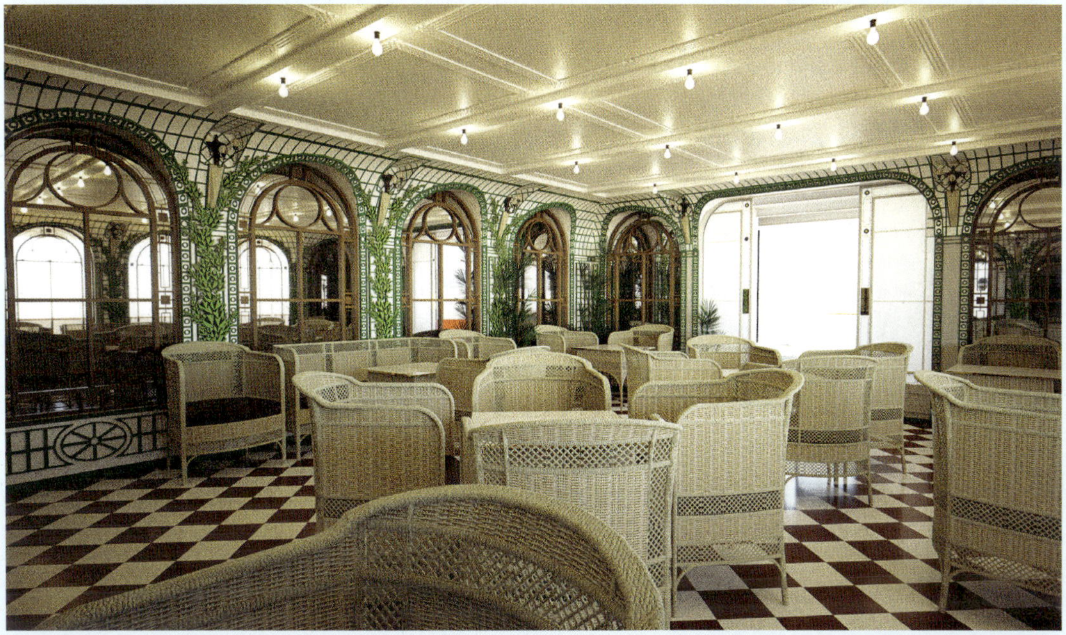

Moving through the revolving door in the Smoking Room, one entered the port-side Verandah Café & Palm Court. Its opposite number on the starboard side was for non-smokers. Free light refreshments were served here, but not meals. The sliding doors aft led out onto the Promenade Deck. (Chris Walker)

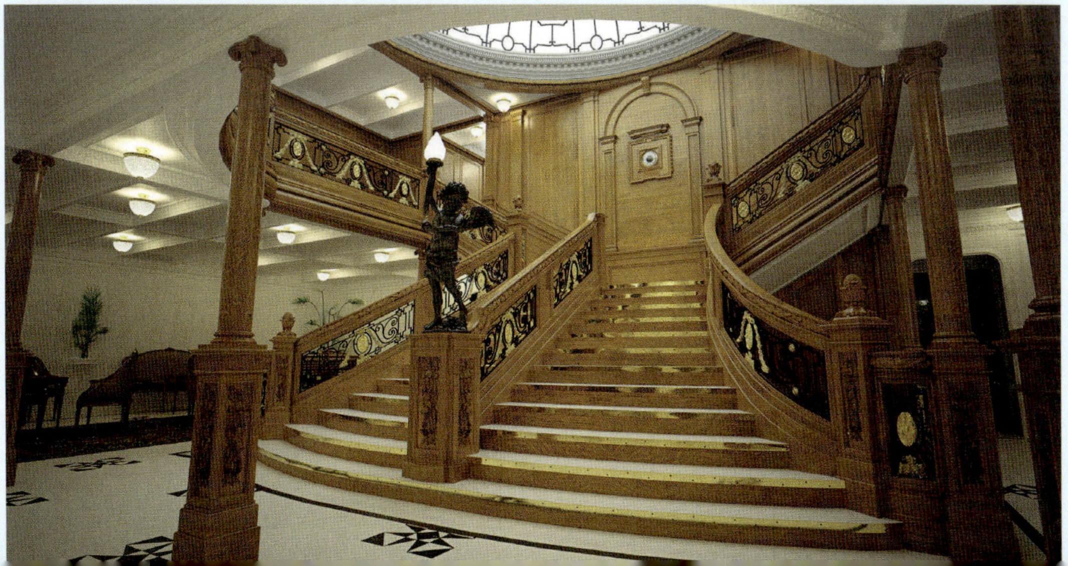

The Aft Grand Staircase, seen from the B Deck Restaurant Reception Room, looking back up to A Deck. The dome here was circular and 18ft 6in diameter. (Chris Walker)

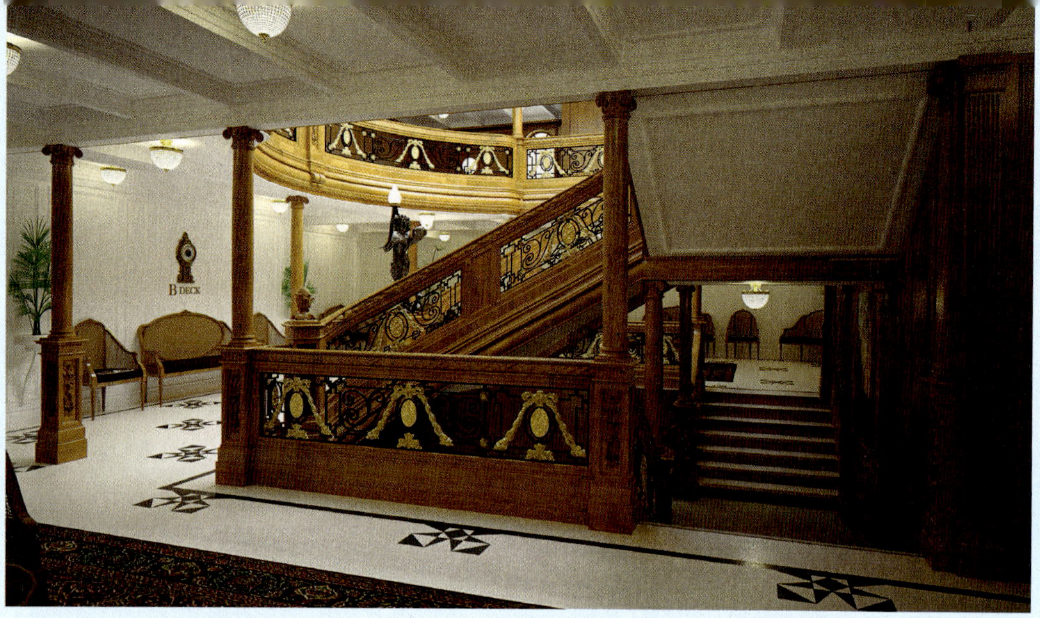


On the starboard side of the Restaurant Reception Room, this view shows the stairs leading down to C Deck, the lowest level of the Aft Grand Staircase. (Chris Walker)

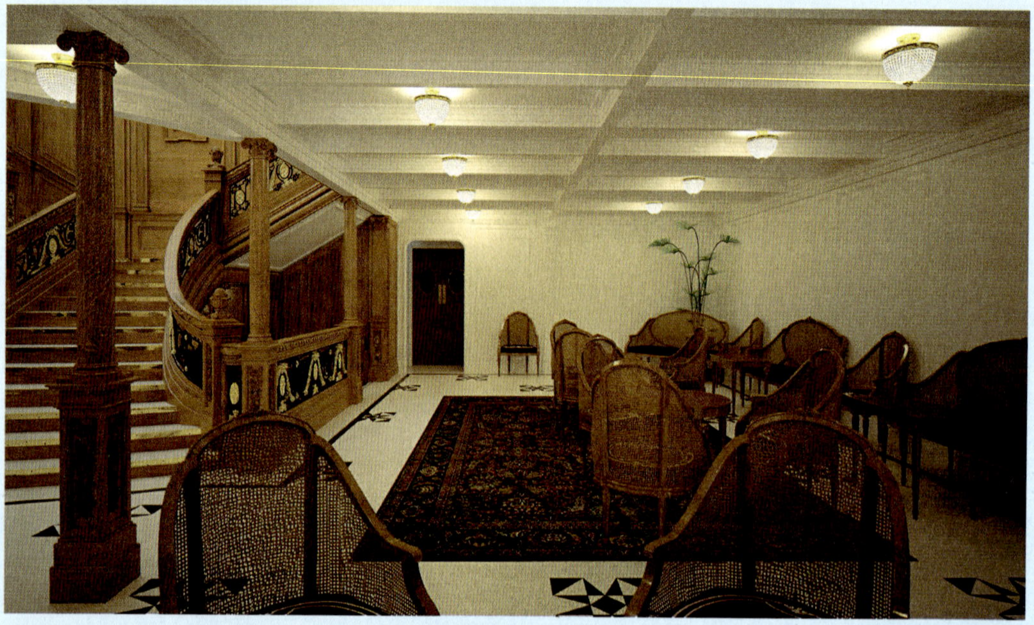

The Restaurant Reception Room, looking forward along the starboard side, was a tremendous improvement over *Olympic*'s layout. It was hoped that this new space would reduce congestion in the popular Reception Room forward on D Deck, and a special trio of musicians played here while passengers socialised. (Chris Walker)

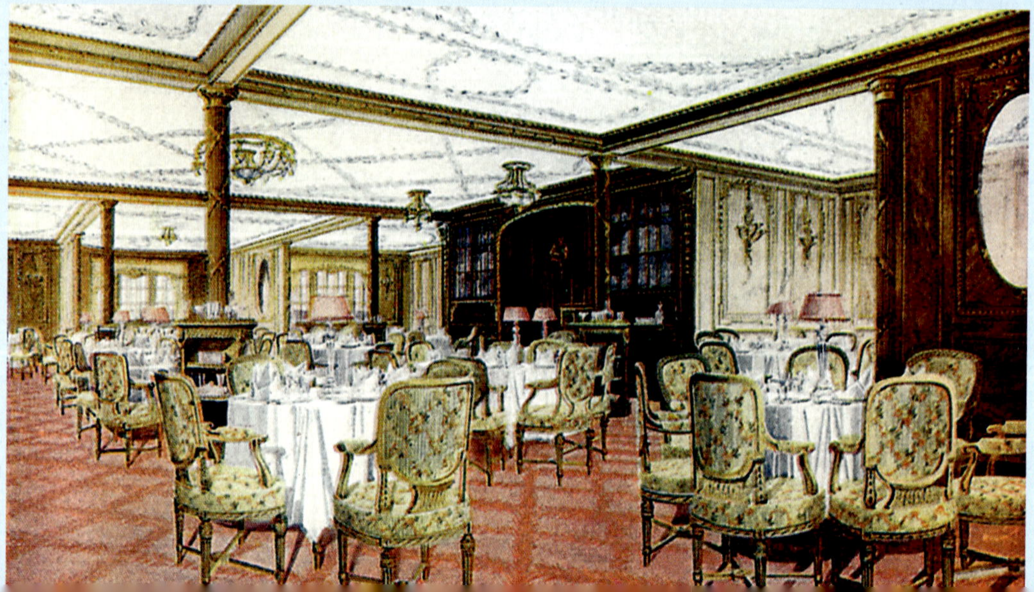

The *à la carte* Restaurant on B Deck was an extra-tariff facility that had proved so popular on *Olympic* that its footprint was expanded on *Titanic*. Rendered in Louis XVI style, it was run by a separate staff under the oversight of Gaspare Gatti. (Authors' Collection)

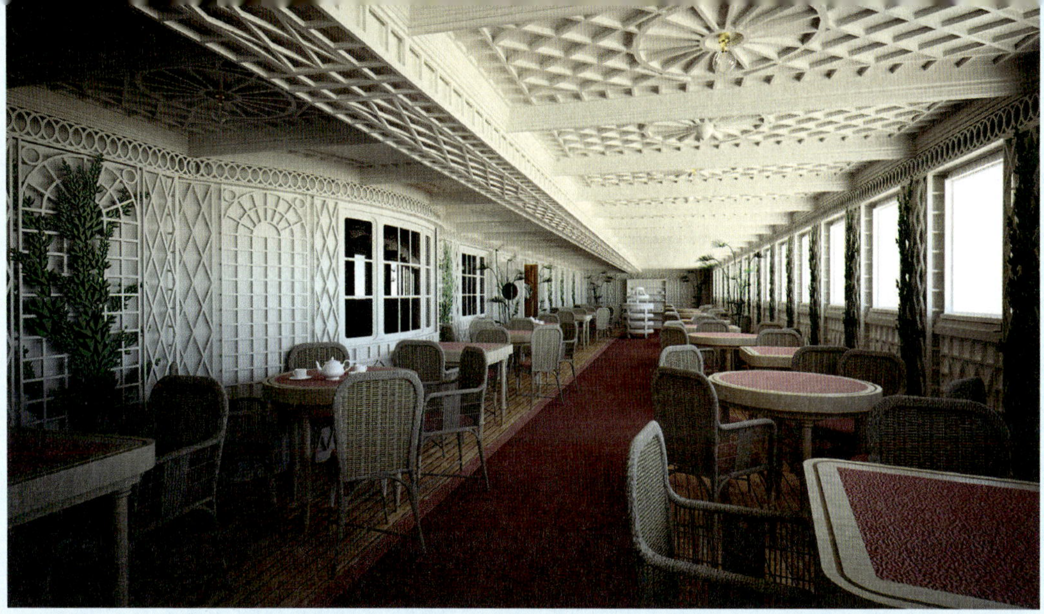

The Café Parisien ran along the starboard side of B Deck, just outside the Restaurant, and was run by the same staff. The side screens gave a splendid sea view and real ivy grew up the trellis along the walls. On the evening of the disaster, a party of guests stayed here until close to 11.30 p.m. (Chris Walker)

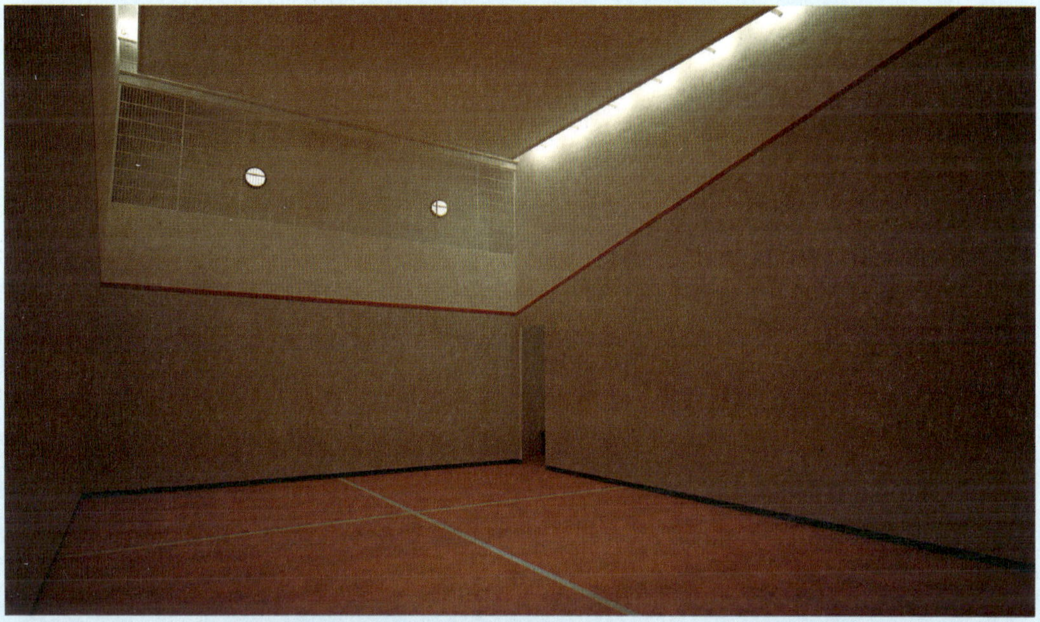

Located deep in the forward end of *Titanic* was the Squash Racquet Court. This view looks up and aft from the court on G Deck toward the spectator gallery above. Passengers could rent balls and racquets from the attendant. (Chris Walker)

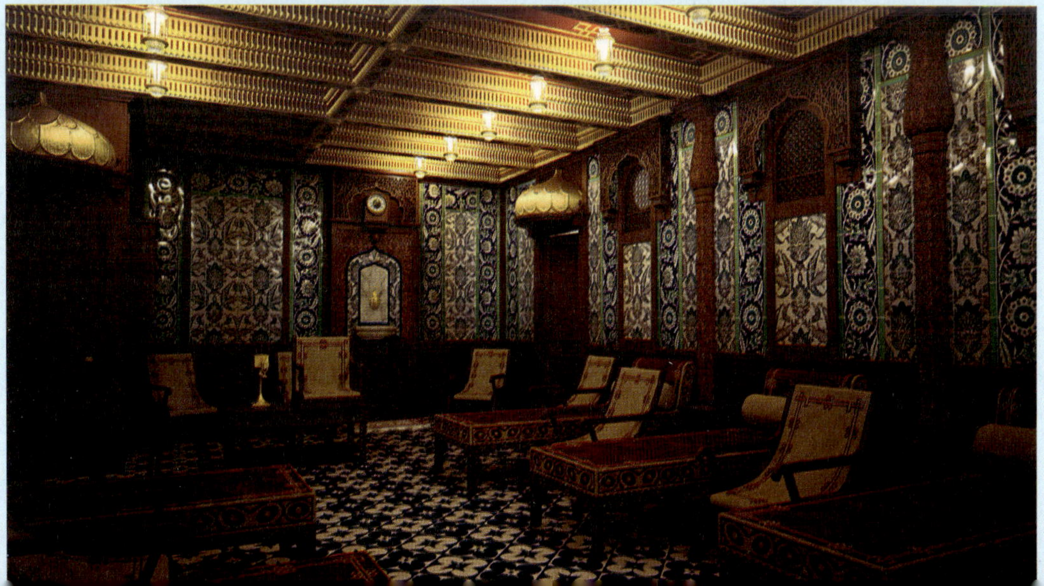

The exotic Turkish Bath Cooling Room was located on the starboard side of F Deck. Although Turkish Baths were all the rage at the time, the experience was not one enjoyed by all who tried it. (Chris Walker)

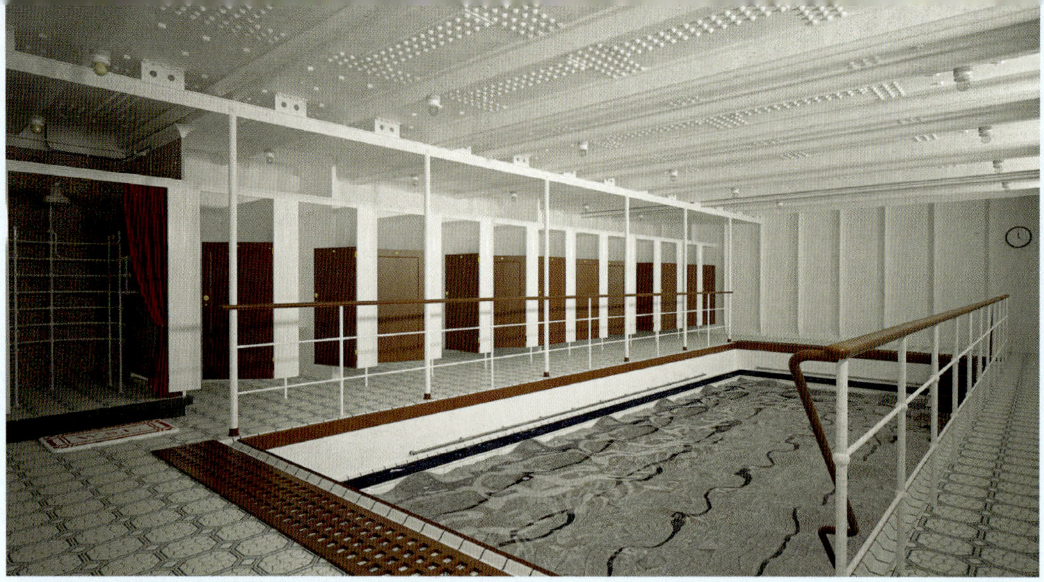

The Swimming Bath was located just forward of the Turkish Bath. The space was something of a new feature on liners but was rather austere when compared with the swimming baths that other liners would feature in the near future. (Chris Walker)

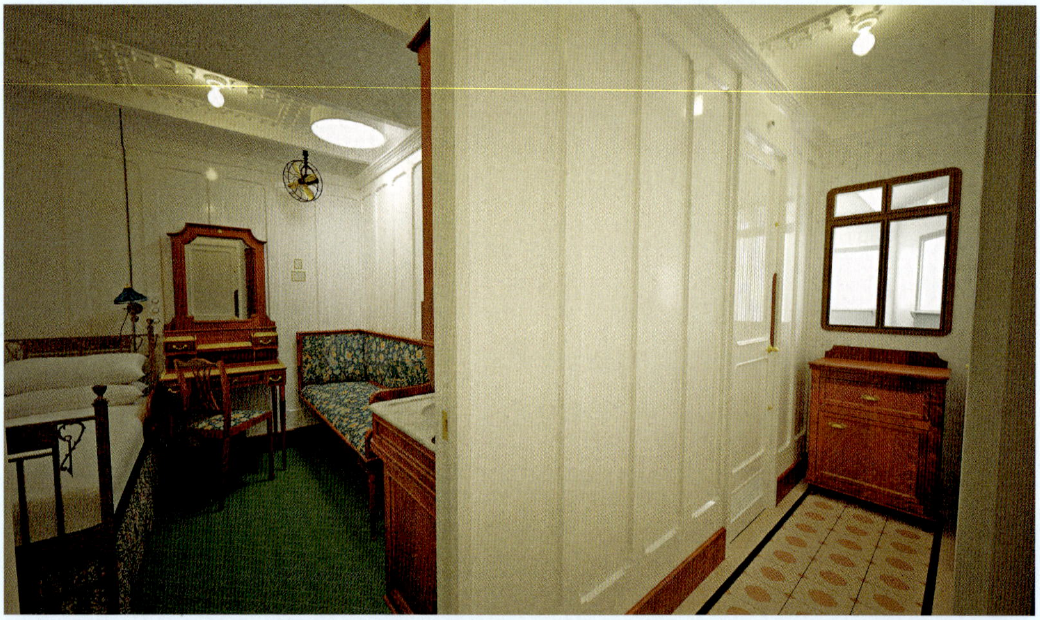

Stateroom A-10, located forward on the port side. Despite the cabin being located inside, extra light was allowed in from the skylights in the base of the Officers' Quarters deckhouse on the Boat Deck. The door to A-12 lies just beyond. (Chris Walker)

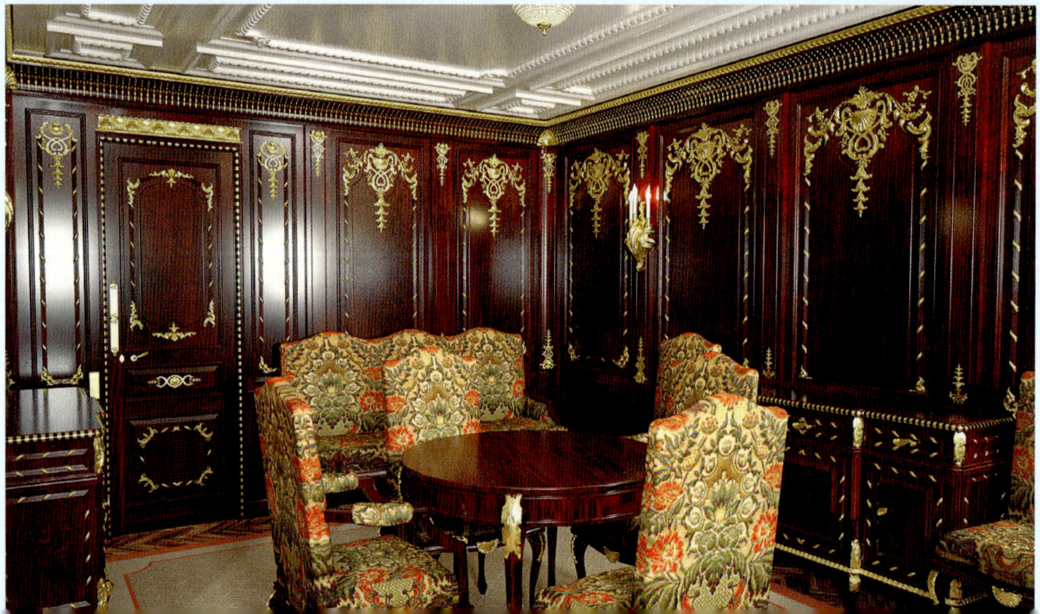

Mr and Mrs Isidor and Ida Straus booked a suite on the starboard side of C Deck, just behind the Purser's Inquiry Office. Their sitting room, numbered C-55, is shown here. (Chris Walker)

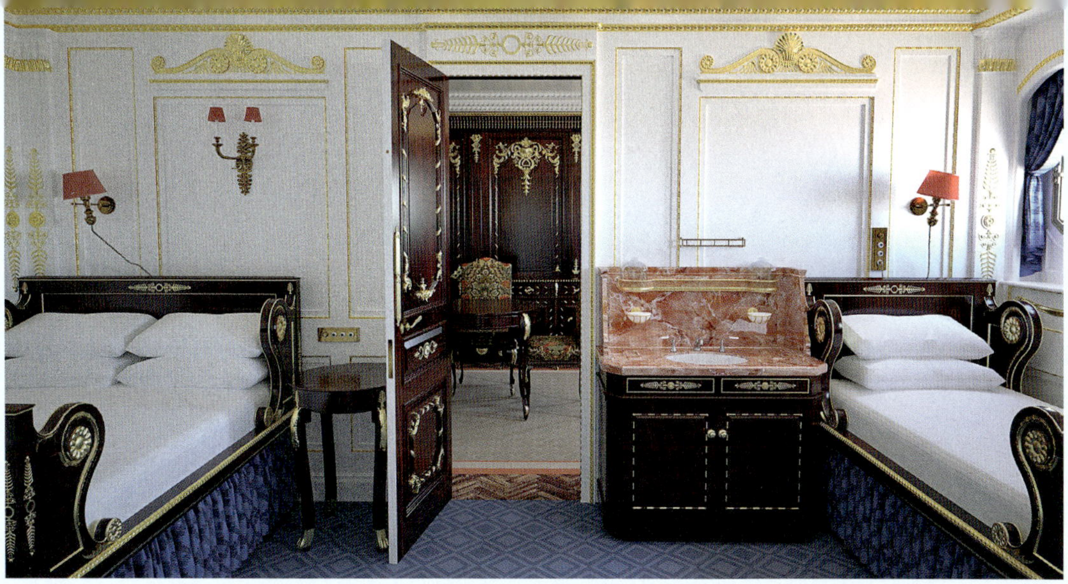

The Strauses' stateroom, C-57, looking through the connecting door back into the sitting room, C-55. (Chris Walker)

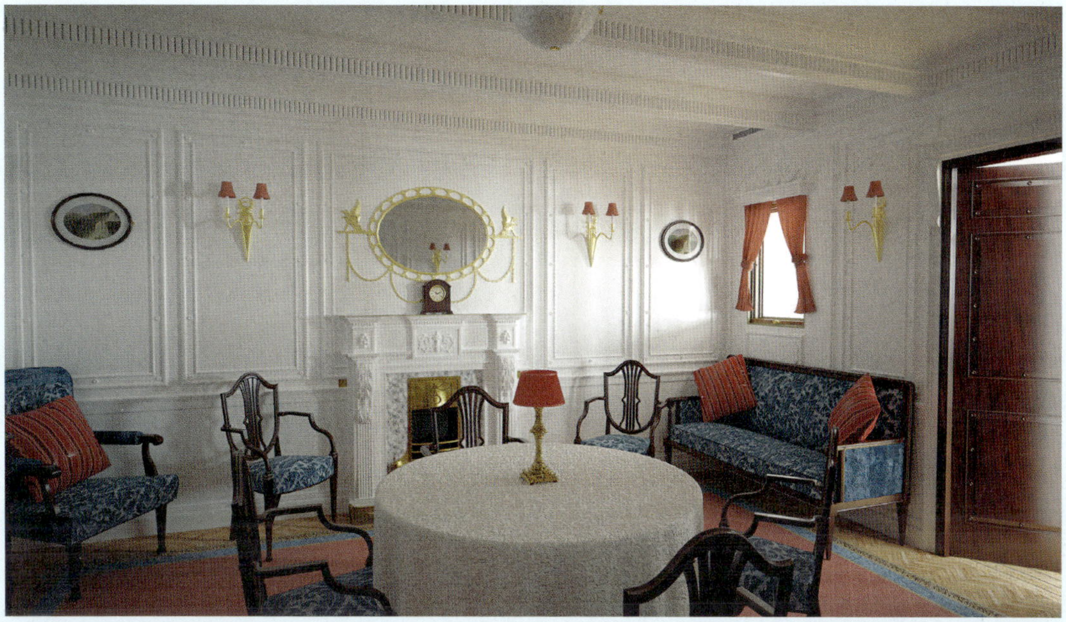

The sitting room for the starboard-side Parlour Suite (B-51–B-55). This suite had its own Private Promenade. It was booked by Mrs Charlotte Cardeza and her son, Thomas. This view brings to life an actual photograph taken from a nearly identical angle aboard *Titanic*. (Chris Walker)

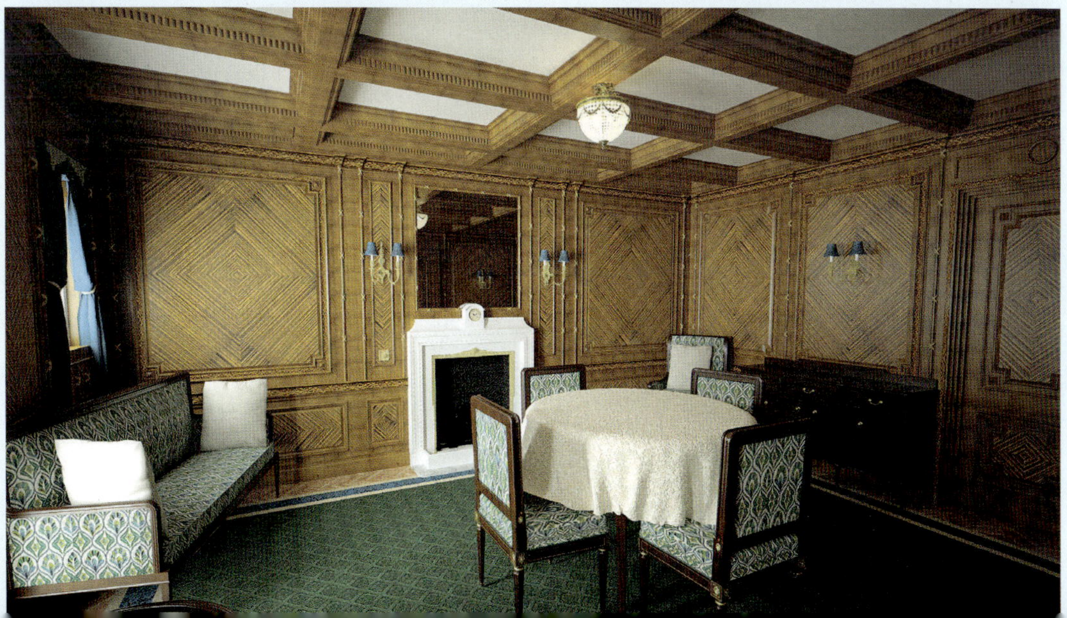

The sitting room of the port-side Parlour Suite (B-52–B-56), which was occupied by J. Bruce Ismay during the maiden voyage. (Chris Walker)

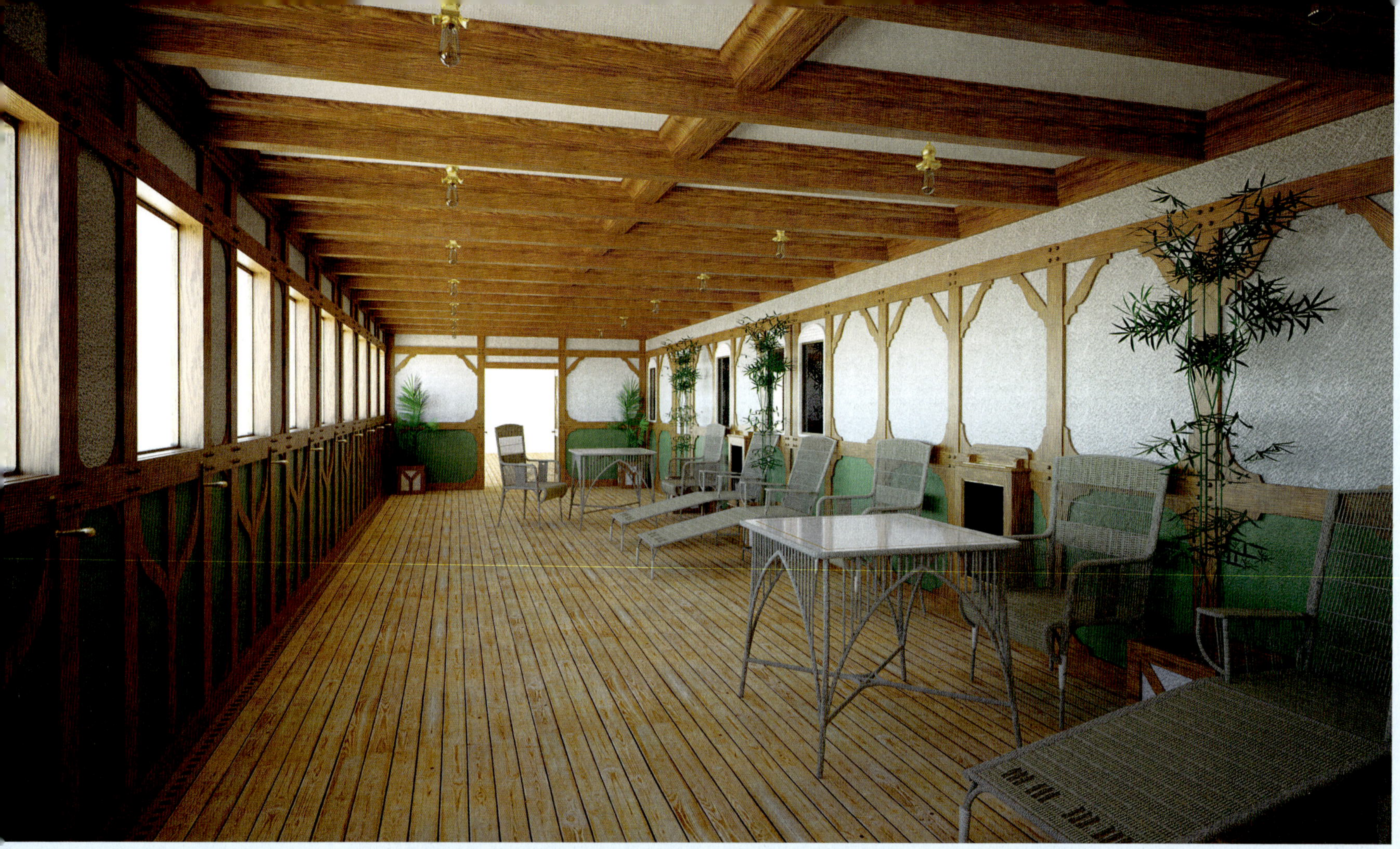

*Above:* The port-side Private Promenade, looking forward. These two Parlour Suites offered amenities unrivalled by any liners then on the Atlantic – even *Olympic*. (Chris Walker)

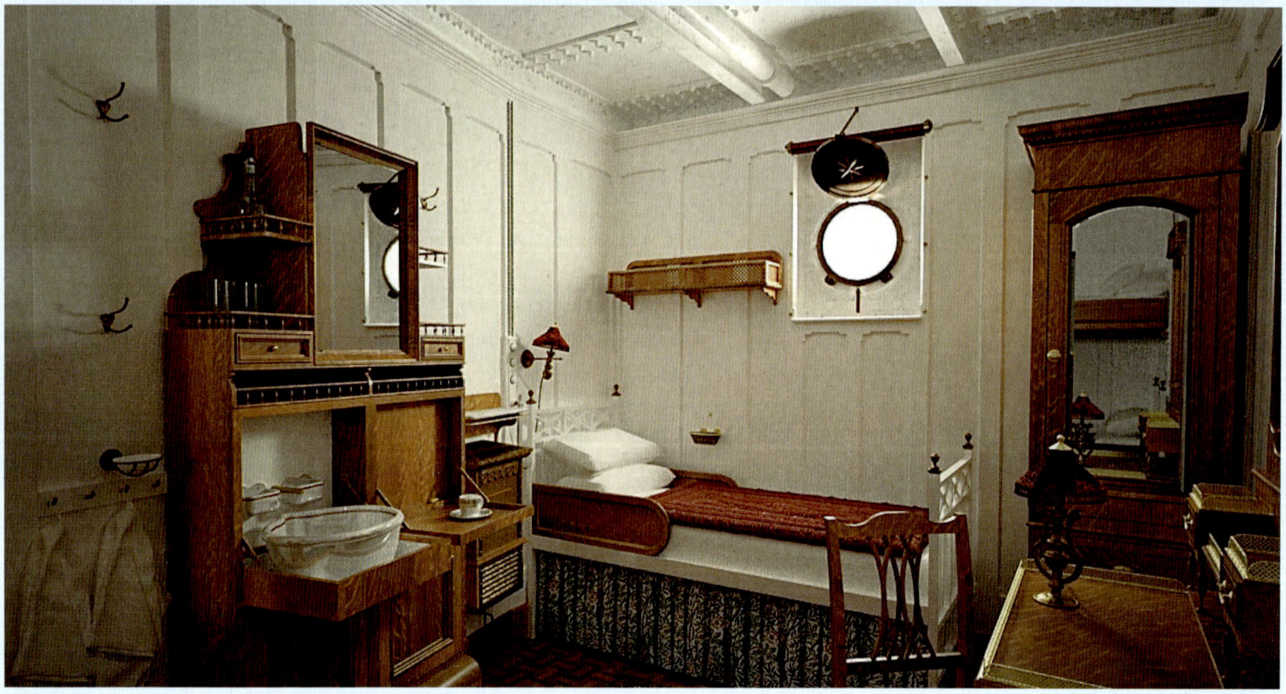

*Right:* Stateroom E-23, booked by Margaret Brown, offered fewer amenities. Most First Class staterooms did not even have private baths or toilet facilities. Notice also the bare, unfinished steel on the ceiling. Yet whether they had booked basic or more luxurious accommodation, all First Class passengers enjoyed the same level of service. (Chris Walker)

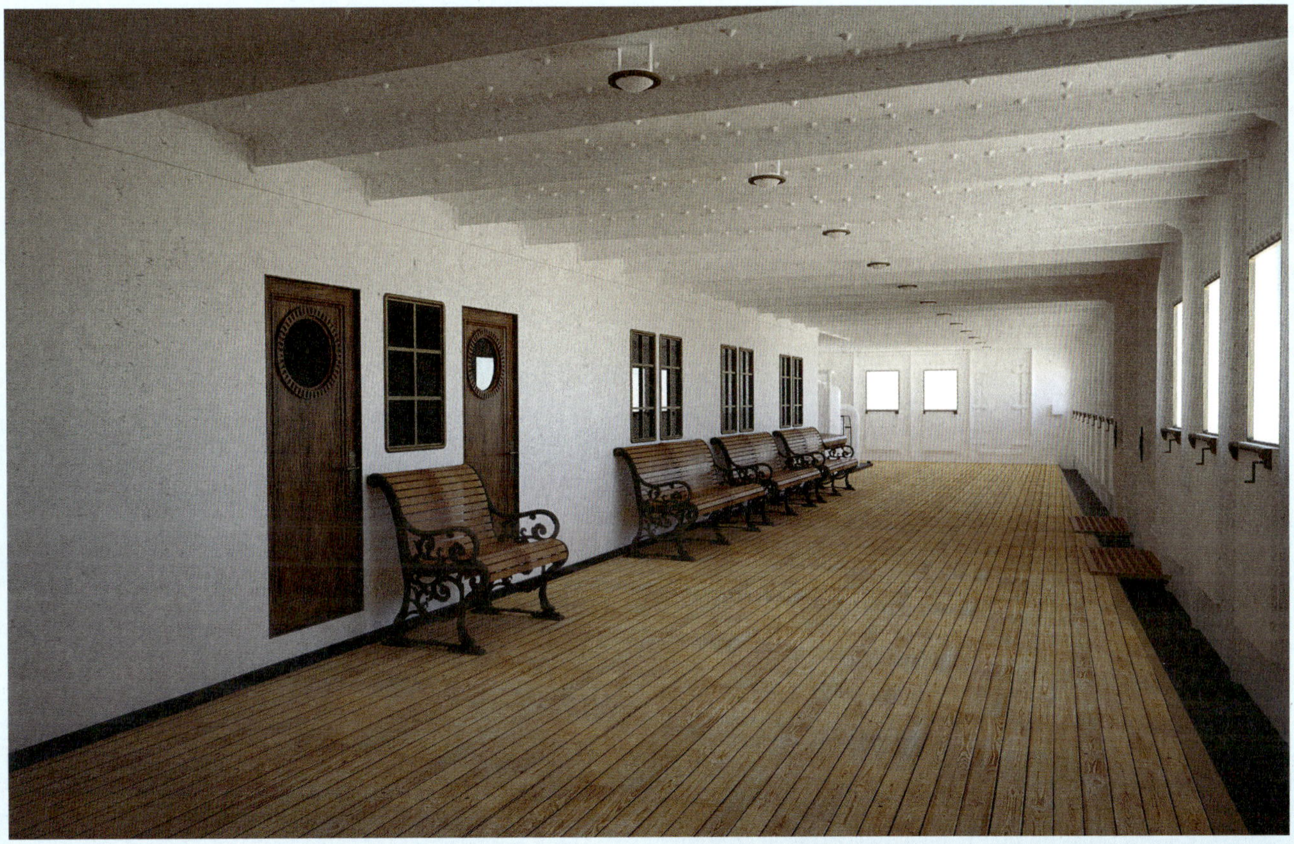

*Left:* The port-side Second Class Promenade on B Deck, looking aft. The shell doors on the right were where passengers boarded in Southampton. (Chris Walker)

*Below:* The Library, or 'Lounge' as it was often called, was on C Deck. On Sunday, 14 April, the Library was quite crowded as passengers sought refuge in warm interior spaces from the chilly temperatures outside. (Chris Walker)

## SECOND CLASS: A TOUR

On 10 April, Second Class passengers boarded from the lower level of a gantry identical to that used by the First Class passengers. It was connected to the shell doors on the port side of C Deck, which opened onto the covered Second Class Promenade, located just forward of the Aft Well Deck.

Second Class accommodations on *Titanic* were not as luxurious as their First Class counterparts, but it was no exaggeration to say that they were the equivalent of First Class on many other ships or, indeed, of many fine hotels ashore. Second Class passenger Bertha Lehmann recalled that she had been in many beautiful hotels in Europe and America but had 'never seen anything quite as beautiful as this boat'. A newspaper reporter who disembarked at Queenstown later recalled that Second Class accommodations on ships were not usually associated with 'lifts and lounges and libraries', but that all these were found on *Titanic*, and he remembered needing reassurance that they had actually left First Class and were touring Second Class.

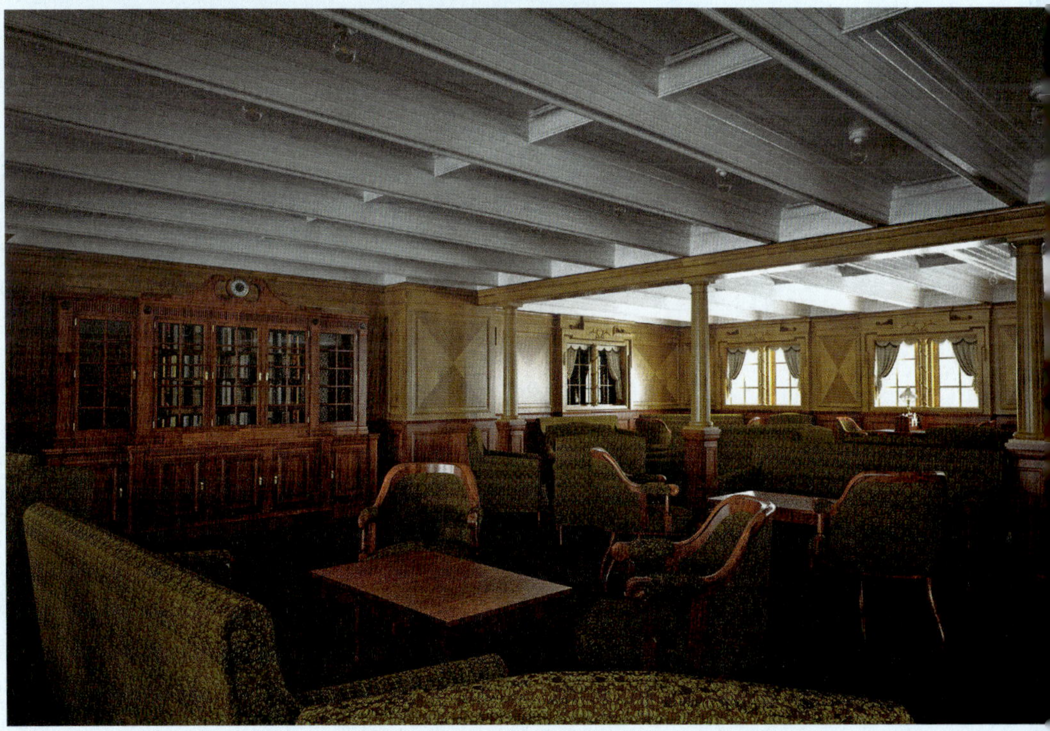

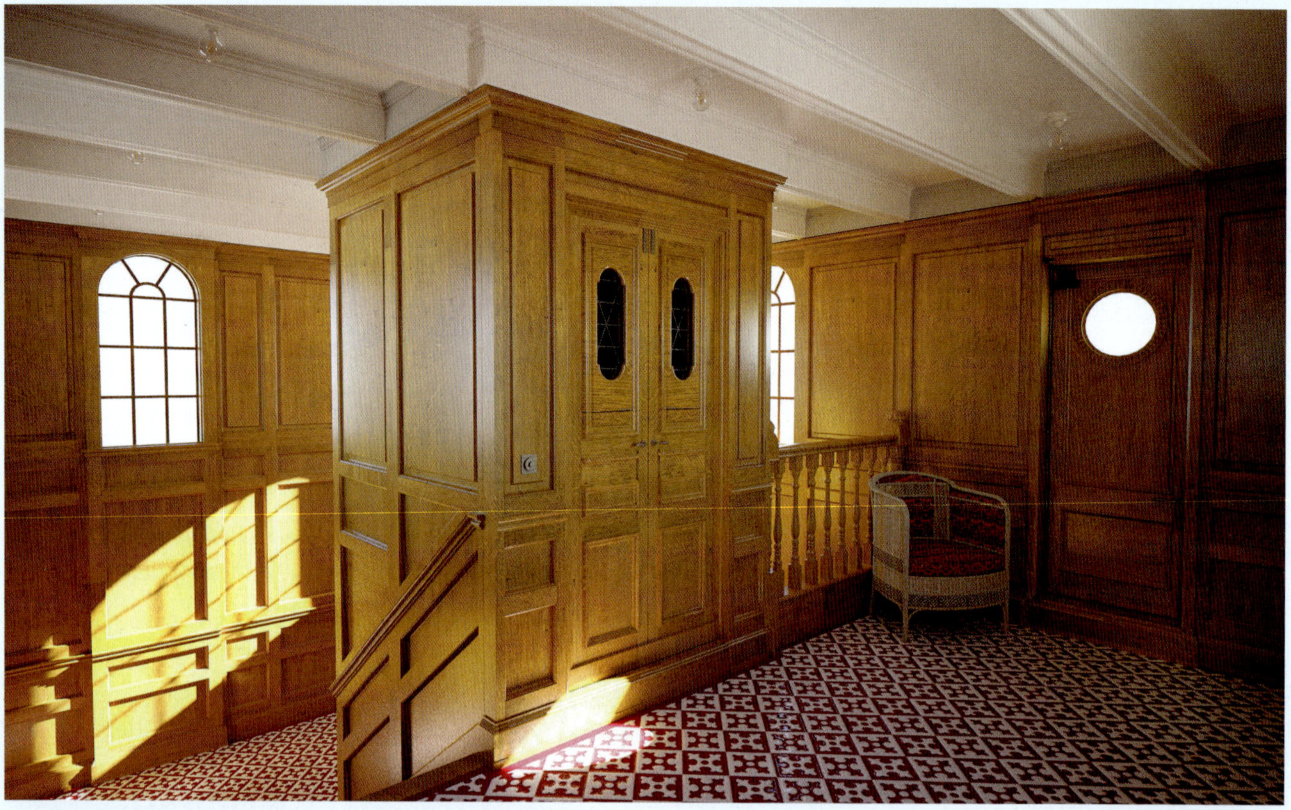

The forward Second Class Entrance ran from the Boat Deck down to F Deck. This is the interior of the Boat Deck level, showing the lift. This was the only way for passengers to access their promenade area on the Boat Deck. (Chris Walker)

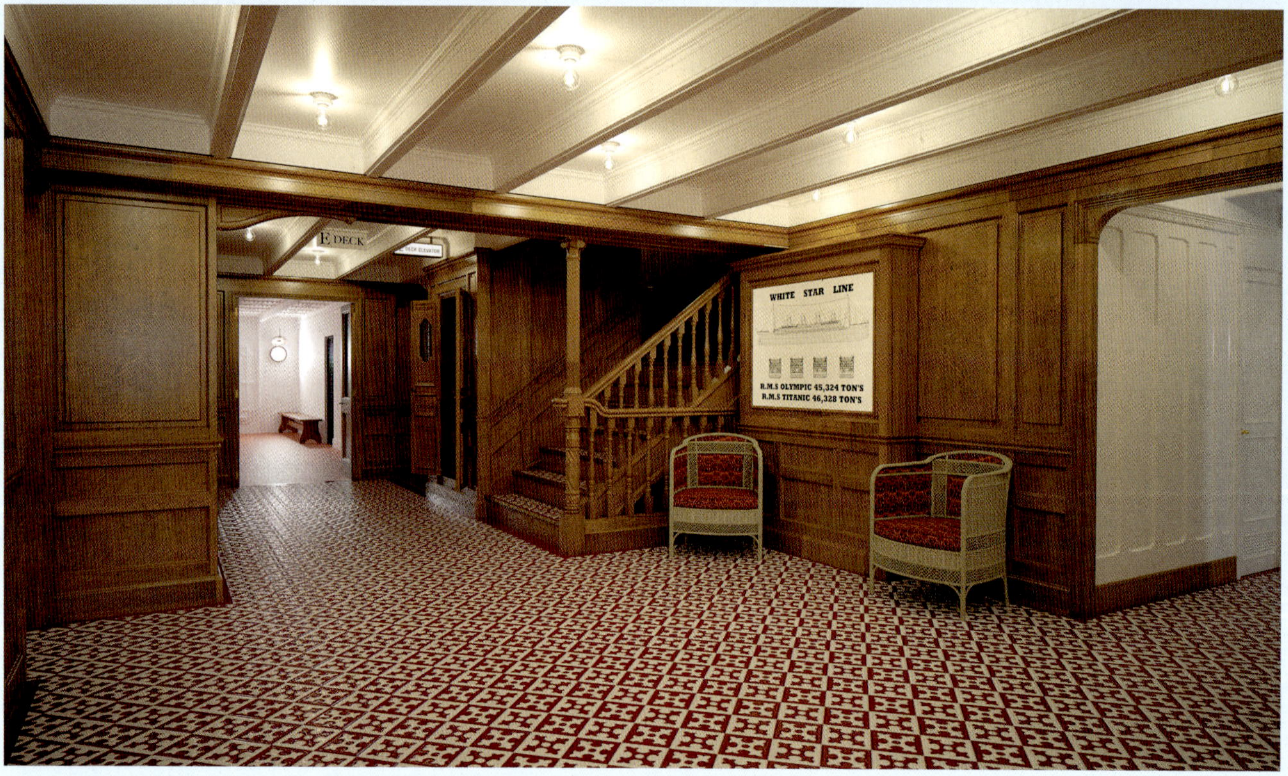


This is the E Deck level of the same forward Second Class Entrance, near a bank of lavatories for men and women. Note the electric sign, one of only three found on the ship. (Chris Walker)

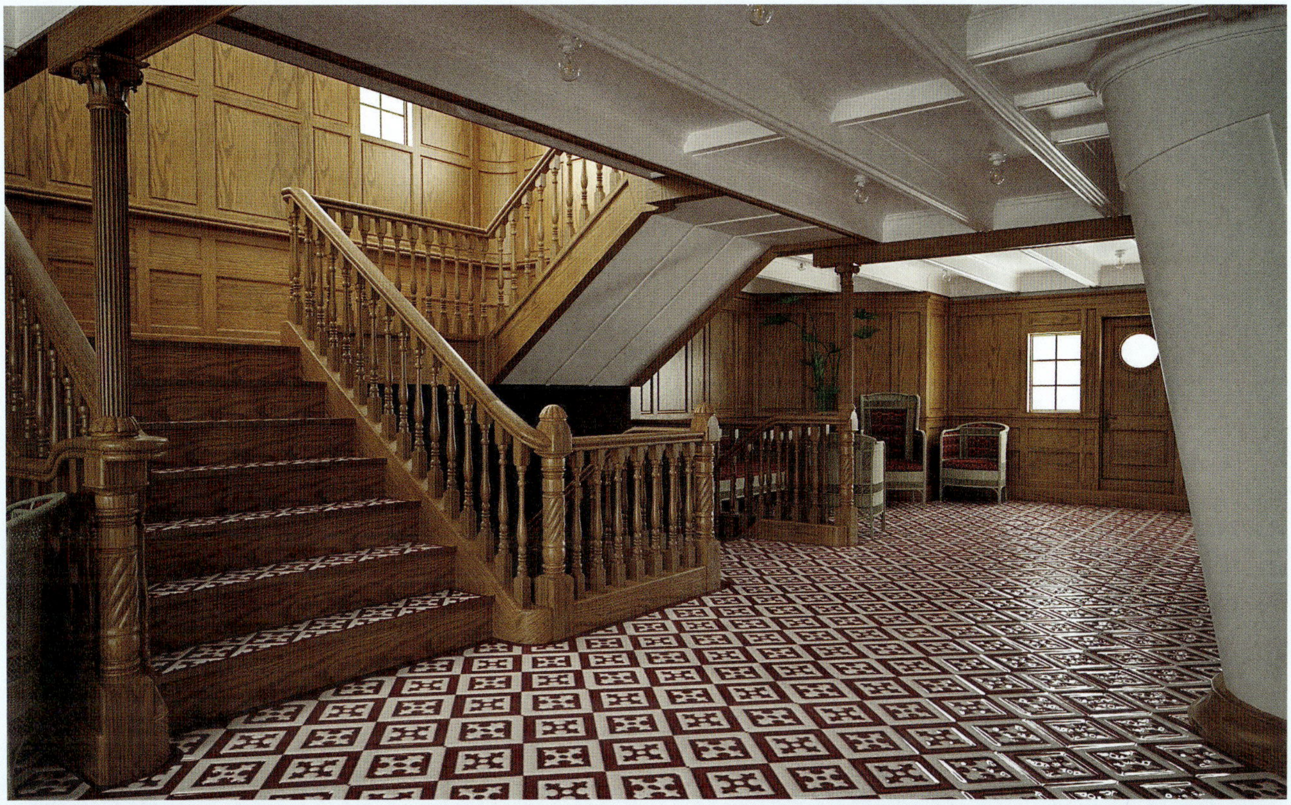

The aft Second Class Entrance on C Deck, astern of the Library. The base of the mainmast is on the right. The door in the distance opens onto the enclosed Second Class Promenade. The band played here each day from 10.00 to 11.00 a.m., 5.00 to 6.00 p.m. and 9.15 to 10.15 p.m. (Chris Walker)

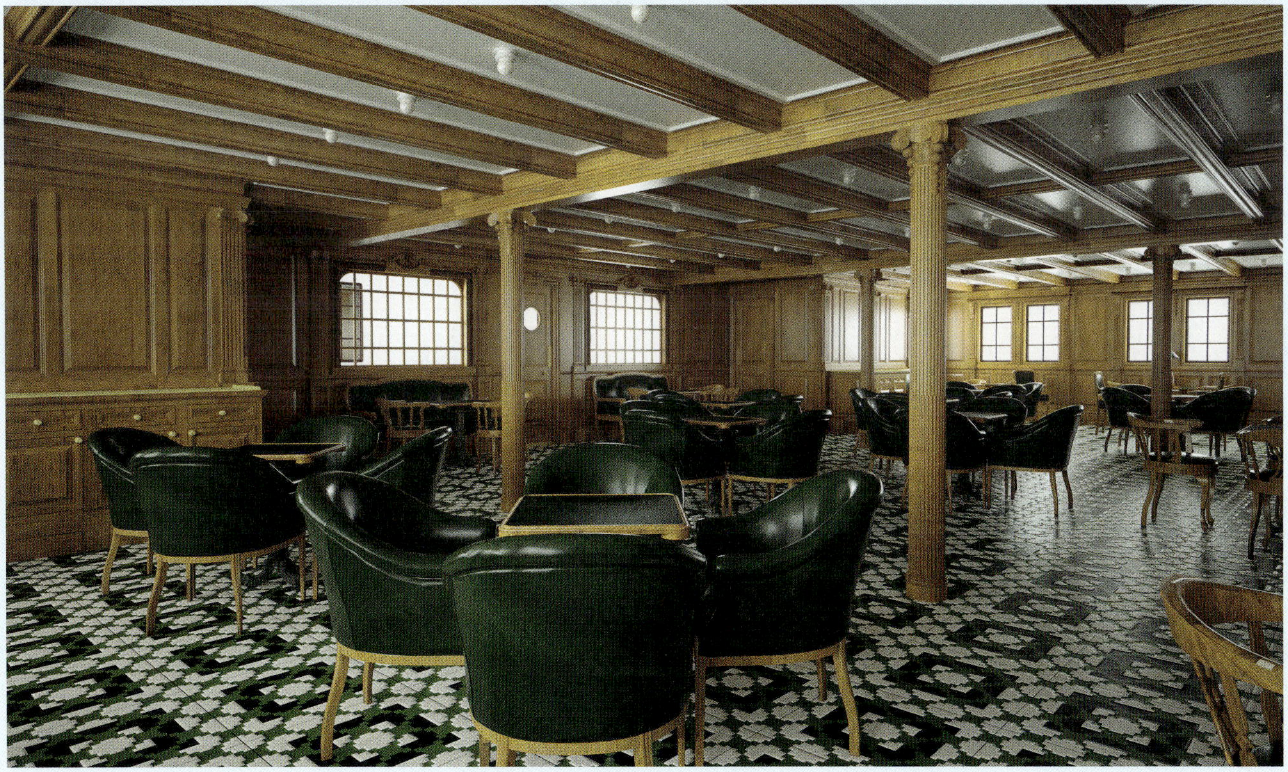

The Second Class Smoking Room was located on B Deck directly above the Library, and was intended solely for the use of gentlemen. The door at the aft end opens onto the open-air Promenade, and the aft Second Class Entrance can be seen outside. (Chris Walker)

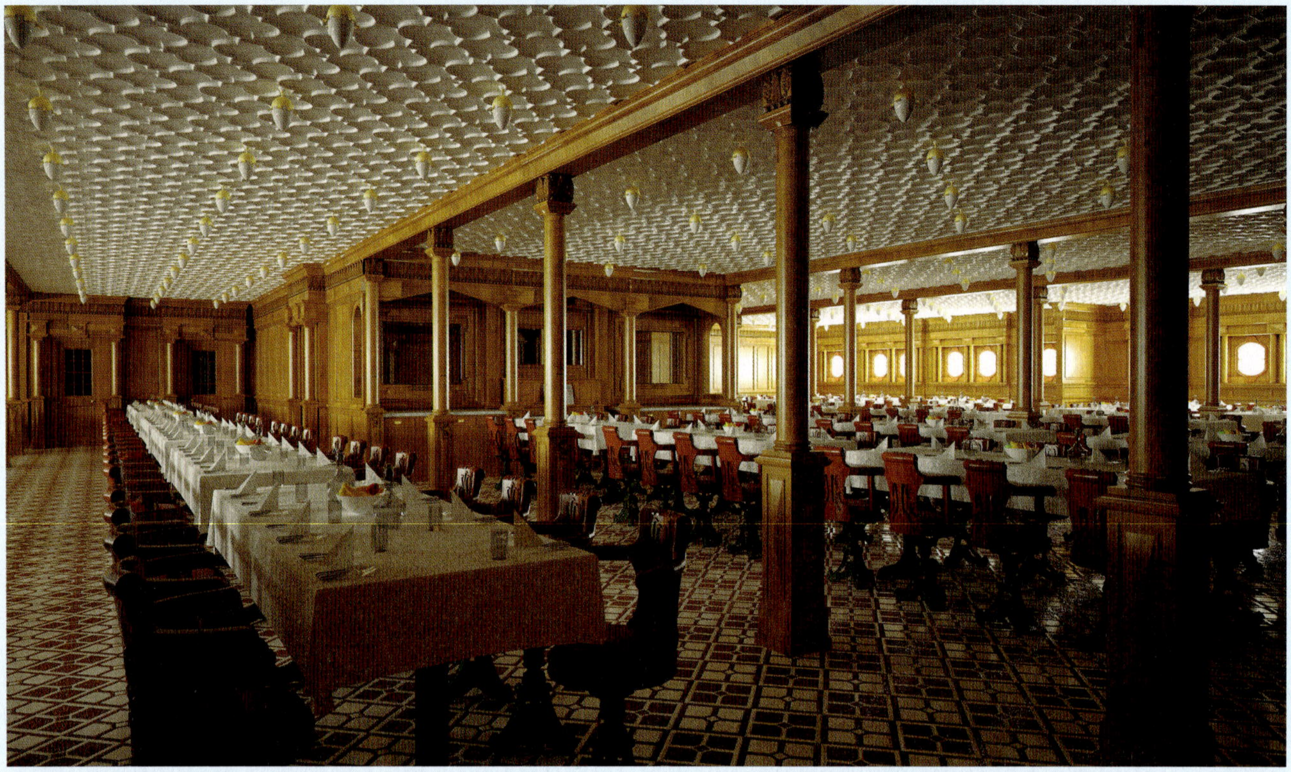

The Second Class Dining Saloon was on D Deck, nestled between the forward and aft Second Class Entrances. It could accommodate up to 394 diners and, unlike the First Class Saloon, the chairs here were bolted to the deck. (Chris Walker)

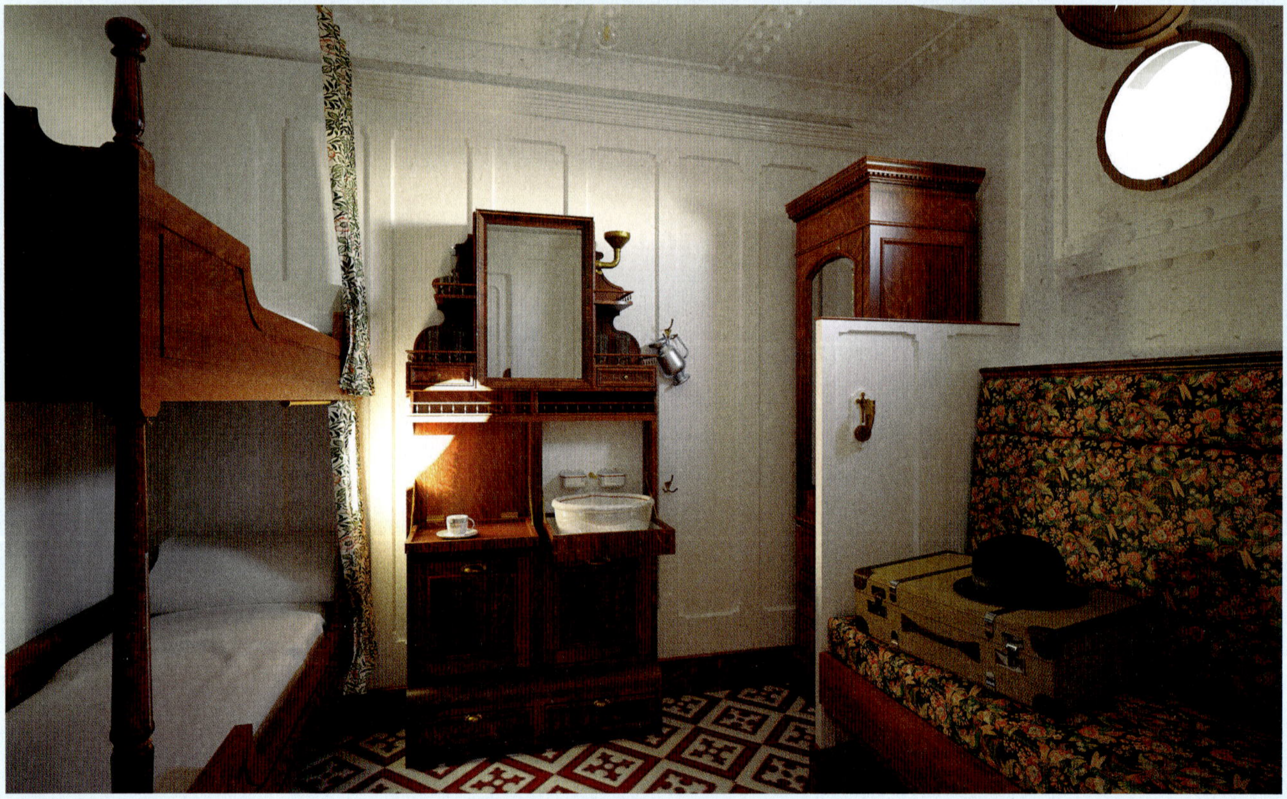

Second Class Cabin E-74 was located on the starboard side near the Reciprocating Engine Casing. It represents the class's superior outside accommodations. It could be used as an alternative First Class cabin, depending on bookings. (Chris Walker)

## THIRD CLASS: A TOUR

In Third Class, also referred to as 'Steerage', accommodations were plain and simple, but they were also clean and comfortable. Today, there is often a tendency to frown upon the concept of Third Class, with its modest facilities. However, modern sensibilities cloud the realities of the era. These facilities had seen tremendous advances since the mid-nineteenth century when passengers were advised to bring their own cutlery for the Atlantic trip, lest they be obliged to rent it on board. Indeed, aboard *Titanic*, conditions for these passengers were better than those found on Second Class in ships of only a few years before, many of which were still in service, and were certainly far above what most of these passengers were used to. Many of these travellers were not even accustomed to such novelties as electricity, clean running water, toilets, or even simple things such as cold, refrigerated milk.

The immigrant trade was very competitive at the time. Far from taking advantage of poor travellers, Atlantic liners of *Titanic*'s era were more interested in providing a top-quality experience for Steerage passengers so that they would recommend passage on the ship to other relatives contemplating a similar journey.

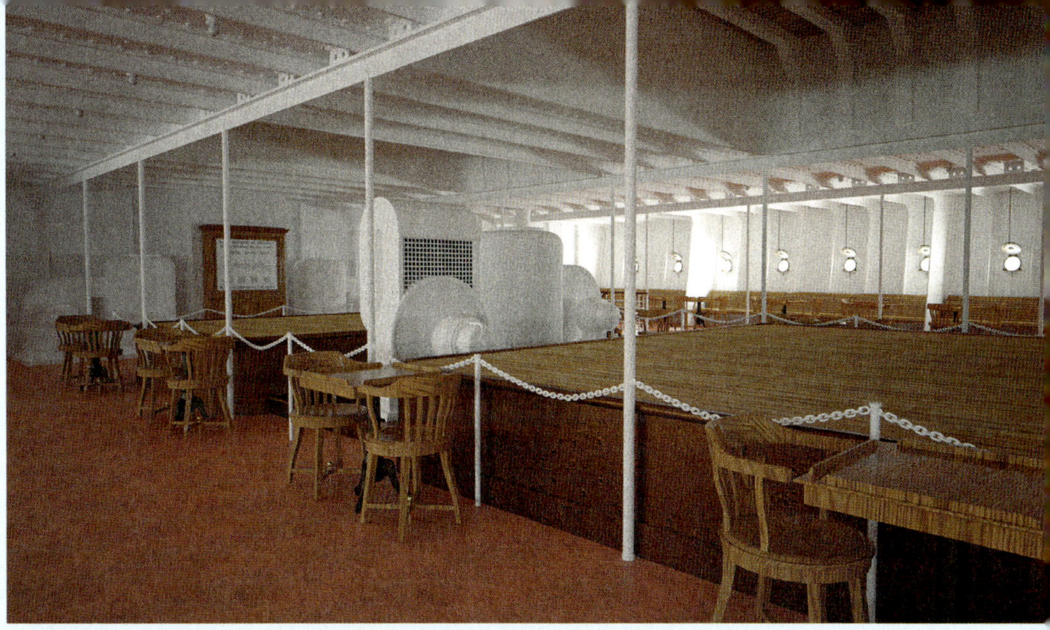

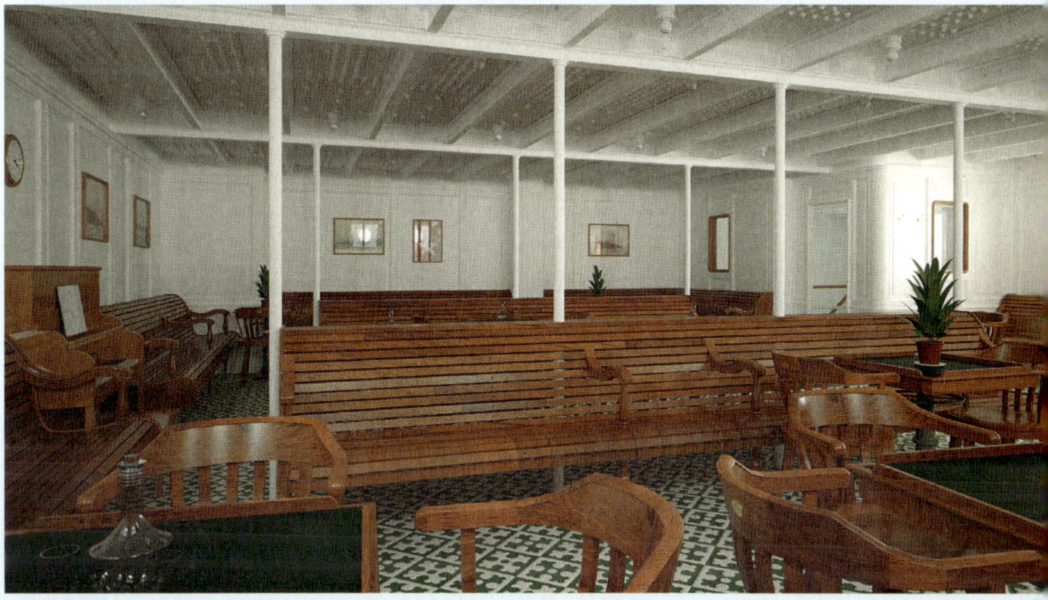

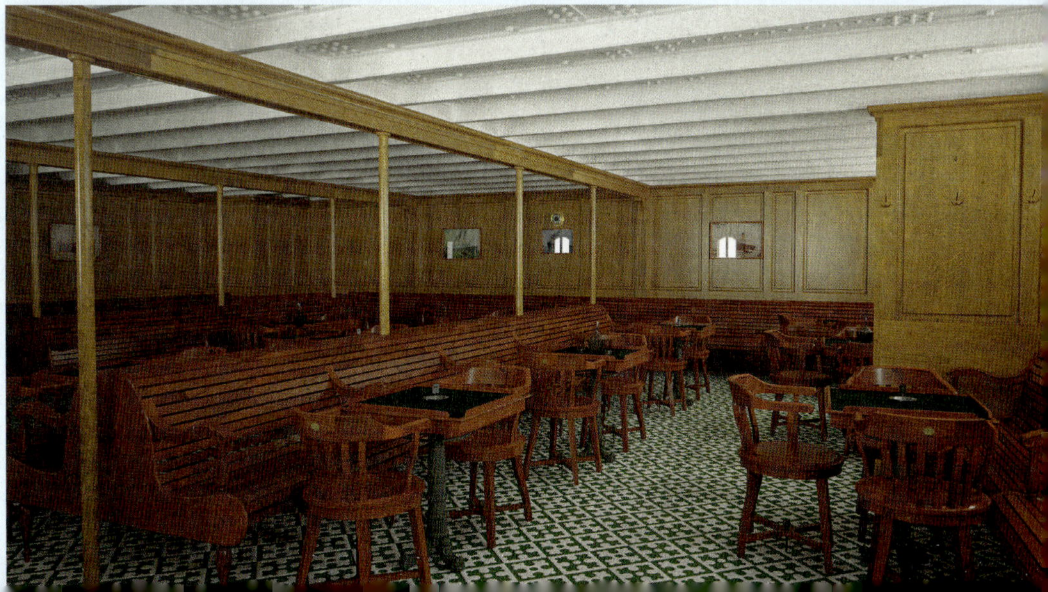

*Top:* The large Third Class Open Space forward on D Deck, beneath the Forward Well Deck, looking aft and to port. Cargo Hatch Nos. 2 and 3 are visible. Passengers could gather in this comfortable space for games or even impromptu dances, while sheltered from the weather. (Chris Walker)

*Centre:* The Third Class General Room was aft on the starboard side of C Deck, beneath the Poop Deck. It featured an upright piano and plenty of seating where passengers could take shelter from an often stormy sea. (Chris Walker)

*Bottom:* The Third Class Smoking Room was a mirror image of the General Room, located on the port side of C Deck. It was intended for the use of gentlemen only. The floor tiles have recently been discovered to have been green rather than red. (Chris Walker)

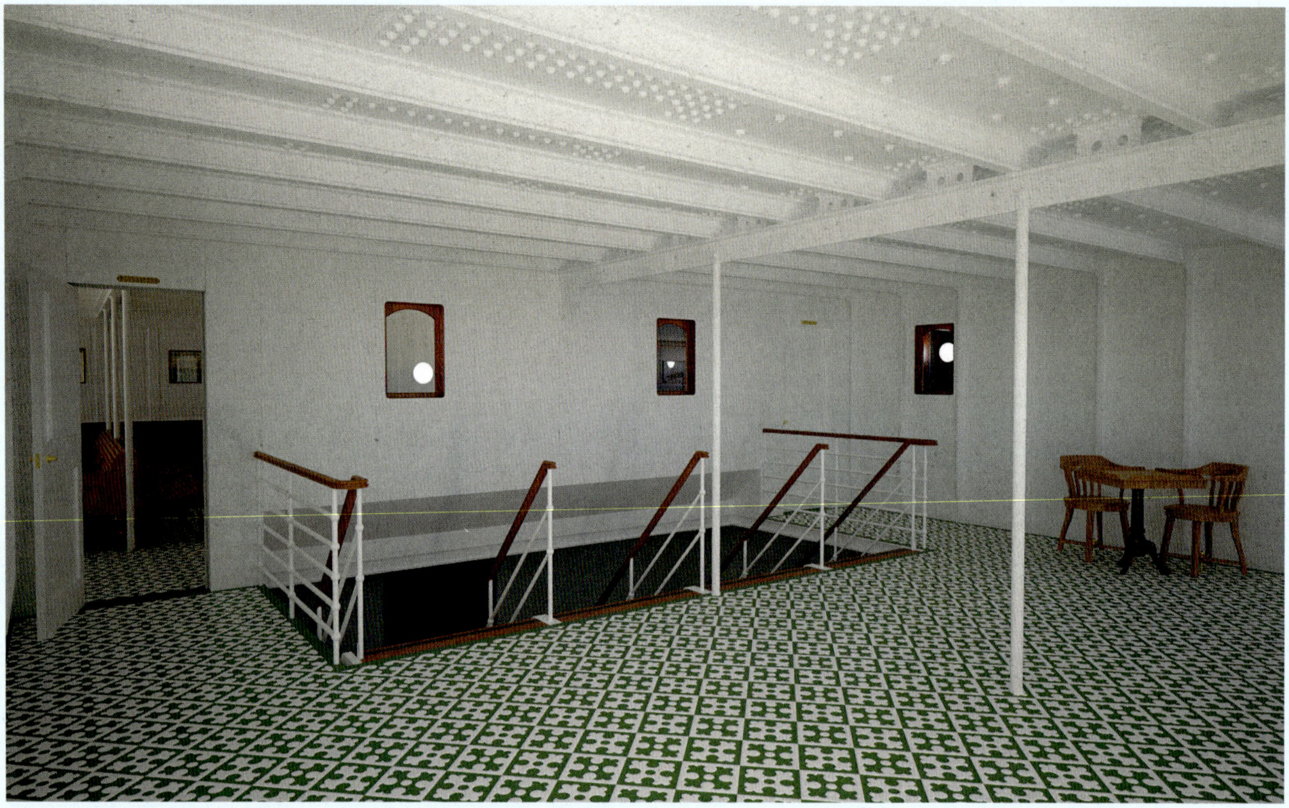

The Third Class Entrance, on C Deck beneath the Poop Deck, gave entry to both the General Room and the Smoking Room. The door to the General Room is open on the left. (Chris Walker)

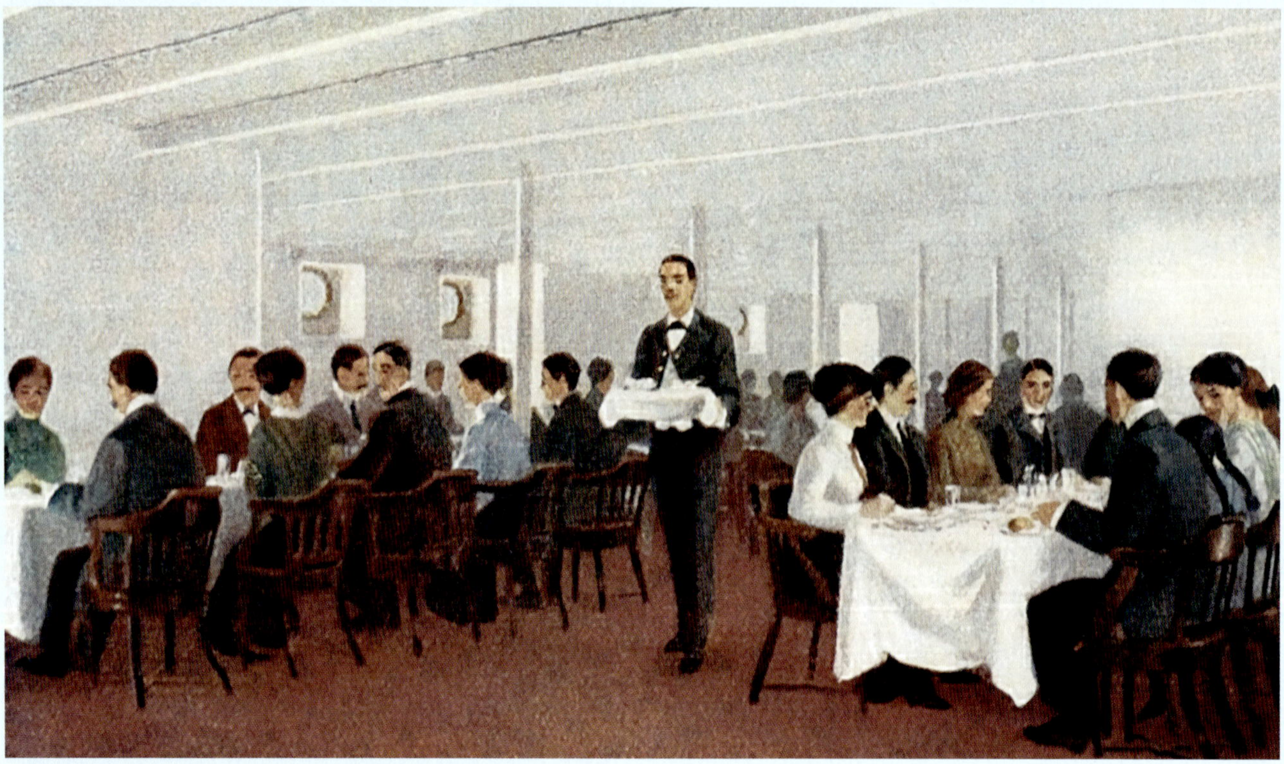

The Third Class Dining Saloon was located on F Deck, and accessed from 'Scotland Road' on E Deck. It was divided into two main areas by a watertight bulkhead. Passengers could enjoy being waited upon, a refreshing change from their ordinary lives. (Authors' Collection)

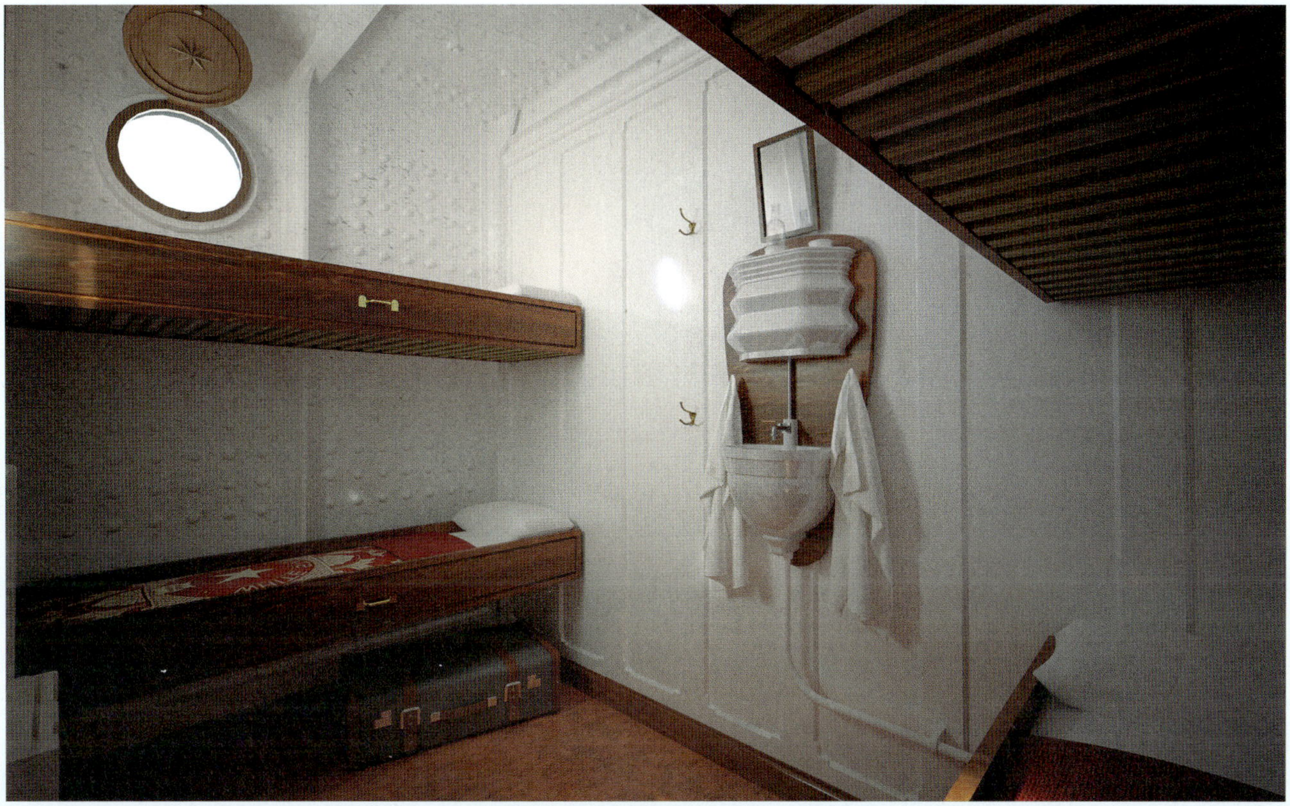

Third Class stateroom E-107 was a comfortable cabin located aft on the port side. Men were primarily housed forward, while women and families would have been accommodated in cabins in this area. (Chris Walker)

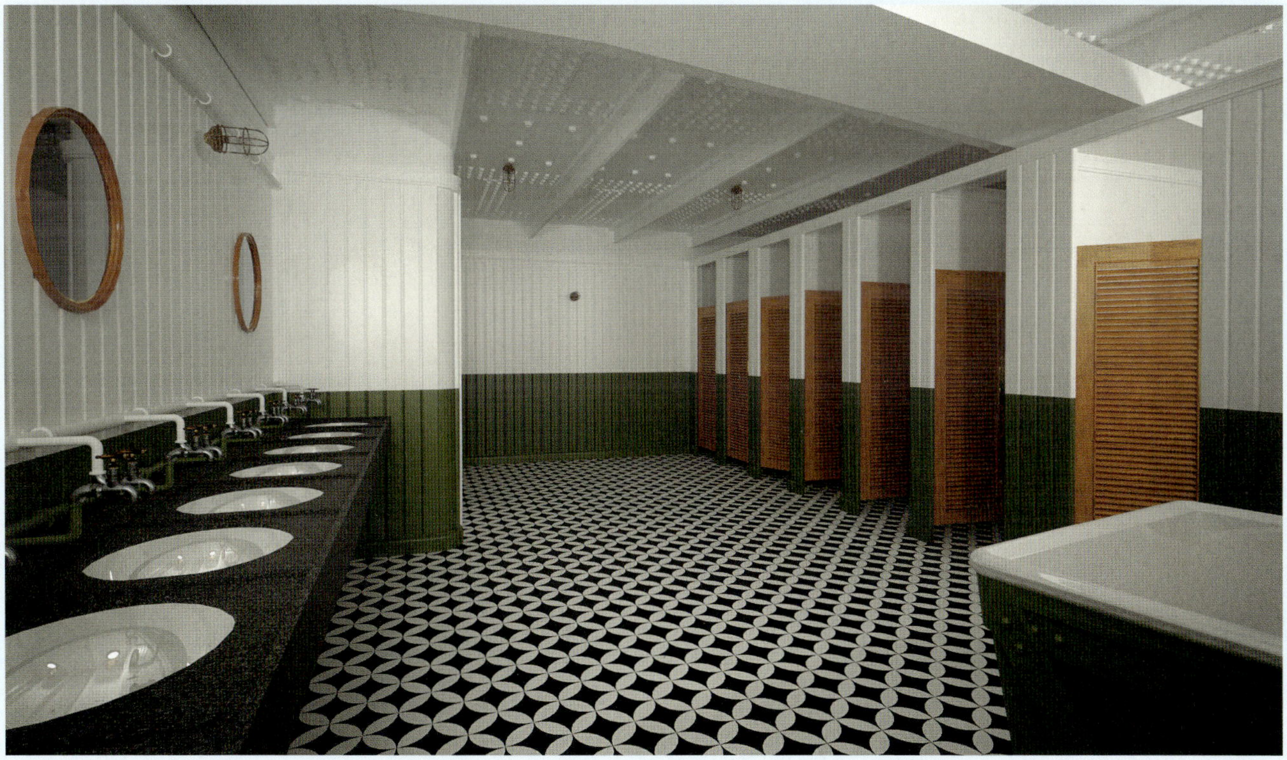

This Ladies' Lavatory was located forward on the starboard side of E Deck, below the Third Class Open Space. (Chris Walker)

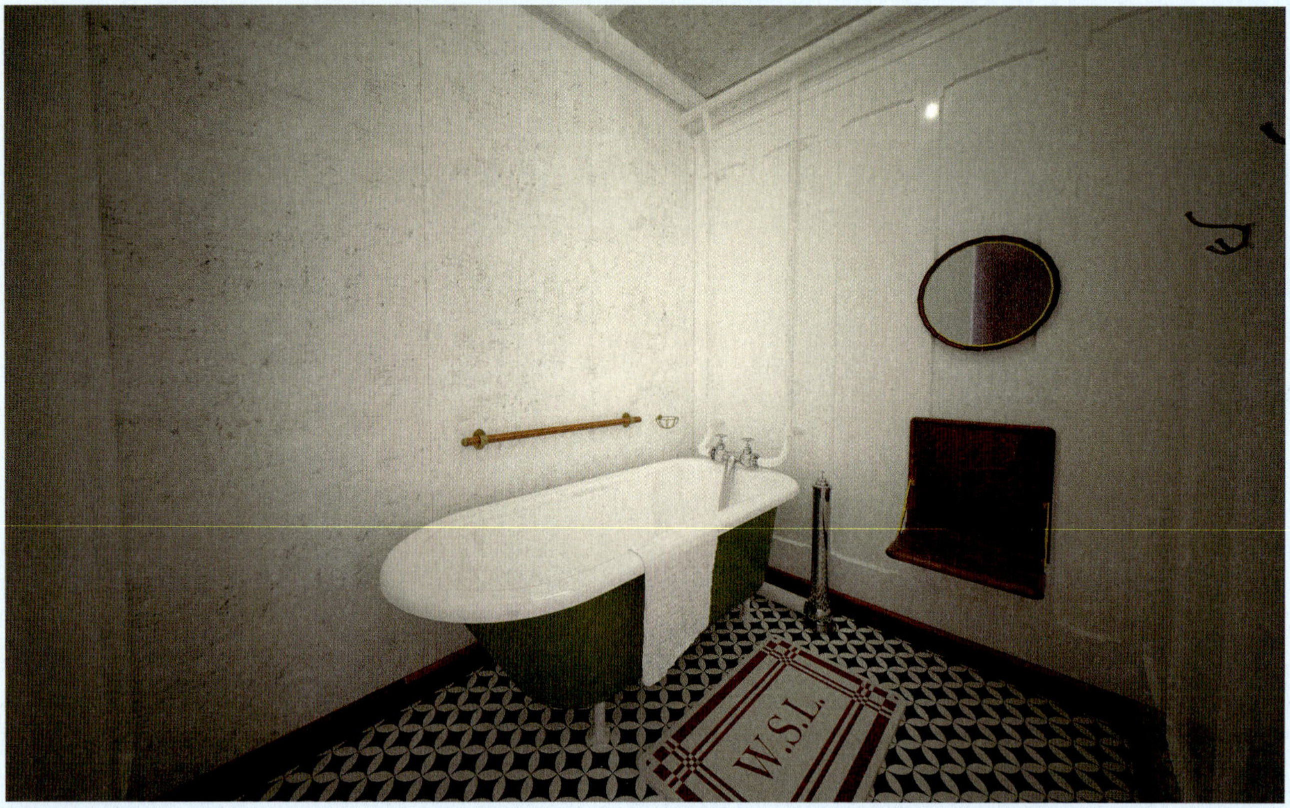



The men's bath on the port side of D Deck, located aft of the Third Class Entrance. It was one of only two tubs in Third Class. However, many people in 1912 would not have thought it odd to go without bathing for a week. (Chris Walker)

## OFFICER AND CREW SPACES: A TOUR

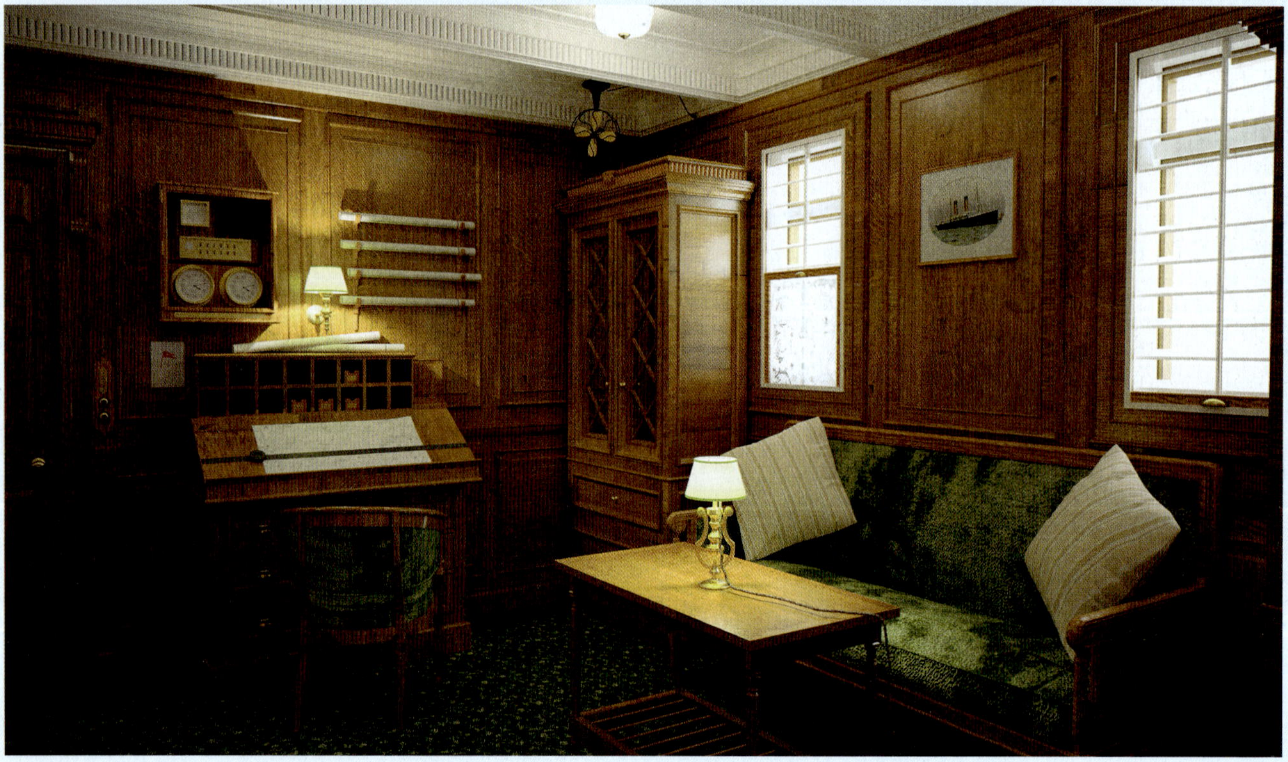

Captain Smith's Sitting Room, located just aft of the Navigating Room. It overlooked the starboard Boat Deck and was the equivalent of some of the best First Class suites. (Chris Walker)

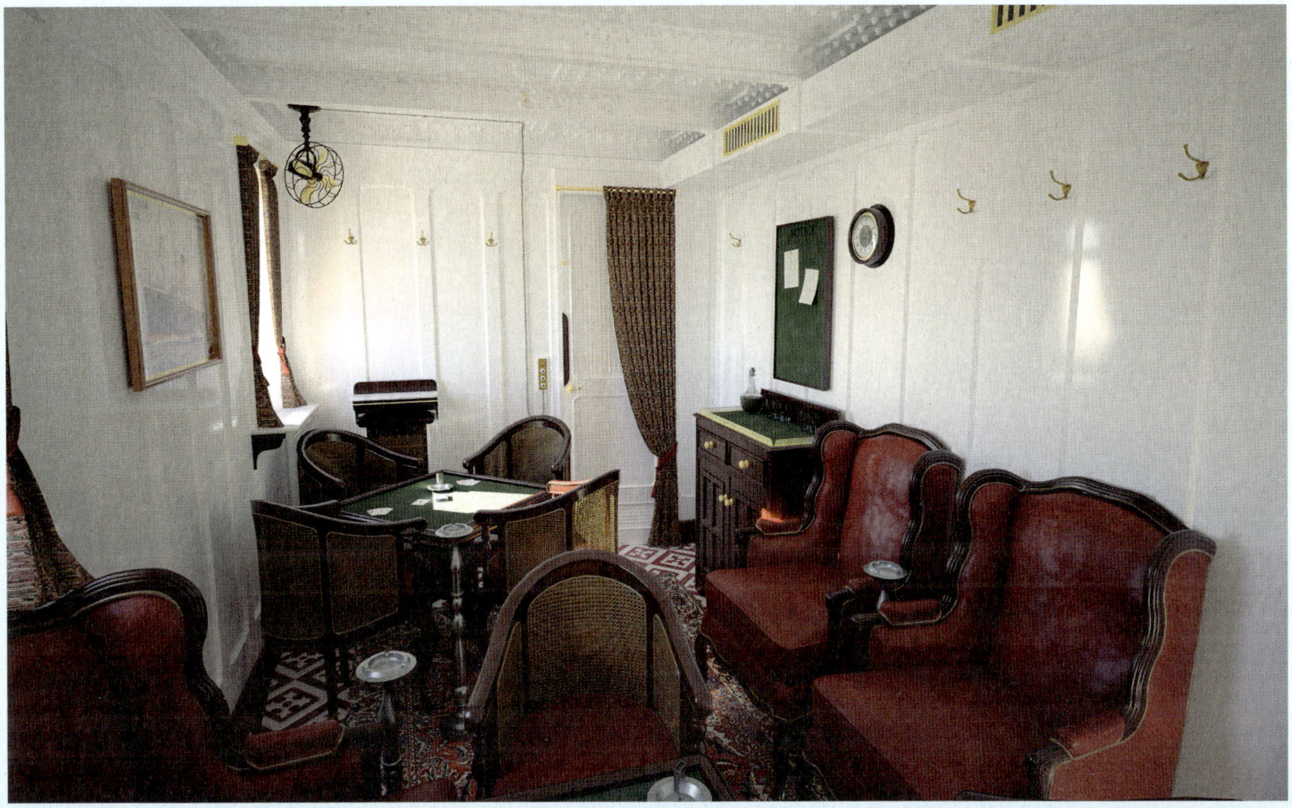

The Officers' Smoke Room on the starboard side of the Boat Deck, just behind the forward funnel. This view looks aft toward the adjoining Pantry. Officers did not have much downtime at sea, but this was a comfortable space in which to relax. (Chris Walker)

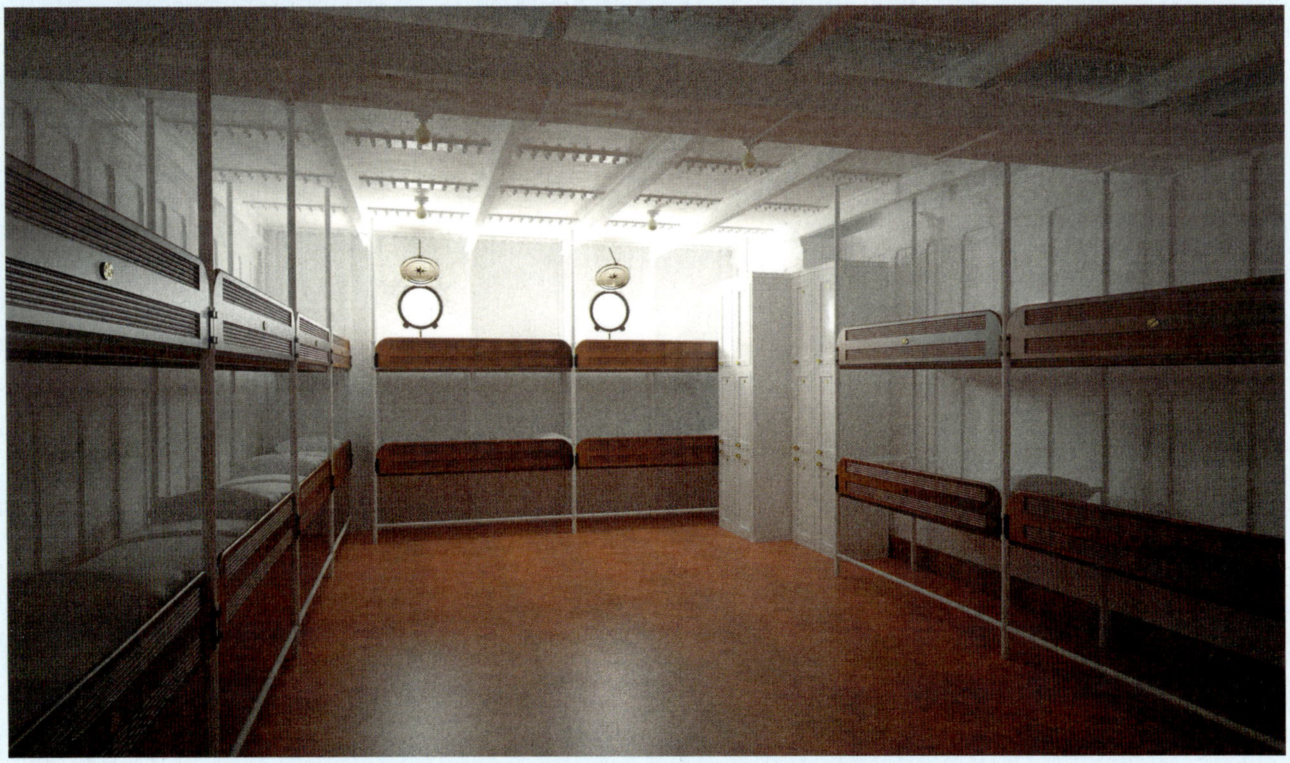

Located on the port side of E Deck, off 'Scotland Road', this dormitory-style room accommodated twenty-two waiters. (Chris Walker)

Although April was not the peak of the travel season on the North Atlantic, bookings were relatively strong. They had been aided by the fact that, because of the coal strike in Britain, IMM had cancelled the sailing of many of its smaller liners, transferring the coal and passengers from each of them to *Titanic* and thus filling many berths that would otherwise have gone empty. Some of those passengers who had been transferred were enthused about the upgrade, while a few others were hesitant about taking the first trip on a ship this big and so thoroughly untested.

## *TITANIC*'S OFFICER CORPS

*Titanic*'s officer corps was composed of a blend of well-seasoned and respected veteran officers, as well as some young but promising up-and-comers. Technically speaking, her first commander was Herbert James Haddock, who took the reins in Belfast on 25 March. He was relieved by Captain Smith on 31 March, and would take Captain Smith's place in command of *Olympic* on Wednesday, 3 April 1912.

Let us now meet the men who were responsible for overseeing *Titanic*. We will meet them in the order in which they joined the ship:

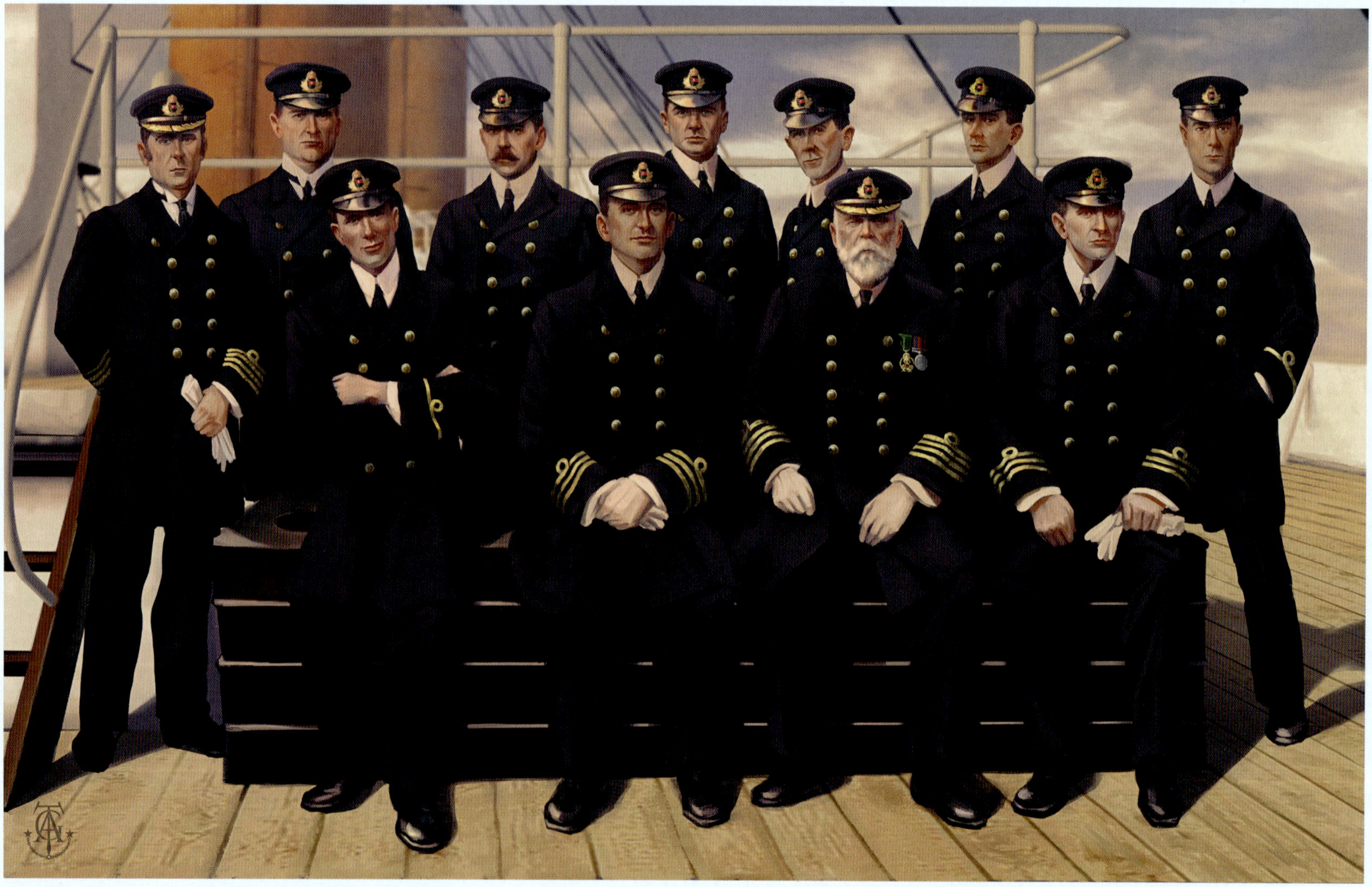

This imaginative view shows all of the officers who ever served aboard *Titanic*. Back row, left to right: Captain Haddock, McElroy, Pitman, Lightoller, Blair, Boxhall, Lowe. Front row, left to right: Moody, Wilde, Smith, Murdoch. (Tatiana Yamshanova)

**Name: William McMaster Murdoch**
**Rank:** Chief Officer (later First Officer)
**Hometown:** Dalbeattie, Dumfries, Scotland
**Residence:** Southampton, Hampshire, England
**Marital status:** Married
**Age:** 39
**Appearance:** 5ft 9in, brown hair (clean shaven), hazel eyes
**Previous ship:** *Olympic*

The ship's three senior officers arrived in Belfast on 20 March 1912, ten days before Captain Smith. Upon arrival, they found their quarters aboard *Titanic* incomplete, and they had to be put up in local hotel accommodations. William Murdoch had been appointed to serve as Chief Officer for the maiden voyage of *Titanic*, whereas on his previous ship, *Olympic*, he had served as First Officer. Murdoch was a Lieutenant in the Royal Naval Reserve, and was also the only Scotsman among the officer corps. A sensible and well-seasoned officer, his actions had prevented a disaster in 1903, while he was Second Officer on the *Arabic*. He narrowly avoided a collision with another vessel by countermanding the helm orders of a superior officer, shoving the Quartermaster out of the way and personally taking control of the ship's helm. He was a veteran of White Star Line's Australian run.

**Name: Charles Herbert Lightoller**
**Rank:** First Officer (later Second Officer)
**Hometown:** Chorley, Lancashire, England
**Residence:** Netley Abbey, Hampshire, England
**Marital status:** Married
**Age:** 37–38 (b. 30 March 1874)
**Appearance:** 5ft 10in, dark hair (previously auburn and clean shaven), blue eyes
**Previous ship:** *Oceanic*

Murdoch's friend Charles Lightoller was to serve as First Officer. Just shy of his 38th birthday upon arrival in Belfast on 20 March, Lightoller had seen an eventful career prior to joining *Titanic*. He had served on a variety of vessels, survived being shipwrecked, and even left the sea for a time to prospect for gold in the Yukon and work as a cowboy. Like Murdoch, Lightoller was a veteran of the White Star Line's Australian run. His sense of humour had occasionally nearly got him into trouble with stricter masters, but he was also a very responsible and hardworking officer.

**Name: David Blair**
**Rank:** Second Officer (did not sail on 10 April)
**Hometown:** Newport, Isle of Wight, England
**Marital status:** Married
**Age:** 37
**Appearance:** 5ft 6in, brown hair (clean shaven), blue eyes
**Previous ship:** *Teutonic*

Rounding off the group of senior officers, 37-year-old David Blair was to serve as Second Officer. Blair's previous posting had been aboard the *Teutonic*. *Titanic* was nearly 300ft longer than *Teutonic*, so the scale of the new vessel must have been hard for Blair to grasp. He seemed quite pleased with the promotion. Blair remained with *Titanic* in Belfast, served as her Second Officer during the trials on 2 April, and stayed on with the ship all the way down to Southampton. During her stay in that port, he continued to serve in that capacity and helped ready the ship for the upcoming trip. He was officially released on 9 April due to the reshuffling of senior officers.

**Name: Herbert John Pitman**
**Rank:** Third Officer
**Hometown:** Sutton Montis, Somerset, England
**Residence:** Castle Cary, Somerset, England
**Marital status:** Married
**Age:** 34
**Appearance:** 5ft 7in (down from 5ft 9in in 1900), dark-brown hair (moustache), blue eyes
**Previous ship:** *Oceanic*

*Titanic*'s four junior officers arrived in Belfast at noon on 27 March 1912, following a gruelling overnight trip from Liverpool. The young but promising group of men immediately reported to Chief Officer Murdoch after arriving aboard *Titanic*. Herbert Pitman would serve as Third Officer. An inch or two shorter than average for the officers who sailed on the maiden voyage, Pitman's signature moustache made him stand out among his mostly clean-shaven fellow officers. He had already been at sea for sixteen years, serving on both the Australian and Japanese routes. On his previous vessel, *Oceanic*, he had served as Fourth Officer.

**Name: Joseph Groves Boxhall**
**Rank:** Fourth Officer
**Hometown:** London, England
**Residence:** Hull, Yorkshire, England
**Marital status:** Married
**Age:** 28
**Appearance:** 5ft 9in, dark-brown hair (clean shaven), dark-brown eyes
**Previous ship:** *Arabic*

Joseph Boxhall came from a family that seemed to have the sea in their blood. By the time he joined *Titanic*, he had already been at sea for thirteen years, despite being just 28 years old. He had been confirmed as a Sub-Lieutenant in the Royal Naval Reserve the preceding October. Another veteran of the Australian run, Boxhall had previously served on *Oceanic* and *Arabic*.

**Name:** Harold Godfrey Lowe
**Rank:** Fifth Officer
**Hometown:** Llanrhos, Conwy, Wales
**Residence:** Barmouth, Gwynedd, Wales
**Marital status:** Single
**Age:** 29
**Appearance:** 5ft 8in, brown hair (clean shaven), brown eyes
**Previous ship:** *Belgic*

Harold Lowe was new to the North Atlantic service, as he had served on the Australian run aboard *Belgic* and *Tropic* prior to his transfer. Lowe was a teetotaller and a no-nonsense man, although he was occasionally a little strong with his language, and sometimes came off a little brusquely. Yet the North Atlantic route was a prestigious service, and so this promotion was no doubt very pleasing to the new Fifth Officer.

**Name:** James Paul Moody
**Rank:** Sixth Officer
**Hometown:** Scarborough, Yorkshire, England
**Residence:** Grimsby, Lincolnshire, England
**Marital status:** Single
**Age:** 24
**Appearance:** 5ft 11in, brown hair (clean shaven), blue eyes
**Previous ship:** *Oceanic*

As it turned out, three of *Titanic*'s officers had come over directly from *Oceanic*. In addition to Lightoller and Pitman, rounding off the officer corps was the promising James P. Moody, just 24 years old. Moody had only received his Master's Licence the previous April. Having endured a harsh winter, he was hoping to spend the summer on the Atlantic aboard the *Oceanic* and was actually initially reluctant to accept the assignment aboard *Titanic* in spite of the prestige of being promoted to a position on White Star's newest and greatest ship.

**Name:** Edward John Smith
**Rank:** Captain
**Hometown:** Hanley, Staffordshire, England
**Residence:** Southampton, Hampshire, England
**Marital status:** Married
**Age:** 62
**Appearance:** Approx. 5ft 8in, white hair and beard, eyes possibly grey or blue
**Previous ship:** *Olympic*

Edward John Smith was White Star's Senior Captain, and the unofficial Commodore of the line. He was thus a natural choice to command *Titanic* on her maiden voyage. He had been with the company since 1880, and had skippered their *Majestic* (1889), two of the 'Big Four', *Baltic* and *Adriatic*, and *Olympic*. A member of the Royal Naval Reserve and a veteran of the Boer War, Smith was popular with crew and passengers alike; indeed, some passengers would only cross the Atlantic on vessels he commanded.

He had taken *Olympic* out on her maiden voyage in June 1911, and had left her to travel up to Belfast, arriving there on Saturday, 30 March 1912, just in time to take *Titanic* out on her sea trials. He took over from *Titanic*'s first Captain, Herbert James Haddock, who had been in charge less than a week. Haddock had overseen the ship at Belfast while the delivery crew was arriving. Haddock swapped positions with Captain Smith, departing Belfast and taking command of the *Olympic* in time for her next crossing. When Captain Smith stepped aboard *Titanic*, his entire senior officer corps was already in place aboard.

There has been great debate on whether this was to be Captain Smith's final crossing of the Atlantic before retiring, or whether he was to stay on until the third ship of the class made her maiden crossing two or two and a half years on but, considering his age and the fact that he was apparently involved in spreading the rumours of his upcoming retirement, it seems very likely that this was actually going to be his final round-trip voyage in command.

**Name:** Henry Tingle Wilde
**Rank:** Chief Officer
**Hometown:** Liverpool, Merseyside, England
**Residence:** Walton, Merseyside, England
**Marital status:** Widowed
**Age:** 39
**Appearance:** 6ft 1in, brown hair (clean shaven), blue eyes
**Previous ship:** *Olympic*

Almost as soon as *Titanic* reached Southampton, news was received of a significant shake-up in the senior officer corps. *Olympic*'s Chief Officer, Henry T. Wilde, 39 years old, had been kept ashore when his ship departed Southampton on 3 April. He had been slated to receive command of another White Star Line steamer, *Cymric*, and had been very excited at the prospect of taking command of his own ship. By the end of March, however, he had learned that the coal strike would dash those plans.

So, why was he kept ashore when *Olympic* sailed on 3 April? It is possible that a transfer to *Titanic* was then in the works. In a letter to his niece and nephew on Sunday, 7 April, he still was not sure that he would be sailing with *Titanic*, although he had been aboard her that day and Friday, 5 April, helping to prepare the ship for her maiden voyage. It was only at 2.00 p.m. on Tuesday, 9 April that Wilde finally received confirmation, in a telegram from White Star's headquarters in Liverpool, that he would be sailing with *Titanic*. He wrote to his children's nanny that he would be back in eighteen days, enclosing £10, and also referring to how busy and unsettled he had been with all of the confusion.

This news was not welcome to many of the other officers aboard – not due to any ill will towards Wilde, but rather because of the ways that the change impacted upon them. Once Wilde arrived on the evening of 9 April, William Murdoch was bumped down to First Officer, Lightoller was bumped from First Officer to Second Officer, and David Blair was bumped from the ship entirely. These demotions were said to be temporary, just for the maiden voyage, and the junior officers retained their original ranks. Yet Blair was upset, writing, 'this is a magnificent ship, I feel very disappointed I am not to make the first voyage'. Lightoller himself later referred to the change as a 'doubtful policy' that threw both he and Murdoch 'out of our stride', and added that, in addition to the temporary demotion, it 'caused quite a little confusion'.

Wilde's wife and twin sons had passed away in December 1910, but he was by no means alone, as he still had four children to look after, and was also close to his sister and her children. On 11 April, while *Titanic* was in Queenstown, Wilde posted three letters. In one to his daughter Jennie, he said that *Titanic* was 'a very fine ship, an improvement on the *Olympic* in many ways', and he added that he hoped she would be able to come down to see the ship, apparently after the liner returned to Southampton. By stark contrast, in a letter supposedly posted at the same time and written to his sister, Wilde allegedly said he didn't like *Titanic* and had 'a queer feeling about it', yet the letter has never come to light for analysis, though it is often quoted.

## SECOND OFFICER BLAIR AND THE MISSING BINOCULARS

Some have erroneously claimed that David Blair's departure from *Titanic* on 9 April actually doomed the vessel. This is because, following the departure from Belfast, *Titanic*'s lookouts found that they did not have any binoculars. Due to this, Blair loaned Lookout George Hogg his own set of glasses.

After arriving in Southampton, Blair came up to the Crow's Nest and told Hogg to lock the binoculars up in the Second Officer's cabin and to return the keys to him. Blair gave this order despite there being a storage box in the Crow's Nest. This was supposedly because binoculars were not left in the Crow's Nest while the ship was in port, to prevent theft. Hogg did as he was told, but before he could return the keys to Blair, he was tapped to do some work in the Forecastle. As a result, he handed the keys to Seaman William Weller, to give them to Blair. Blair then accidentally took them with him when he left the ship.

Following Blair's departure from the ship, nobody knew where the binoculars were and, despite repeated requests, none were provided to the lookouts during the voyage. The conspiracy minded among us would have readers believe that if only Blair had remained aboard the ship, the lookouts would have had binoculars on the night of the disaster, thus spotting the iceberg in time to avoid the collision. However, this is nothing but hyperbole. While contemporary opinions vary, evidence shows that binoculars were not universally provided to lookouts, and many lookouts and officers felt that binoculars were not necessary, or even particularly useful.

Using binoculars tended to narrow a lookout's field of vision dramatically and thus led to missing other hazards. While the lookouts who testified at the disaster inquiries liked having binoculars handy, their testimony was nearly unanimous that having them was unnecessary for initially spotting objects such as icebergs. Binoculars were only really useful in looking at an object after it was initially sighted. Indeed, modern tests made with binoculars at night on open bodies of water have demonstrated that using them to detect objects is highly ineffective.

A far more vital factor in reducing the ability of the lookouts to see well that night would have been the icy wind that stung their eyes, causing them to sting and tear up, and that was a factor that had nothing to do with binoculars.

While Blair's missing binoculars make for a dramatic story and catchy headlines, they simply did not have any effect on the disaster.

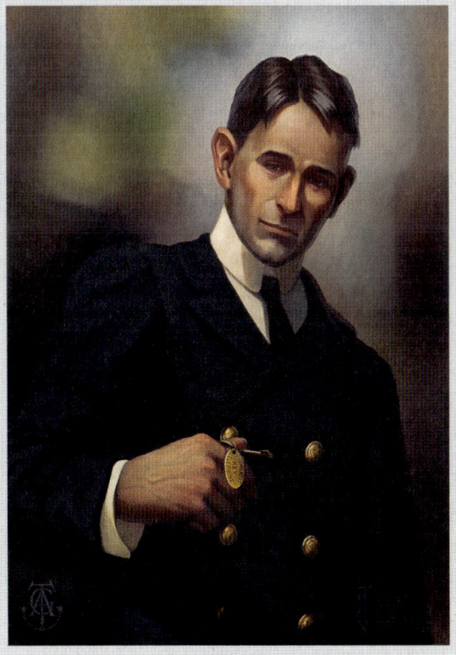

*Titanic*'s first Second Officer, David Blair, holding the infamous key 'that sealed *Titanic*'s fate'. Blair forgot to return this key when he left the ship, though the mistake certainly did not doom *Titanic*. (Tatiana Yamshanova)

Late that morning, the ship's officers made their official status reports to Captain Smith and the Board of Trade representatives that *Titanic* was ready to sail. Pilot George Bowyer boarded yet again, preparing to take the new ship out of port for the first time, and at 11.45 a.m. came the cries of, 'All ashore!'

Stragglers still aboard who were not staying for the trip disembarked, the gangways were retracted, and then the massive whistles that stood proudly on the forward two funnels blasted three times, signalling imminent departure. A last few crew members who were running late from the pub tried to make it aboard, and only some managed to make it onto the ship while others were turned aside at the gangway. In the end, *Titanic* cast off a few minutes after noon, and the tugs began to ease her away from the dock. Once she was positioned, the tugs cast off and her engines were started. *Titanic* was underway.

Very quickly, the maiden voyage almost ended in disaster. Just after engaging her engines, *Titanic* drew parallel with two smaller liners tied up to the piers; the suction from her engines pulled the SS *New York* from its moorings and it began to drive toward *Titanic*'s port stern. Passengers aboard *Titanic* braced for what seemed an inevitable impact and *Titanic*'s engines were quickly stopped and the port engine rapidly reversed in an ingenious but vain attempt to stop a calamity. Just when all seemed lost, a single tug, *Vulcan*, got a line on the errant *New York* and stopped its advance toward *Titanic*. Some said there was as little as 4ft separating the ships.

This very rare original photograph shows *Titanic* proceeding down Southampton Water on the afternoon of 10 April 1912. (Günter Bäbler Collection)

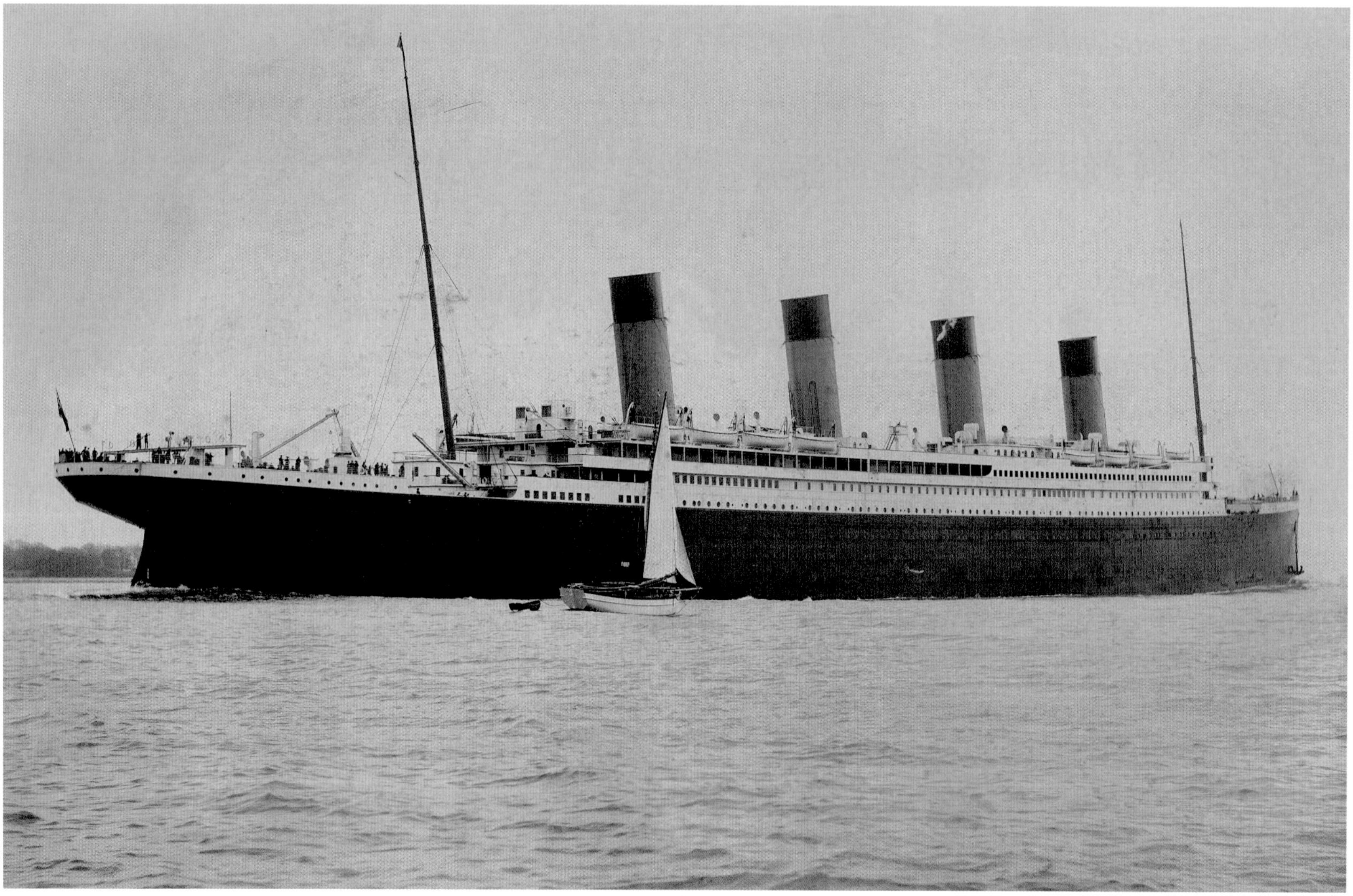

*Titanic* backed up the channel, while *Vulcan* and other tugs moved *New York* to a safer mooring, and while they firmly secured the other moored liner, White Star's own *Oceanic*. *Titanic* moved downriver some distance before stopping again to offload some standby crew members who would not be needed for the trip, and only then did *Titanic* manage to get underway in earnest, now thoroughly behind schedule. Whereas she should have reached open water at the Nab Light Vessel, off the east side of the Isle of Wight, at about 2.35 p.m., it seems she did not pass that point until shortly after three o'clock.

Ahead of her lay a journey of about 65 miles across the English Channel, toward her first port of call – Cherbourg, France. The trip took roughly three and a quarter hours.

Many of *Titanic*'s most noteworthy or memorable passengers were to board at Cherbourg: the Astors, Margaret Brown, Edith Rosenbaum, Sir Cosmo and Lady Duff Gordon, the Thayer family, tennis star Karl Behr, and many others. Because *Titanic* was too big to tie up directly at a pier, the new White Star tenders *Nomadic* and *Traffic* would instead take these passengers, their luggage and the mails bound for Ireland and America out to *Titanic* once she came to a halt. The big ship crossed the breakwater and dropped anchor at 6.30 p.m., about twenty minutes before sunset, with her starboard side facing south, toward the city.

Passengers aboard *Nomadic* and *Traffic*, many of whom were not accustomed to waiting for much of anything, had been cooling their heels waiting for the tardy ship to arrive while they tried to stay warm, but a thrill ran through the group as *Titanic* appeared. The two little tenders started out to meet her across the choppy harbour. *Traffic* arrived first, tying up alongside with the mail and 102 Third Class passengers.

*Nomadic* took nearly half an hour to arrive alongside *Titanic*, and her passengers were awed to see the liner's lights coming to life as they approached. She tied up shortly after sunset, about 7.00 p.m., and soon 172 First and Second Class passengers were climbing aboard *Titanic*.

Those in First Class entered via the gangway doors on D Deck, which gave entry to the Reception Room. Passengers who had been aboard *Titanic* from Southampton watched through the windows of the Dining Saloon as the newcomers made their way aboard and began to find their way around the enormous vessel. Meanwhile, Thomas Andrews looked down from the decks with pride on the *Nomadic* and *Traffic*. His firm had built both of them just the previous year.

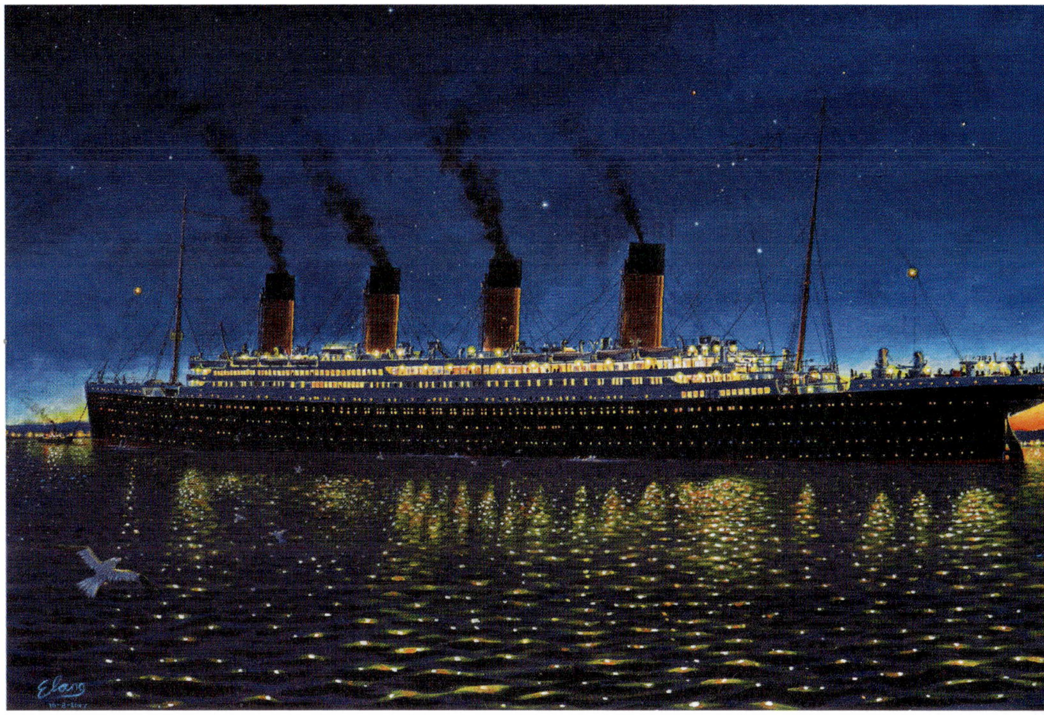

*Titanic* anchored in Cherbourg Harbour at dusk on 10 April. As day turned to night, the ship's lights presented boarding passengers with a fantastic spectacle as they approached *Titanic*. (Elang Erlangga)

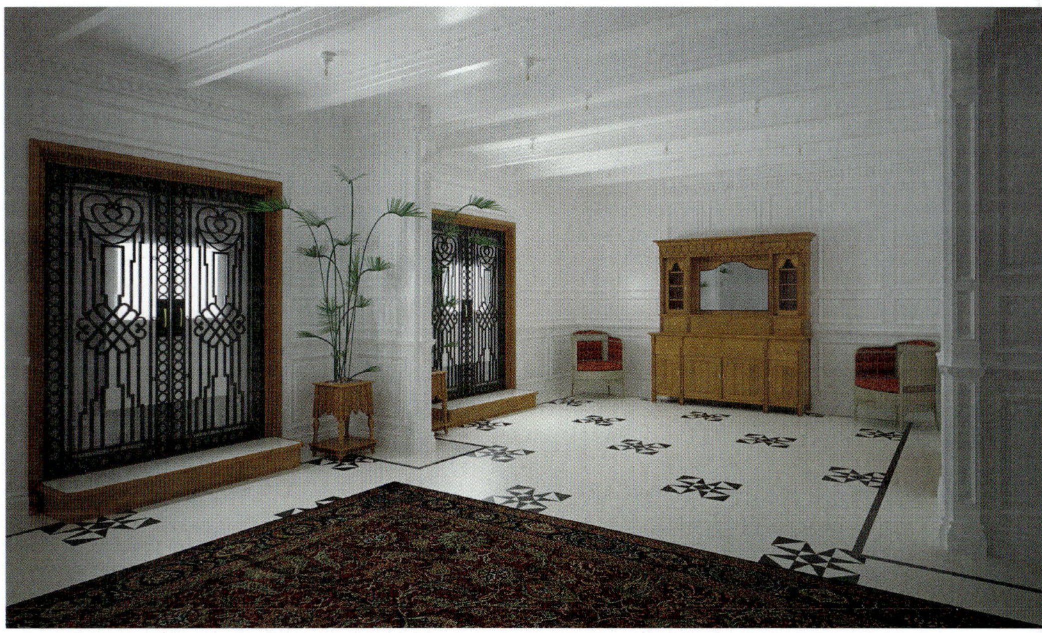

First Class passengers boarding from *Nomadic* that evening came through the gangway doors in the starboard Entrance on D Deck, just off the Reception Room. Passengers at dinner in the Saloon remembered watching them come aboard. (Chris Walker)

*Titanic* weighed anchor at 8.10 p.m. and set course for Queenstown, Ireland, steaming through the night at just over 20 knots. In each class, passengers began to work out how to find their way around *Titanic* and settled into their accommodation. Even those who were accustomed to travel at sea found *Titanic* bewildering in her sheer size, and neither was the crew impervious to this effect. For those who had transferred from *Olympic*, of which there were many, *Titanic* was at least not unfamiliar in scale; however, even for these, there were so many subtle differences between the two ships that some clearly had trouble getting accustomed to *Titanic*. For crew or officers who had come over from smaller White Star ships, *Titanic* was a completely new sort of creature to master and operate.

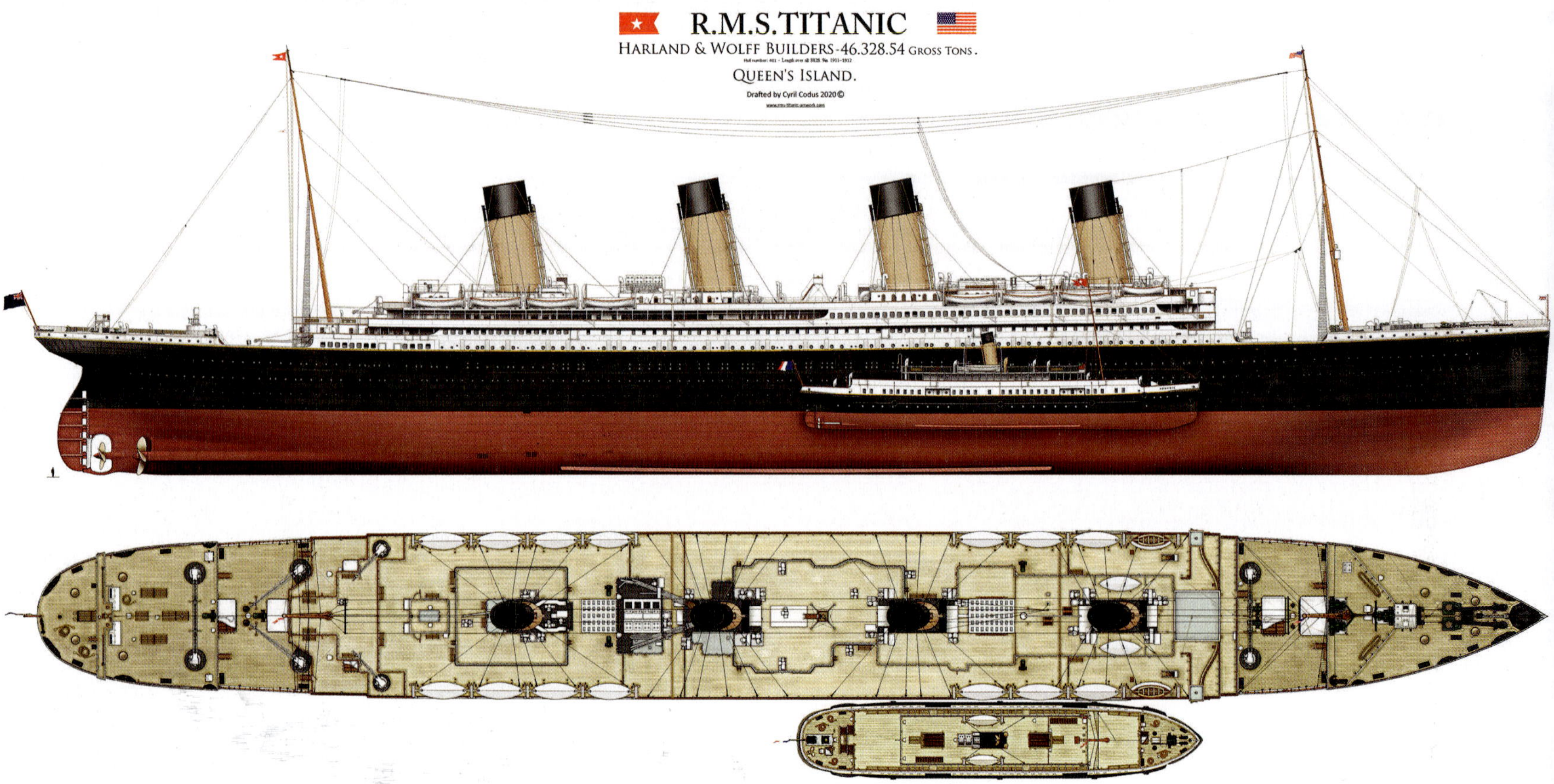


This profile and overhead view shows the tender *Nomadic* alongside *Titanic*, as she was in Cherbourg. (Cyril and Lionel Codus)

## REBUILDING A FIRST CLASS STATEROOM

*Titanic*'s First Class accommodation offered a wide variety of styles and price points for potential passengers to choose from. Beyond budgetary limitations or a preferred location aboard the ship, there really was something for every taste.

First Class Stateroom B-58 was among one of *Titanic*'s finest offerings. It was rendered in Louis XVI style, with damask silk wall coverings inset into white-and-gilt panels and a luxurious brass bed. Although the ship featured a second stateroom with similar decor (C-63), B-58 was an offering not found on *Olympic*. It was located on the port side of the ship, tucked directly behind the port-side Parlour Suite – B-52, -54 and -56, occupied by Bruce Ismay, with its Private Promenade.

The 'footprint' of B-58 actually formed part of the spacious B Deck First Class Promenade on the *Olympic*. A door on its aft wall could connect this stateroom to B-60, a stateroom of similar size but rendered in Queen Anne style. B-58 had been largely completed by January 1912, with some minor work continuing into February; certain items like the quilt and cushions arrived in March and the completed stateroom was photographed by Robert Welch early that same month.

During *Titanic*'s maiden voyage, B-58 was booked by 50-year-old Hélène Baxter, and her two grown children, 27-year-old Mrs Mary Hélène Douglas and 24-year-old Quigg Baxter, stayed in the adjoining B-60. The party paid nearly £250 for the two rooms, which would equate to over £29,000 (or about US$41,000) in 2021. The party boarded at Cherbourg, France, and they were returning to their home in Montreal, Québec, after a European tour. Quigg Baxter also booked accommodations in Stateroom C-90 for his lover, Berthe Antonine Mayné, a Belgian entertainer who he apparently intended to marry, and who was travelling under the assumed name of Madame de Villiers. Mrs Baxter and her daughter survived, as did Berthe Mayné, but Quigg Baxter did not.

*Titanic* aficionado Alexandre de Araújo Nunes was fascinated by the broad array of architectural periods represented aboard *Titanic* and undertook to recreate Stateroom B-58 in full size. Interestingly, it could also double for *Titanic*'s Stateroom C-63, since other than the fact that the two rooms had a slightly different size and layout, they were outfitted rather identically. C-63 is not known to have been occupied during the maiden voyage.

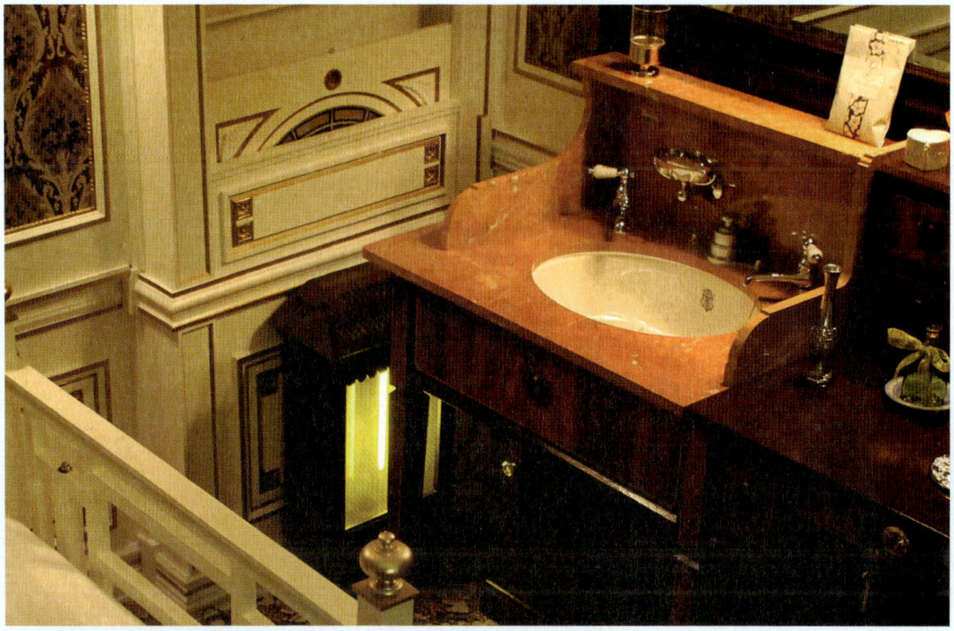

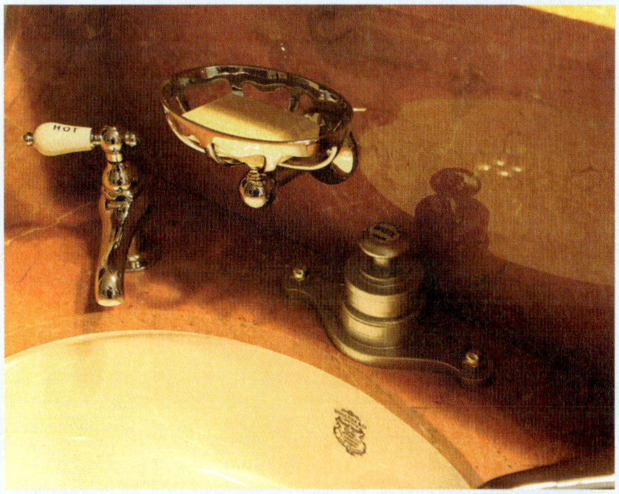

*Above:* An electric heater in recreated Stateroom B-58. Similar heaters were put to good use throughout the maiden voyage by many passengers, who found the ship bitterly cold. A washbasin with running water sits beside it. (Alexandre de Araújo Nunes)

*Left:* A close-up of the washbasin shows the attention to detail in this recreation. Note the Doulton & Co. logo in the sink basin as well as the nickel-plated taps and hardware, all a close match to *Titanic*'s original washbasins. (Alexandre de Araújo Nunes)

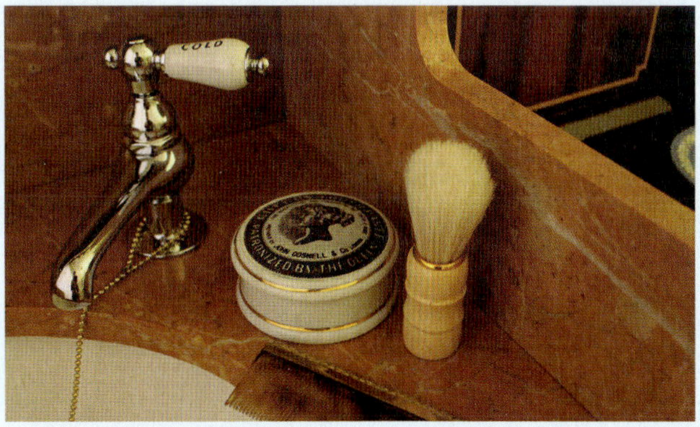

On the right side of the washbasin is a porcelain jar for cherry-flavoured toothpaste from John Gosnell & Co., said to be used by Queen Victoria herself. Many such jars are known to have been supplied to *Titanic*'s passengers, apparently by the company. (Alexandre de Araújo Nunes)

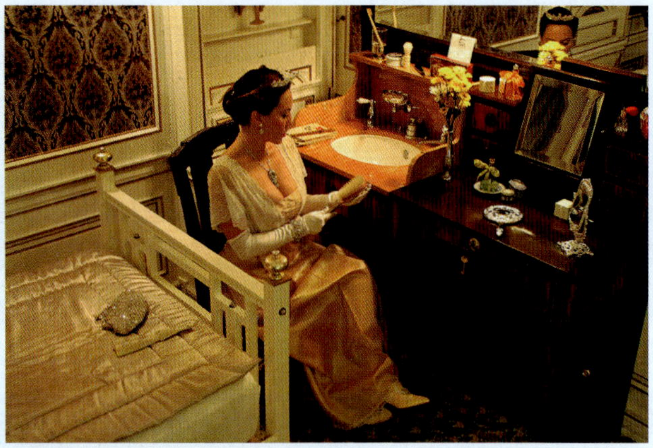

Dressing for dinner was de rigueur in First Class, except on the first and last nights at sea. Here, a woman dons her finest attire before heading down for supper. (Alexandre de Araújo Nunes)

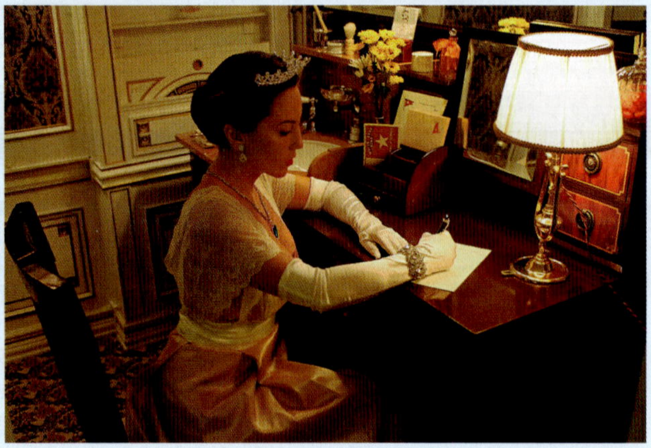

A lady dashes off a letter on White Star stationery to friends or family before going down to dinner. The letter would be deposited in the post office boxes made available for passengers and would actually be mailed when the ship reached port in New York. (Alexandre de Araújo Nunes)

Working from original photographs and plans, and meticulously rendering each detail into wood, plaster and iron, he skilfully crafted even the minutest details. These included the beautiful slide-down stained-glass privacy windows that sat just inside the outer ports – round ports have been included to represent C-63, while with the privacy screens closed and the outer ports obscured, the room easily passes for B-58 with its square ports. Damask inserts are set into the white-and-gilt wall panels, just like the originals, and a Doulton & Co. bowl in the washstand is nearly identical to those found and recovered from *Titanic*'s wreck site. Even the wires for the portable lamps look authentic to the originals, with their twisted braiding.

Then there are the finishing touches: Vinolia Otto toilet soap, known to have been supplied to First Class passengers to offer a 'higher standard of toilet luxury and comfort at sea'. A small jar of John Gosnell & Co. Ltd cherry toothpaste rests on the washstand. There is also a selection of White Star Line stationery and promotional materials on the writing desk.

This talented artisan then brought the scene to life with a model dressed in period garb. This recreated stateroom shows what many of *Titanic*'s First Class passengers experienced on their first night at sea as they settled into their staterooms.

Overnight and in the early morning of Thursday, as *Titanic* turned north-west toward Ireland, she began to take the Atlantic swell on her port side. It was at this point that some of the passengers who were less accustomed to ocean travel began to notice the ship was moving with a steady roll. While those familiar with ships marvelled at how steady *Titanic* was, some of those who had not yet gained their sea legs began suffering ill effects.

The visit at Queenstown (now called Cobh), Ireland, was brief, lasting just over two hours from the time she dropped anchor to the time she departed. She anchored approximately 2 miles offshore and, as in Cherbourg the previous evening, transferred mail and passengers by means of two small tenders, *America* and *Ireland*. The immigrants who boarded here were Irish, some of the last members of the great westward migration from that country that had begun over half a century earlier.

One First Class passenger, Francis Browne, who was then studying to become a Jesuit priest, disembarked, taking with him the photos he had captured while on board *Titanic*. Today, these offer spectacular glimpses of life aboard the ship during the first day of her maiden voyage. Local reporters also clambered over the ship during her brief stay, taking a number of now-famous photos.

*Titanic* raised anchor just after 1.30 p.m. ATS (Apparent Time Ship), and just less than twenty-five minutes later, at 1.55 p.m., she passed Daunt's Rock, the official starting point of her transatlantic crossing. She had on board 2,208 persons: 324 First Class, 284 Second Class, and 709 Third Class passengers, and a crew of 891.

The crossing ahead, comprising 2,891 miles, looked to be a triumph of all the hard work, labour, skill and forethought that White Star and Harland & Wolff had put into creating *Titanic*. Instead, the first signs of trouble would come with a break in the rhythm of the ship's motion some eighty-four hours and eighteen minutes after she passed Daunt's Rock, at 11.40 p.m. on Sunday, 14 April.

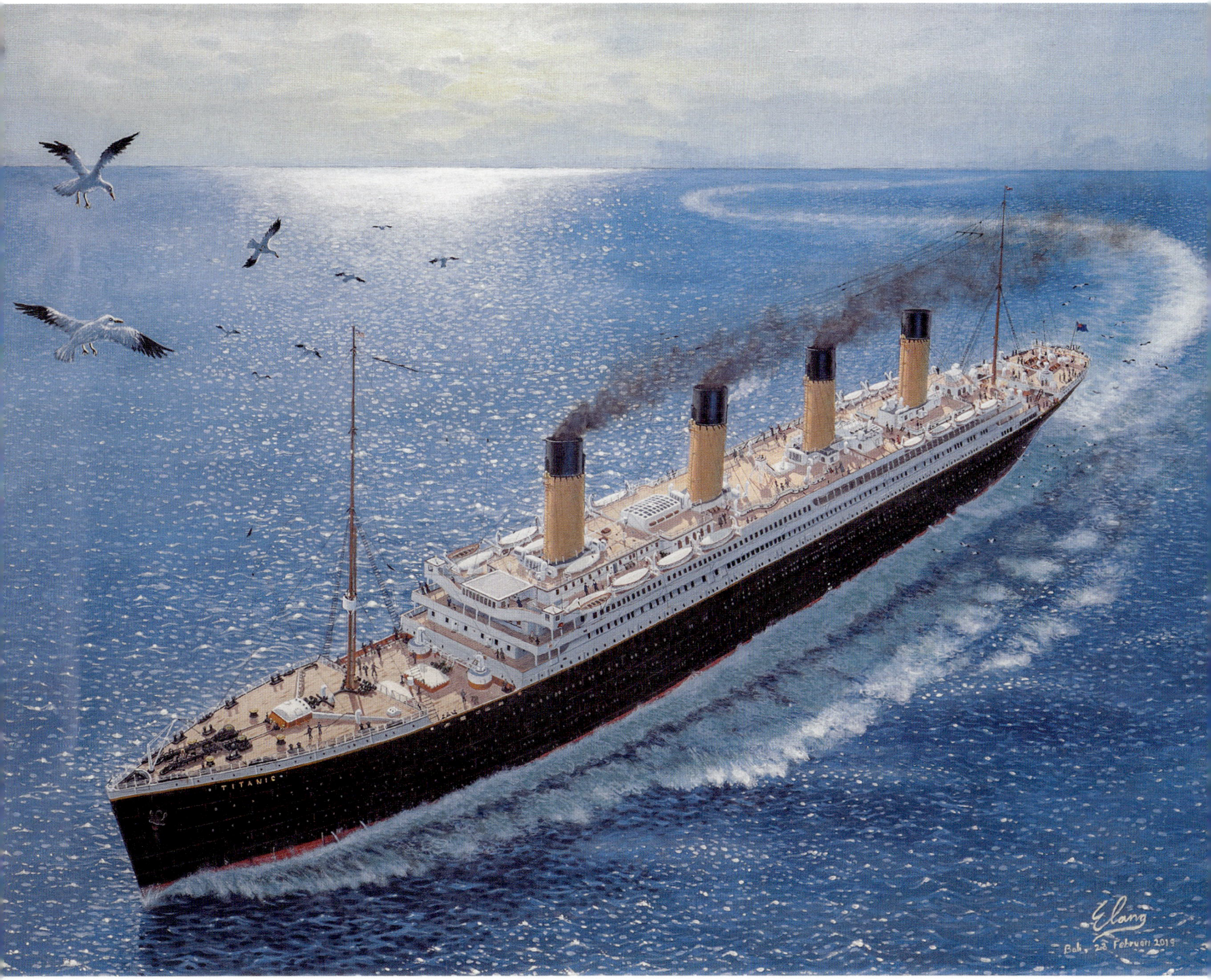


Early in the morning of Thursday, 11 April, as *Titanic* steams for Queenstown, headed north from the Lizard and Land's End, she steers a serpentine course to adjust her compasses. (Elang Erlangga)

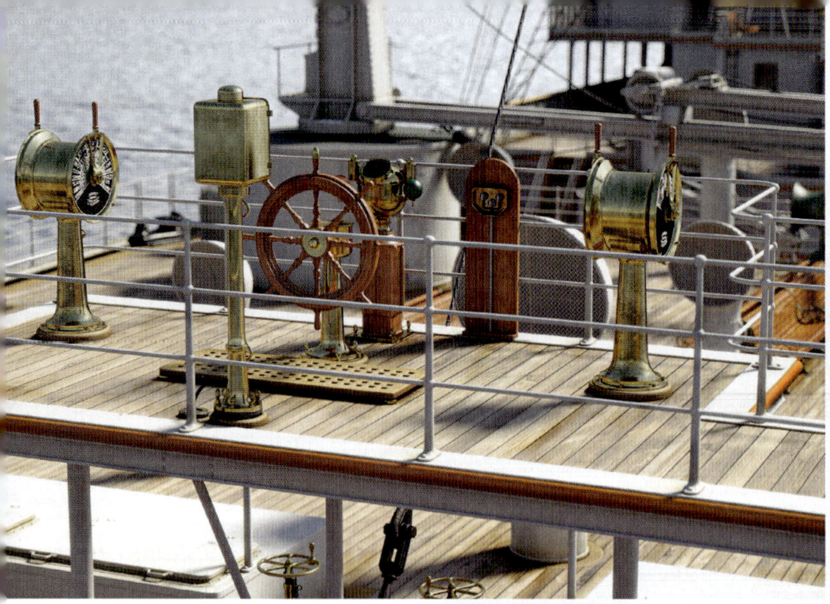

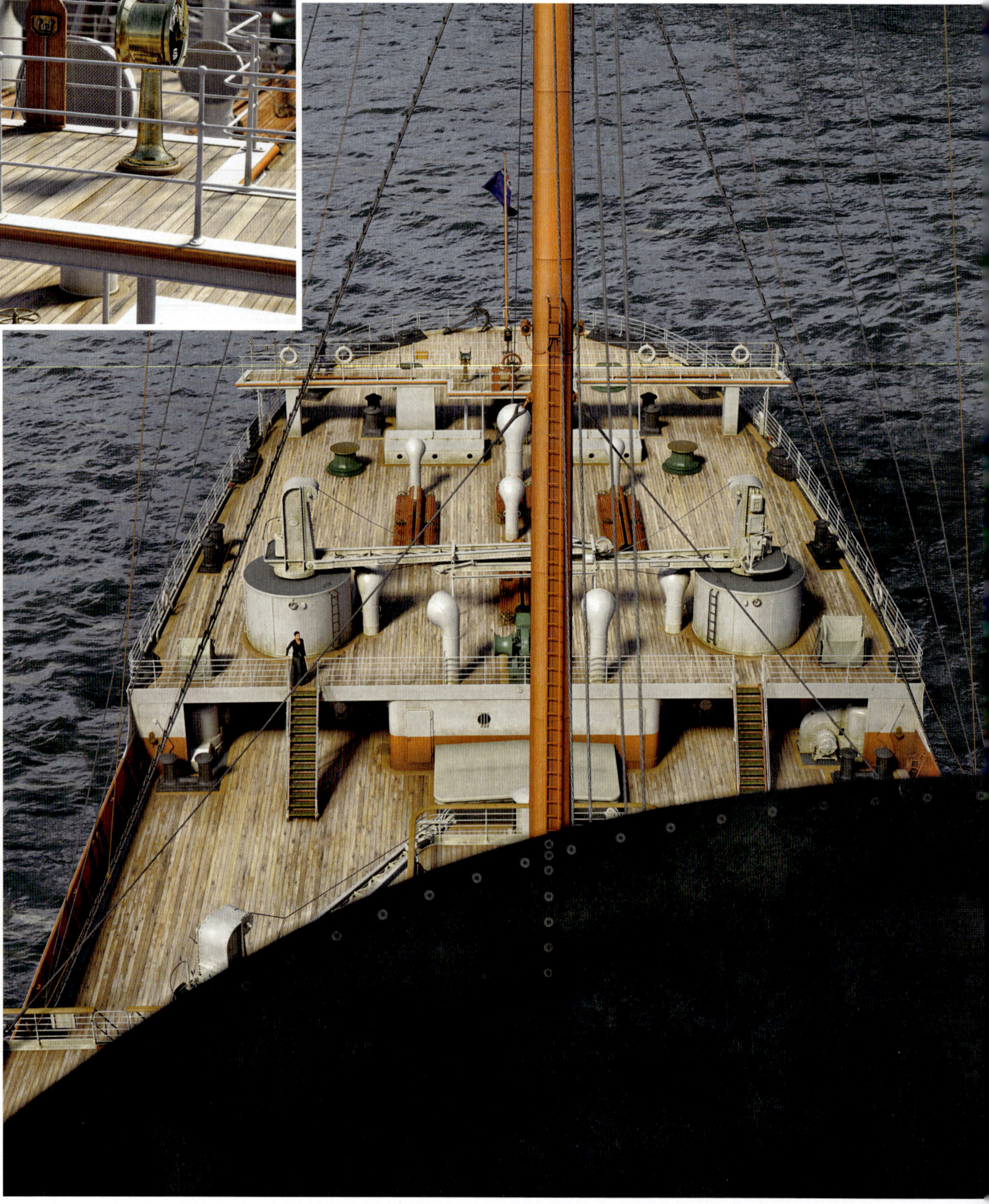

*Above:* This view shows the equipment on the Aft Docking Bridge, which sat above the Poop Deck at the stern of the ship. It featured an emergency helm, a telephone, an engine order telegraph and a steering/manoeuvring telegraph, and was typically manned by a Quartermaster throughout the trip. (Vasilije Ristović)

*Right:* While *Titanic* was anchored in Queenstown, a stoker poked his head above the rim of the aft funnel and was photographed from the tender *America*. This view shows the stoker's perspective from that vantage point, looking out over the aft decks. (Vasilije Ristović)

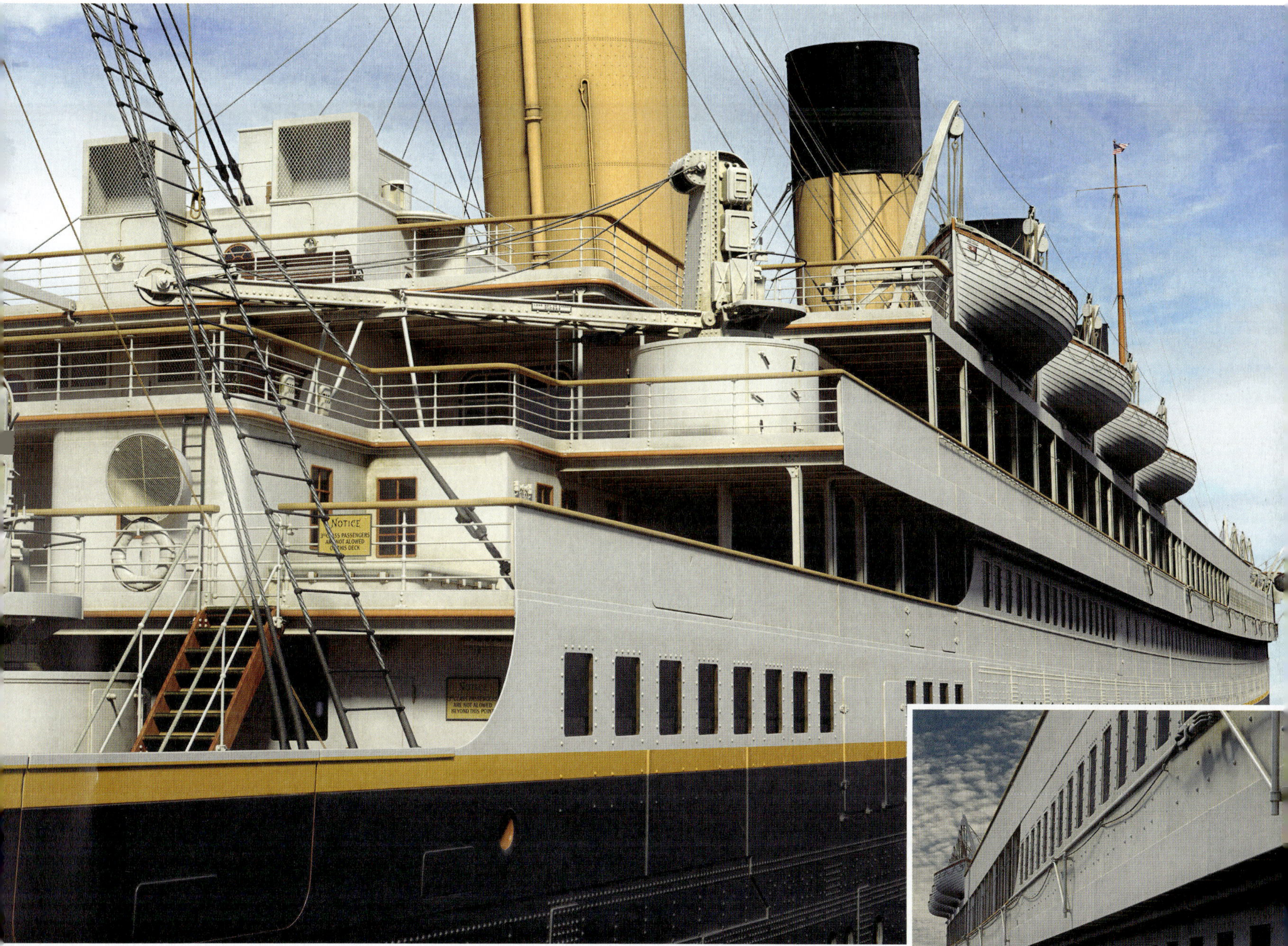


*Above:* Looking forward along *Titanic*'s starboard side, we see the aft lifeboats have been cranked out slightly over the edge to make more room for Second Class passengers to stroll along the Boat Deck. (Vasilije Ristović)

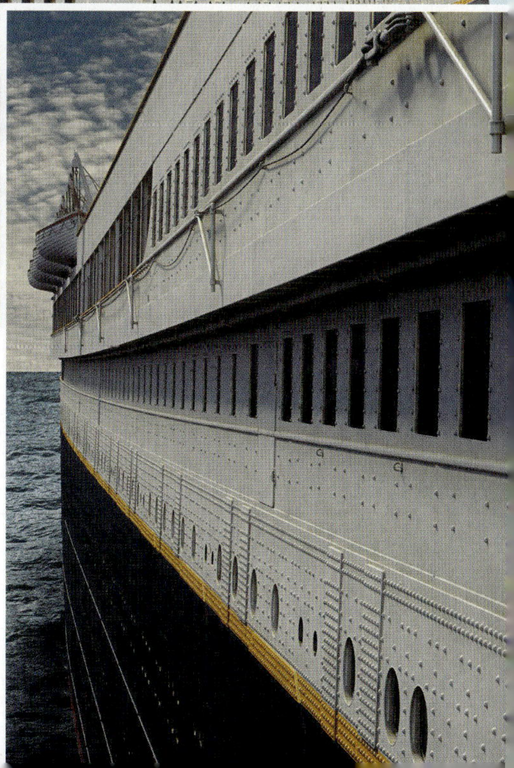

*Right:* The side of *Titanic* towers above the water as she pauses in Queenstown Harbour. (Vasilije Ristović)

## KEEPING TRACK OF *TITANIC*'S MAIDEN VOYAGE

In this section, we will outline all of the known times, locations, distances, speeds, engine revolutions and time differences to shore for the maiden voyage in one succinct place. We will also show how *Titanic*'s maiden voyage would likely have finished, straight through to her anticipated arrival time in New York. Before we begin, let us discuss several technical points and terms:

### Time Differences

Just as today, precisely five hours separated clocks in Southampton, England, and New York, USA. We will refer to British time (Southampton, Liverpool, London) as Greenwich Mean Time (GMT) and time in New York City, where *Titanic* would dock, as New York Time (NYT).

### Apparent Time Ship (ATS)

This was the time being carried by shipboard clocks. *Titanic*'s officers would each afternoon calculate the ship's anticipated geographic location when the sun reached its zenith in the sky (local noon) the next day, and work this to compute an overnight clock change so that shipboard clocks would read 12.00 at local noon the next day. They were highly skilled at this, but any minor clock adjustments – if any, they usually totalled only a minute or two – would be made in the morning hours. The ships did not change time zones in one-hour increments as most ships do today. While steaming westbound, days were longer, and while steaming eastbound, days were shorter. As *Titanic* was steaming westbound on her maiden voyage, we will refer only to clock 'setbacks'.

Notices were posted in the companionways every evening to inform passengers of the overnight clock change. Many passengers adjusted their personal timepieces before going to bed so that their watches would read the correct shipboard time when they woke up and, for example, went to breakfast.

For crew members in the Deck and Engine Departments, the clock setbacks were as evenly divided between the two overnight watches as possible, and crew members often adjusted their timepieces before retiring so that they would agree with ATS when they next went on watch. The first of these adjustments would be carried out to delay the ship's midnight and the time to go off duty for the majority of crew members serving on the 8.00 p.m.–12.00 midnight watch. The second would be done shortly after the new watch came on duty, placing a similar amount of time into the beginning of their 12.00 midnight–4.00 a.m. watch.

### Revolutions Per Minute (RPM)

These refer to the rate at which *Titanic*'s reciprocating engines, and thus her wing propellers, were turning.

Speeds are given in nautical miles per hour. To convert a speed given in knots to land miles per hour, multiply by 1.150779; to convert knots to kilometres per hour, multiply by 1.852. Distances are given in nautical miles.

### Speed Through the Water

'Speed through the water' tells the speed at which the ship was being pushed through the water by her propellers. This speed does not refer to actual geographic progress, however, as that could be dramatically affected (positively or negatively) by wind, weather and currents.

### Speed Made Good

This refers to actual average speeds made between fixed geographic locations (such as noontime positions).

### WEDNESDAY, 10 APRIL (PART 1): THE CROSS-CHANNEL TRIP

**Weather:** High of about 48°F, mostly overcast with occasional sun, windy.
**Distance in open water:** Approx. 65 nautical miles.
**Speed and RPM:** 20.4 knots through the water, 68rpm.
**Anchor dropped in Cherbourg:** 6.30 p.m.

**Notes:** *Titanic* was scheduled to sail at noon, but she cast off a few minutes late. She was further delayed by the near collision with SS *New York*, and by a second stop to offload unnecessary standby crew to a tug. These delays seem to have added up to setting her back on her schedule by about three-quarters of an hour.

Once she had cleared Southampton waters at the Nab Light Vessel, she made the cross-Channel trip to Cherbourg of about 65 miles and dropped anchor there at about 6.30 p.m. Working backwards, in rough terms, from the anchor drop point and time, we can estimate that she likely passed the Nab and entered open water between 3.00 and 3.30 p.m. She encountered a moderate chop in the Channel that gave smaller ships a ride, but most reported that *Titanic* felt very stable.

In Cherbourg, many passengers waiting to board *Titanic* recalled that it was cold and rainy, with quite a chop even in the harbour; many aboard the tenders for the half-an-hour ride out to *Titanic* were reportedly chilled, 'uncomfortable' or 'actively ill'.

### WEDNESDAY, 10 APRIL (PART 2): CHERBOURG TO QUEENSTOWN

**Anchor weighed:** 8.10 p.m. ATS/GMT (NYT +5 hrs).
**Speed and RPMs:** Up to 20.9 knots through the water, worked up from 68 to 70rpm overnight.
**Overnight weather:** Clearing skies, cool to cold, moderate winds.
**Overnight clock setback:** 25 minutes (to match local Queenstown time).

### THURSDAY, 11 APRIL (PART 1): CHERBOURG TO QUEENSTOWN

**ATS:** (GMT –25 min, NYT +4 hrs 35 min).
**Distance travelled:** Approx. 320 nautical miles.
**Speed and RPMs:** 20.9 knots through the water, 70rpm.

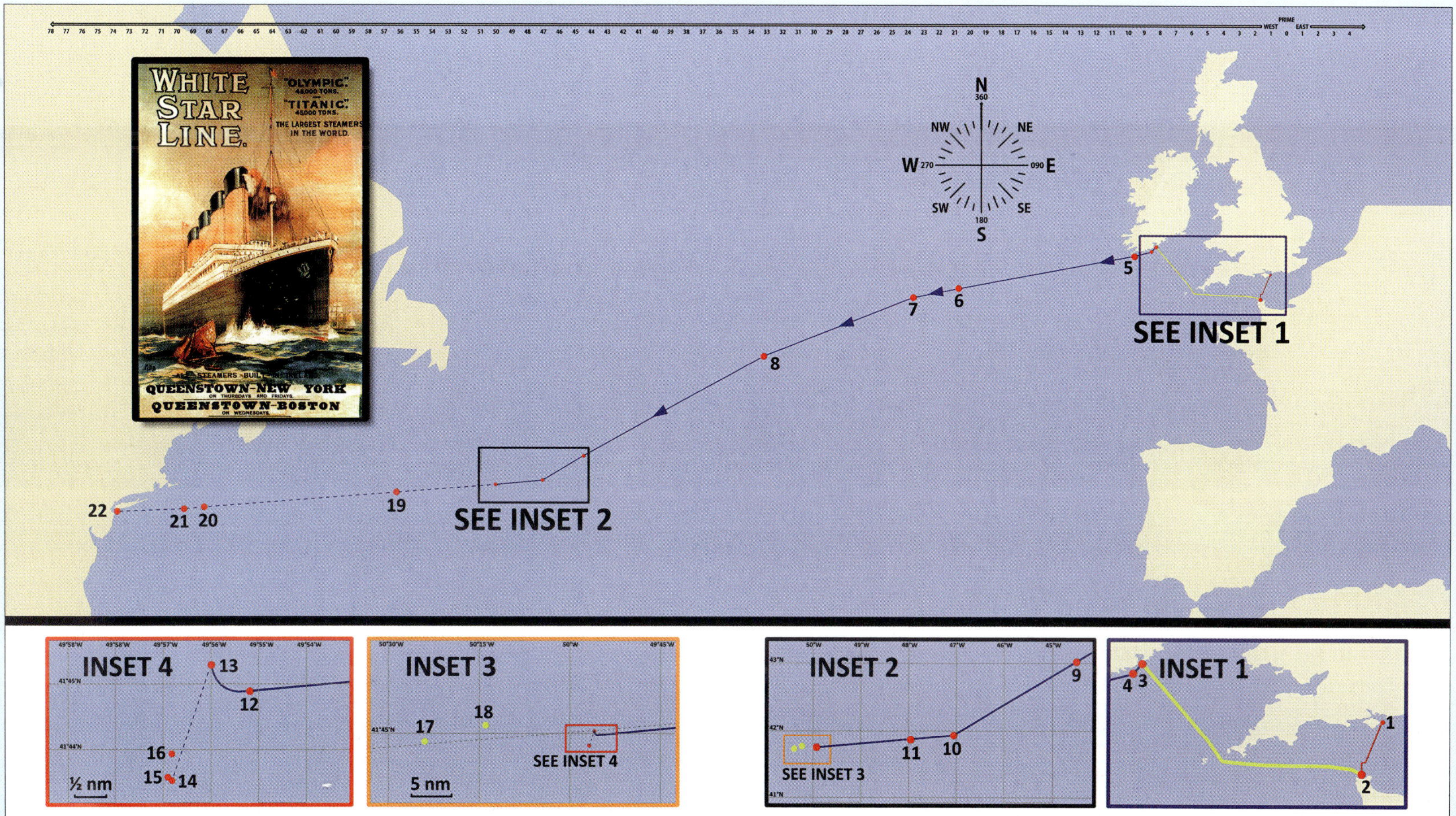


1. 10 April: Entrance to English Channel after leaving Southampton
2. 10 April: Cherbourg, France
3. 11 April: Queenstown, Daunt's Rock (51° 43′N, 08° 16′W)
4. 11 April: Turning point off the Old Head of Kinsale (51° 33′N, 08° 32′W)
5. 11 April: Turning point off of Fastnet Light (51° 23′N, 09° 36′W)
6. 12 April: noon position (50° 06′N, 20° 43′W)
7. 12 April: 5.36 p.m. position (49° 45′N, 23° 38′W)
8. 13 April: noon position (47° 22′N, 33° 10′W)
9. 14 April: noon position (43° 01.8′N, 44° 31.4′W)
10. 14 April: 5.50 p.m. ATS turning location (41° 55.9′N, 47° 04.4′W)
11. 7.30 p.m. celestial fix: 41° 52.5′N, 47° 58′W (est)
12. 11.40 p.m. collision point: 41° 44.9′N, 49° 55.2′W (est)
13. Final stopping point: 41° 45.3′N, 49° 56′W (est)
14. Center of boiler field: 41° 43.5′N 49° 56.8′W
15. Stern Section: 41° 43′ 35″N, 49° 56′ 54″W
16. Bow Section: 41° 43′ 57″N, 49° 56′ 49″W
17. Captain Smith's SOS location: 41° 44′N, 50° 24′W
18. Fourth Officer Boxhall's SOS location: 41° 46′N, 50° 14′W
19. 15 April: noon position (41° 22′N, 56° 16′W) (est)
20. 16 April: noon position (40° 40′N, 68° 22′W) (est)
21. Turning point off Fire Island Light Vessel: 40° 35′N, 69° 36.5′W
22. Entrance to Ambrose Channel, New York: 40° 28′N, 73° 50′W

This chart shows details of *Titanic*'s maiden voyage from Southampton, past the evening of Sunday, 14 April, through to New York. (Authors' Collection)

**Weather during the morning:** Cool to cold, windy, mostly sunny, 'glorious'.
**Weather at Queenstown:** About 50°F, 15–20 knot NW winds, cloudy.
**Anchor dropped:** 11.30 a.m. ATS (11.55 a.m. GMT) (15.75 hours after weighing anchor in Cherbourg).
**Anchor weighed:** Just after 1.30 p.m. ATS, just after 1.55 p.m. GMT.

**Notes:** *Titanic* dropped anchor in Queenstown precisely 15.75 hours after weighing anchor in Cherbourg the night before. This means that she averaged approximately 20.3 knots in terms of 'speed made good' between those two points, over a course of approximately 320 miles. The ship took the Atlantic swell on her port side during the morning as she steamed north-west, but reportedly behaved well, with a steady roll that only a few noticed.

### THURSDAY, 11 APRIL (PART 2): TRANSATLANTIC CROSSING

**Passing Daunt's Rock Light Vessel:** 1.55 p.m. ATS, 2.20 p.m. GMT.
**Location:** 51° 43'N, 08° 16'W.
**Speed and engine RPMs:** 20.9–21.4 knots through the water, 70–72rpm.
**Afternoon weather:** Skies clearing, winds decreasing to light. Overnight, winds increased to a headwind of about 15 knots, with an easy but noticeable roll aboard.
**Overnight clock setback 11–12 April:** 59 minutes.

### FRIDAY, 12 APRIL

**ATS:** GMT -1 hour, 24 min, NYT +3 hours, 36 min.
**Noon position:** 50° 06'N, 20° 43'W.
**Nautical miles steamed since passing Daunt's Rock (23 hrs, 4 min):** 484.
**Average speed made good:** 20.98 knots.
**Weather:** 'Beautiful' morning sun, seas calm, headwinds continue; increasing clouds and some showers in the afternoon. High approximately 60°F.
**Overnight clock setback 12–13 April:** 49 minutes.

**Notes:** *Titanic*'s position at 5.36 p.m. ATS (7.00 p.m. GMT, 2.00 p.m. NYT) reported as 49° 45'N, 23° 38'W, some 114.6 miles from her noon position; she had averaged only 20.46 knots in terms of speed made good in those 5 hours, 36 minutes.

### SATURDAY, 13 APRIL

**ATS:** GMT -2 hours, 13 min, NYT +2 hours, 47 min.
**Noon position:** 47° 22'N, 33° 10'W.
**Nautical miles steamed since noon, 12 April (24 hrs, 49 minutes):** 519.
**Average speed made good since previous noon:** 20.91 knots.
**Weather:** Broken cloud cover in the morning, high of about 60°F. Steady roll continued; some reported that the ship briefly encountered fog before the weather cleared again. Scattered showers overnight 13–14 April; ongoing headwinds 15–20 knots.
**Overnight clock setback 13–14 April:** 45 minutes.

**Notes:** The reported position at 5.36 p.m. ATS on Friday, 12 April shows us that in the 19.21 hours after that position, the ship averaged 21.05 knots up to noon on Saturday, bringing the average for those 24 hours, 49 minutes up to the still-disappointing 20.91 knots reported at noon on Saturday. At noon on 13 April, engine rpms were reportedly 75 (speed through the water approximately 22.15 knots).

### SUNDAY, 14 APRIL

**ATS:** GMT -2 hours, 58 mins, NYT +2 hours, 2 mins.
**Noon position:** 43° 01.8'N, 44° 31.4'W.
**Nautical miles steamed since noon on 13 April (24 hrs, 45 mins):** 546.
**Average speed made good since previous noon:** 22.06 knots.
**Weather:** Morning showers until around daybreak, with a morning swell. Few to no clouds, but colder than the preceding day. Lifeboat drill cancelled due to morning winds.

### FROM NOON ON SUNDAY

From *Titanic*'s noontime position on 14 April, there was a lot of mystery surrounding her speeds and locations up to the collision. Her two transmitted distress co-ordinates were 20.9 and 13.4 miles west of the location she actually stopped at, and this led to great confusion about her final hours' progress. However, after learning the ship's actual position in 1985, we have been able to refine our understanding of her speeds and courses after noon on Sunday.

*Titanic* was to make a turn of roughly 25 degrees to starboard that evening at 'The Corner' point (42° N, 47° W), to line her up with New York. Although myths that the turn was greatly delayed have persisted, it seems that she was only out of position by roughly 5 miles when she made the course adjustment.

**Afternoon/evening weather:** Temperatures dropped steadily through the afternoon from a high of 48–50°F. The wind subsided at around sunset (approximately 6.52 p.m. ATS). Cold, clear, with sea becoming as smooth as glass. Estimated actual alter course location: 41° 55.9'N, 47° 04.4'W (at 5.50 p.m. ATS), 5.2 miles south-west of 'The Corner'.
**Estimated collision point (11.40 p.m.):** 41° 44.9'N, 49° 55.2'W.
**Estimated stopping point:** 41° 45.3'N, 49° 56'W.
**Average speed made good since noon:** 22.16 knots (258.4 miles).
**Measured distance through the water from noon on 14 April to the time of the collision at 11.40 p.m. ATS:** 260 nautical miles. Average speed through the water was 22.29 knots carrying 75–76rpm on the engines.
**Intended overnight clock setback 14–15 April:** 47 minutes (GMT -3 hours, 45 min, NYT +2 hours) (not carried out).

### THE REMAINDER OF THE VOYAGE

Because *Titanic*'s officers navigated in part based on precise timekeeping, we can take the amount of time for *Titanic*'s overnight on 14–15 April clock adjustment of 47 minutes and derive the co-ordinates they expected to reach by noon on Monday, 15 April: 41° 22'N, 56° 16'W. That run would have been about 544 miles at an average speed made good of close to

22 knots. At that point, *Titanic* would have steamed about 2,096 miles of her 2,891-mile course to the entrance of the Ambrose Channel at New York, leaving only 795 miles remaining.

The rest is easily estimated, although not accounting for anything that would have adversely affected her progress, like fog or poor weather. If she neither slowed nor accelerated – the latter a distinct possibility as there was talk of a high-speed run on Monday, which would likely have begun after reaching the noon position – *Titanic* would have reached a noon position on Tuesday, 16 April of about 40° 40'N, 68° 22'W, covering approximately 546 miles since noon on Monday. This would have placed her at the entrance of the Ambrose Channel – the official end point for the transatlantic crossing – a mere 249 miles later, at about 10.54 NYT that evening (3.54 a.m. Wednesday, 17 April GMT), giving a crossing time of five days, thirteen hours and thirty-four minutes at an average speed of 21.65 knots – a very respectable crossing time and speed.

This calculation also bears out what Bruce Ismay and many others on board *Titanic* were expecting: to arrive in New York late on Tuesday rather than Wednesday. Based on the length of time that it took *Olympic* to reach her pier once she passed the Ambrose Light Vessel on her maiden voyage, *Titanic* would not likely have tied up at Pier 59 until roughly 6.30 a.m. on Wednesday morning. First and Second Class passengers would have had the option of breakfasting aboard before disembarking.

## *TITANIC* ABLAZE – FIRE DOWN BELOW?

It has been known since 1912 that *Titanic* had a coal bunker fire during the maiden voyage, and detailed testimony on it was given in both the American and British inquiries. The fire was discovered in the forward coal bunker of Boiler Room No. 5, between Boiler Room Nos. 5 and 6. We do not know precisely when the fire started, but Fireman Charles Hendrickson testified it was discovered before *Titanic* left Belfast; if this is true, it was burning for quite some time. Yet the two senior surviving officers later reported they had not been informed about the fire, indicating it was not a serious issue. Indeed, at the time, coal bunker fires, though not an everyday occurrence, were not uncommon.

Hendrickson said he was tasked with helping remove coal from the bunker as early as the First Watch out of Southampton. Leading Fireman Frederick Barrett said that he was tasked with heading up a team of between eight and twelve men who tried to extinguish the fire during the maiden voyage. According to both Barrett and Hendrickson, the fire was extinguished by Saturday, 13 April. At that point, they were able to examine the bulkhead behind the fire, determining that it had been 'red hot', and had also been 'dented a bit'.

On the night of the disaster, the iceberg damage stretched along the starboard side of Boiler Room No. 6, just past the transverse bulkhead separating it from No. 5, and stopped just inside the starboard bunker where the fire had been. Later on, there was a rush of water in that space that took the lives of two engineers, from which Fireman Barrett only barely managed to escape.

Over the years, this apparent coincidence has led to a number of conspiracy theories on the matter, but the coal bunker fire was a subject that had become rather forgotten by passing enthusiasts in recent years. In 2017, a new British television programme claimed that a 'smudge' seen in two rather newly unearthed photos of *Titanic* in Belfast on 2 April 1912 was evidence of a conflagration that had seriously weakened the ship's hull. It was further claimed that the fire was deliberately covered up so as not to delay the maiden voyage, that in trying to extinguish it the ship had to be sped up until it caused the collision, and that when the damage from the fire was subjected to stresses due to the collision and flooding it led to a significant collapse of the ship's internal structure. In essence, the show claimed, the fire doomed *Titanic* and her passengers and crew.

The programme was a massive publicity success and, worldwide, the media picked up on the story and turned it into a phenomenon. Casual *Titanic* enthusiasts, who were not accustomed to hearing about the coal bunker fire, were enthralled. However, the show made a consistent string of claims that have been easily disproved by historians. In reality, the 'smudge' that was allegedly fire damage was nowhere near the location of the actual coal bunker fire; also, other photos of the same area of *Titanic*'s hull taken later on show absolutely no damage or distortion to the hull plating in the area of the smudge. The ship's Swimming Bath, which was directly over where the fire was, showed no evidence of becoming a steaming-hot cauldron from the alleged conflagration below it. The truth is that the 'smudge' seen in the two photos was a reflection or a trick of light playing across the hull plates; similar 'markings' can be seen in photographs of *Olympic* when viewed from similar angles.

The coal bunker fire was not a raging conflagration. It was extinguished on 13 April and played a very little role in the disaster.

What caused the rush of water that killed two men, then? Barrett himself had closed the bunker hatch to the damaged coal bunker after the collision, but the bunker and its door were never intended to be watertight. Apparently, while the pumps had kept the water out of Boiler Room No. 5, it had built up within the bunker to the point where the door gave way, causing the rush of water. The fire had nothing to do with it, and certainly did not cause or accelerate the sinking.

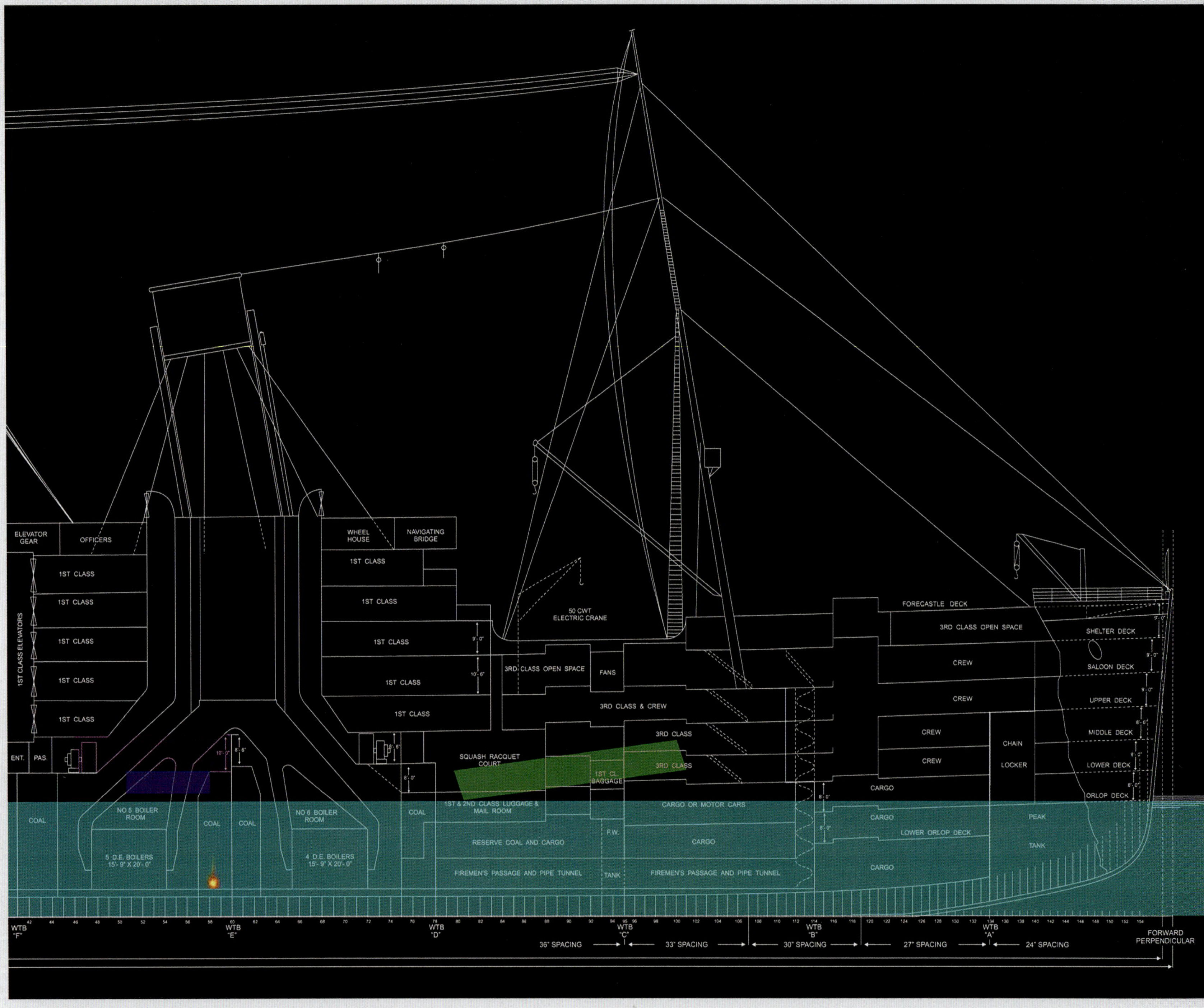


The location of the 'smudge' (green), the bunker fire (red) and the swimming pool (purple). (Authors' Collection, diagram by Steve Hall based on plans by Bruce Beveridge)

## THE MAGNETA CLOCK SYSTEM

In order to help keep shipboard clocks in harmony with each other as the clocks were changed on a daily basis, *Olympic* and *Titanic* were outfitted with a special Magneta clock system. How did this system work? While shore-based Magneta clocks were pendulum-controlled, such a set-up would not work aboard a moving ship, so the seagoing master clocks were similar to a chronometer in function. Each 'master' could control up to twenty-five 'slave' or 'secondary' clocks as long as each was less than 12in in diameter. As *Olympic* and *Titanic* needed more than twenty-five clocks, two masters were installed in the Chart Rooms of each ship, just behind the Wheelhouse, each driving up to twenty-five secondary units located throughout the ship.

The secondary clocks worked on a low-voltage charge sent from the master unit. Every sixty seconds the master clock sent a charge to move the secondary clock's hands one full minute at a time. This also meant that none of the secondary clocks had a second hand. By utilising these high-tech devices and placing secondary units throughout the ship in passenger and crew spaces, this greatly minimised the chance for any confusion about shipboard time as clocks were adjusted at sea. It also reduced the number of clocks that would have to be manually adjusted every day by stewards.

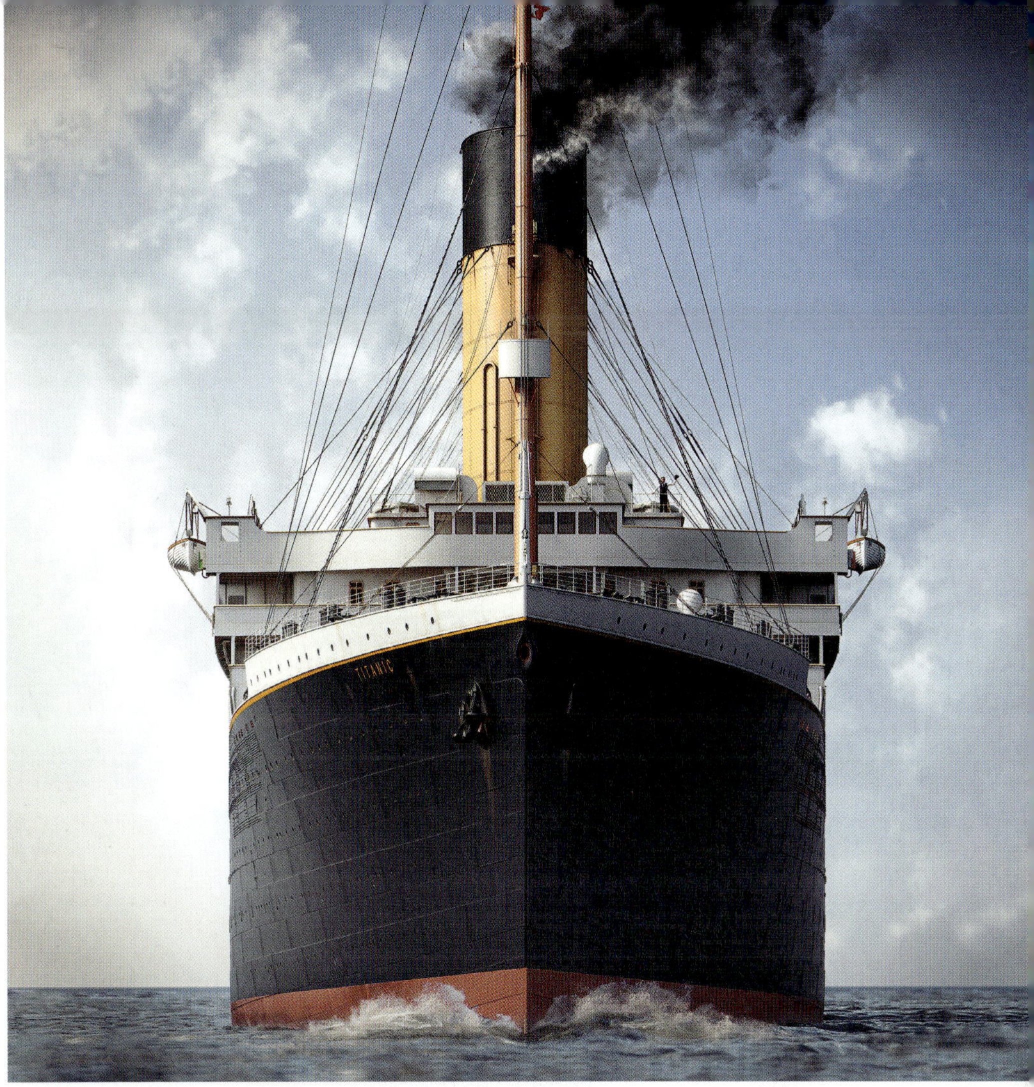

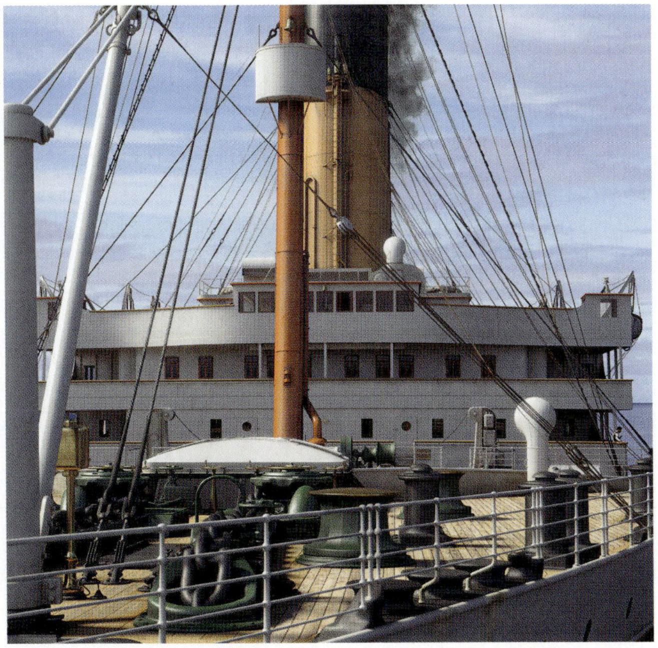

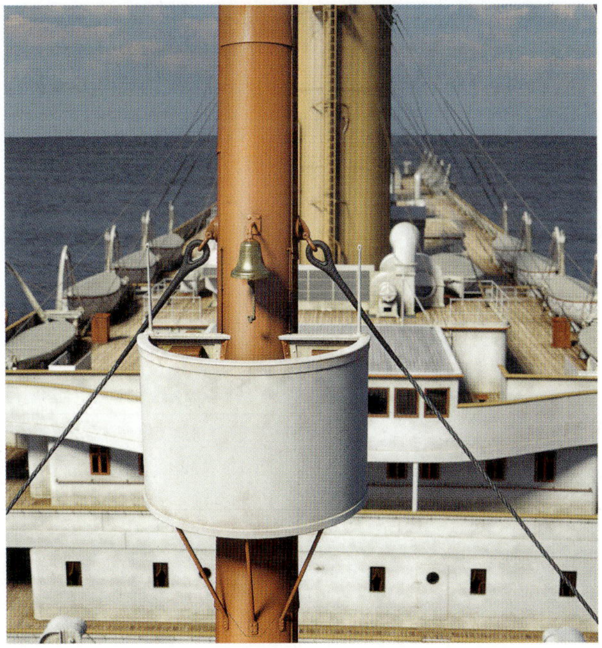

*Above: Titanic*'s clean lines become apparent in this bow-on view. (Vasilije Ristović)

*Far left:* Coming up over the port bow, we see the Bridge and superstructure from the Forecastle. (Vasilije Ristović)

*Left:* A detailed view showing the Crow's Nest and looking aft along the upper decks. (Vasilije Ristović)

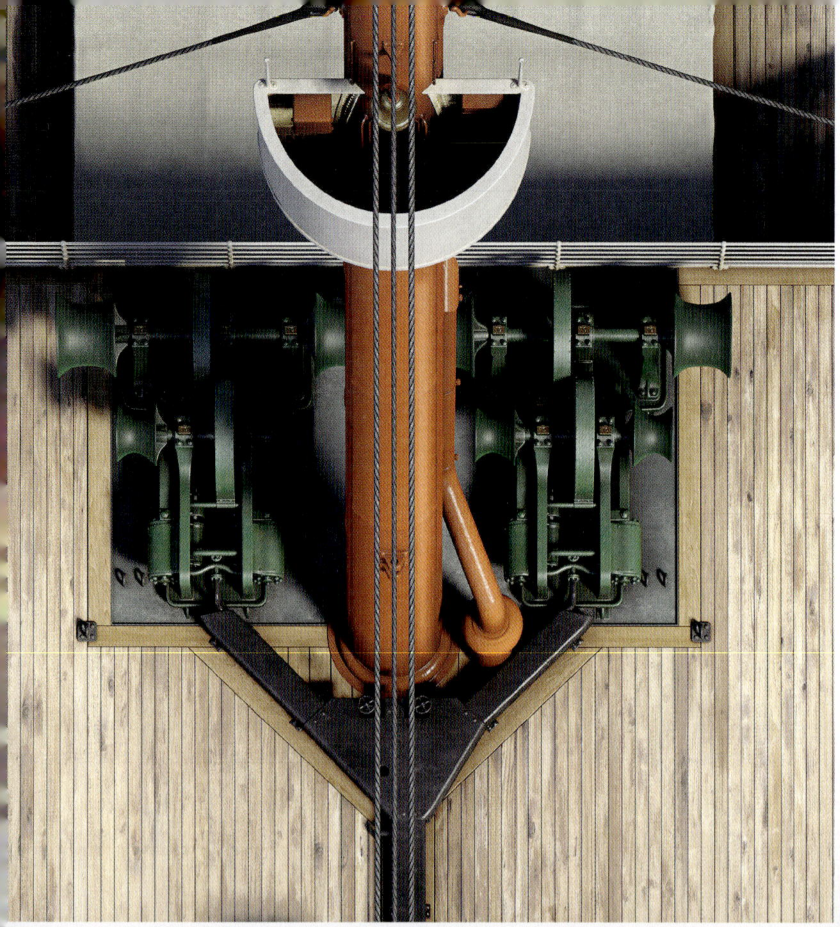

*Left:* Looking directly down into the Crow's Nest. On the starboard side is the box that contained the phone to the Bridge. The brass bell hangs amidships, and the opening through which the lookouts entered and exited the nest is clearly visible. The box on the right was for storage. (Vasilije Ristović)

*Below:* The Navigating Bridge, the nerve centre of *Titanic*, looking to port. The ship's wheel was connected to the one in the Wheelhouse and would turn in conjunction with that wheel. (Vasilije Ristović)

*Opposite:* A fine portrait overlooking the forward portion of *Titanic* from above the funnels. Note the location of Collapsibles A and B on the roof of the Officers' Quarters beside the forward funnel. Emergency Boat No. 1 remains uncovered at sea for quick use. (Vasilije Ristović)

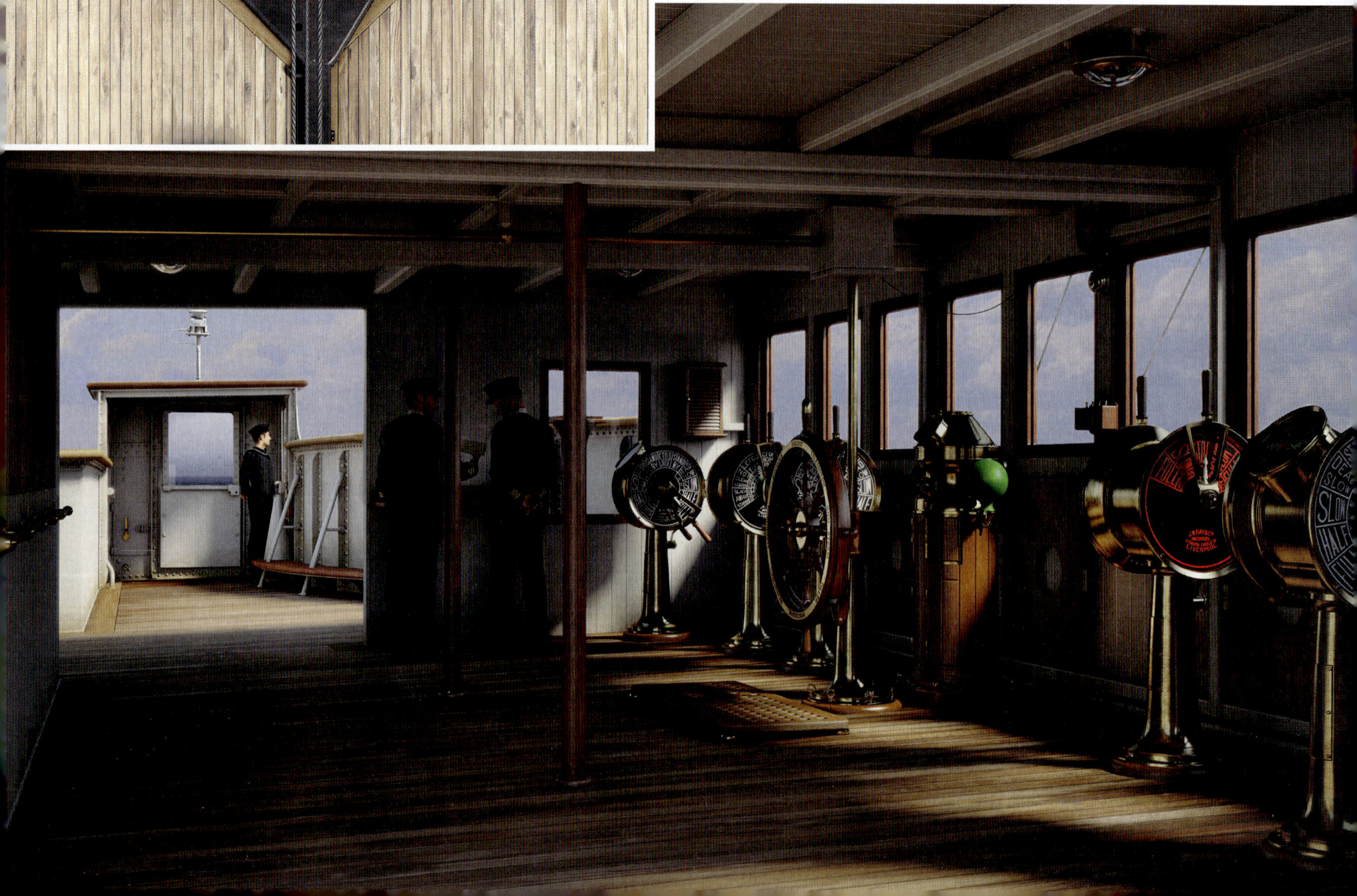

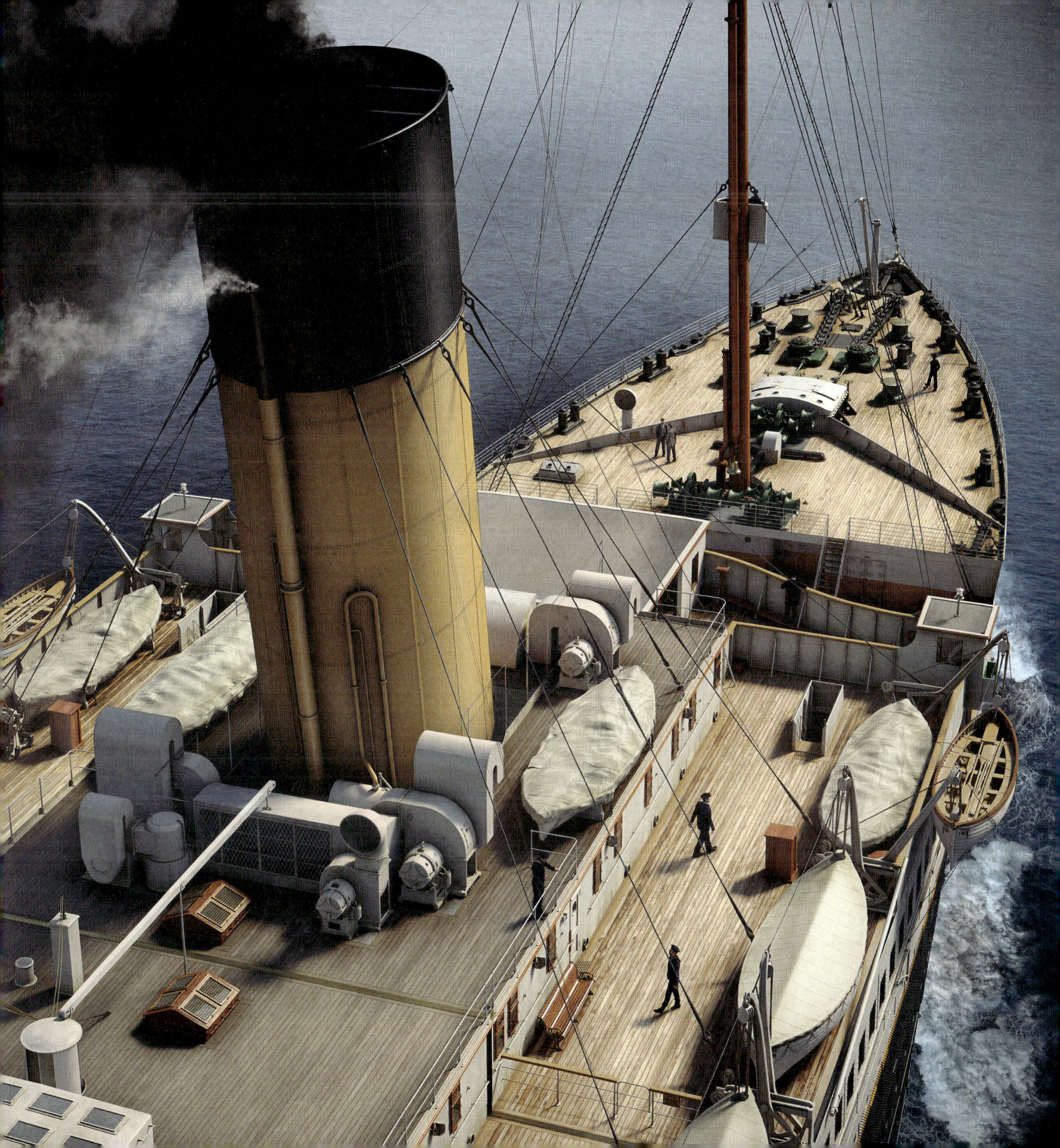

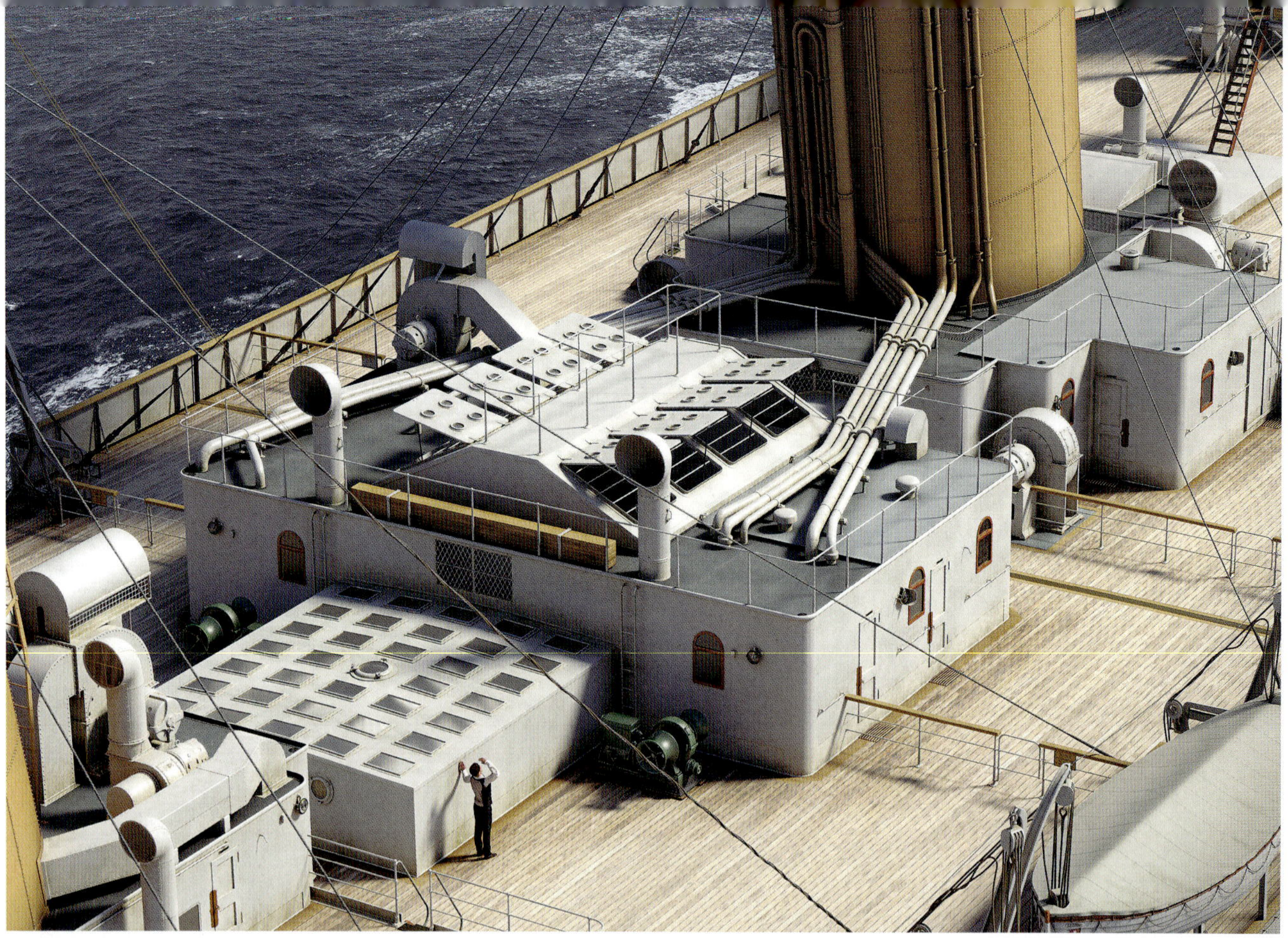

Just aft of the No. 3 funnel is the Reciprocating Engine Skylight and the cover over the Aft Grand Staircase dome. Boat No. 9 sits on the lower right extremity. Between the railings on the starboard side was the Engineers' Promenade. (Vasilije Ristović)

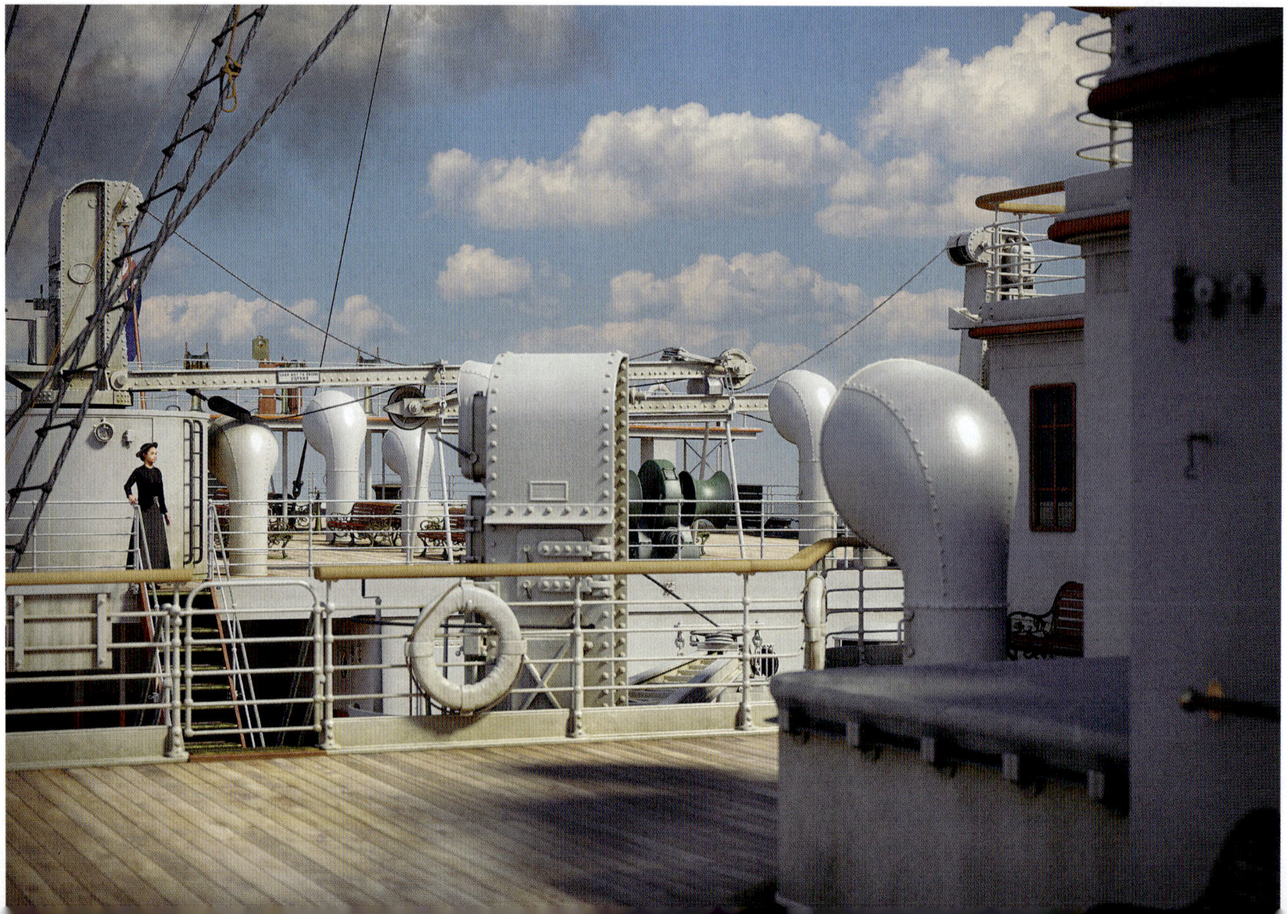

Looking aft from the Second Class Promenade on B Deck toward the Poop Deck. Note starboard Cargo Hatch No. 4 on the right and the gate on the left, which led down to Third Class Open Space on the Aft Well Deck. (Vasilije Ristović)

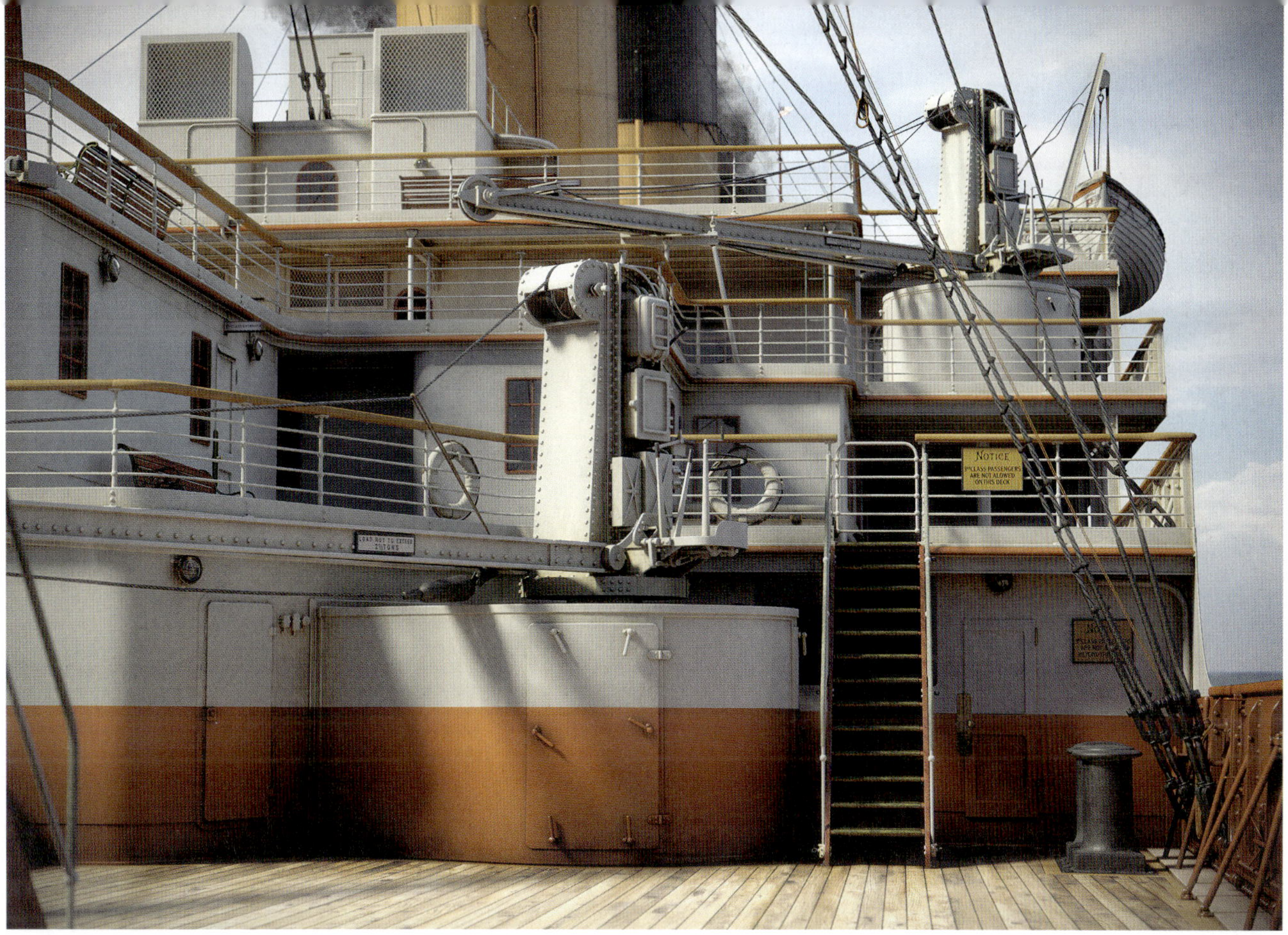

Looking forward on the starboard side of the Aft Well Deck. Second Class passengers could enjoy B Deck while First Class passengers had the use of A Deck, above. Then, on the Boat Deck above that, Second Class passengers could look down on those in First Class. (Vasilije Ristović)

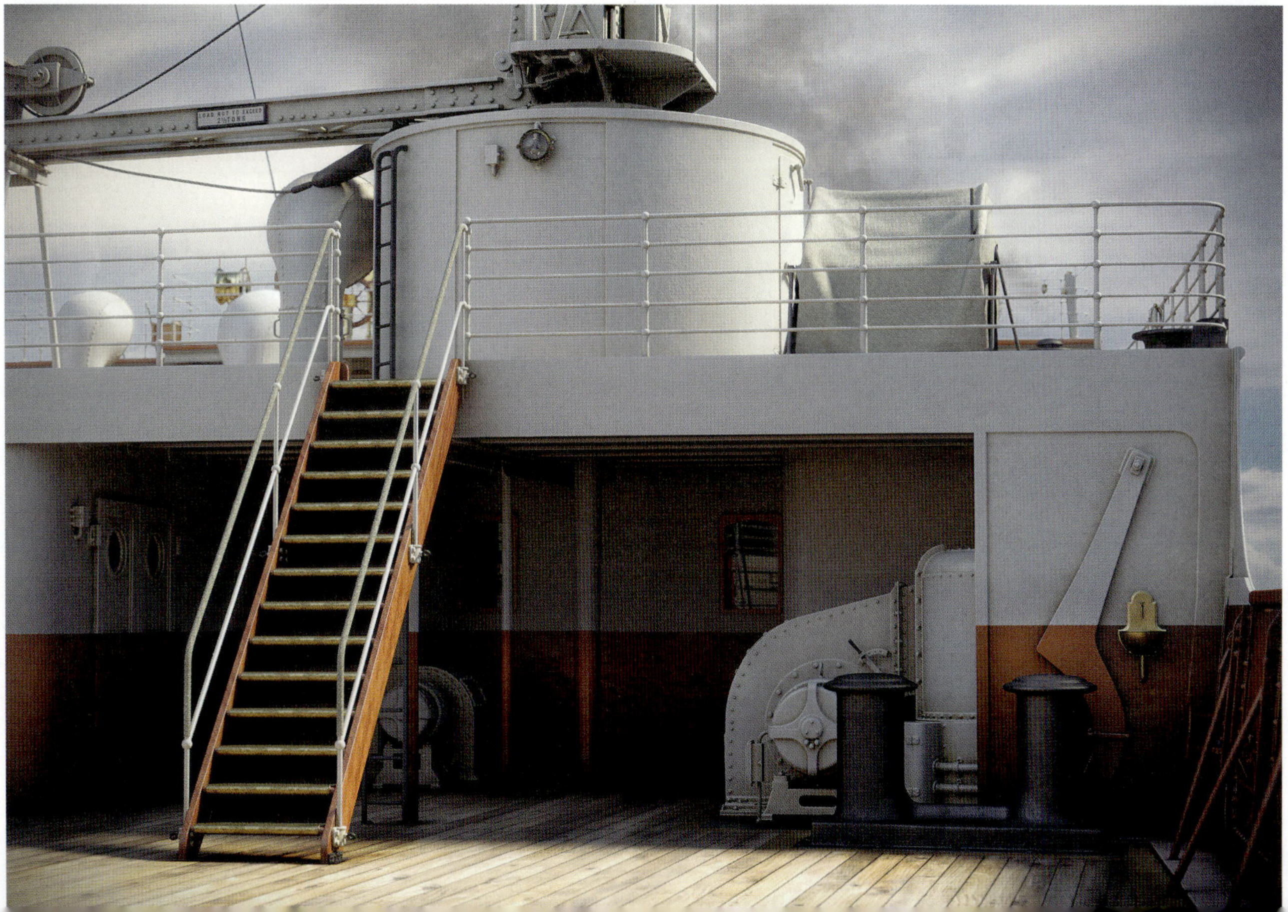

Looking aft on the port side of the Aft Well Deck toward the Poop Deck. A drinking fountain is tucked into the right corner next to the large spanner, while on the left, the double doors give access to the Third Class Entrance. (Vasilije Ristović)

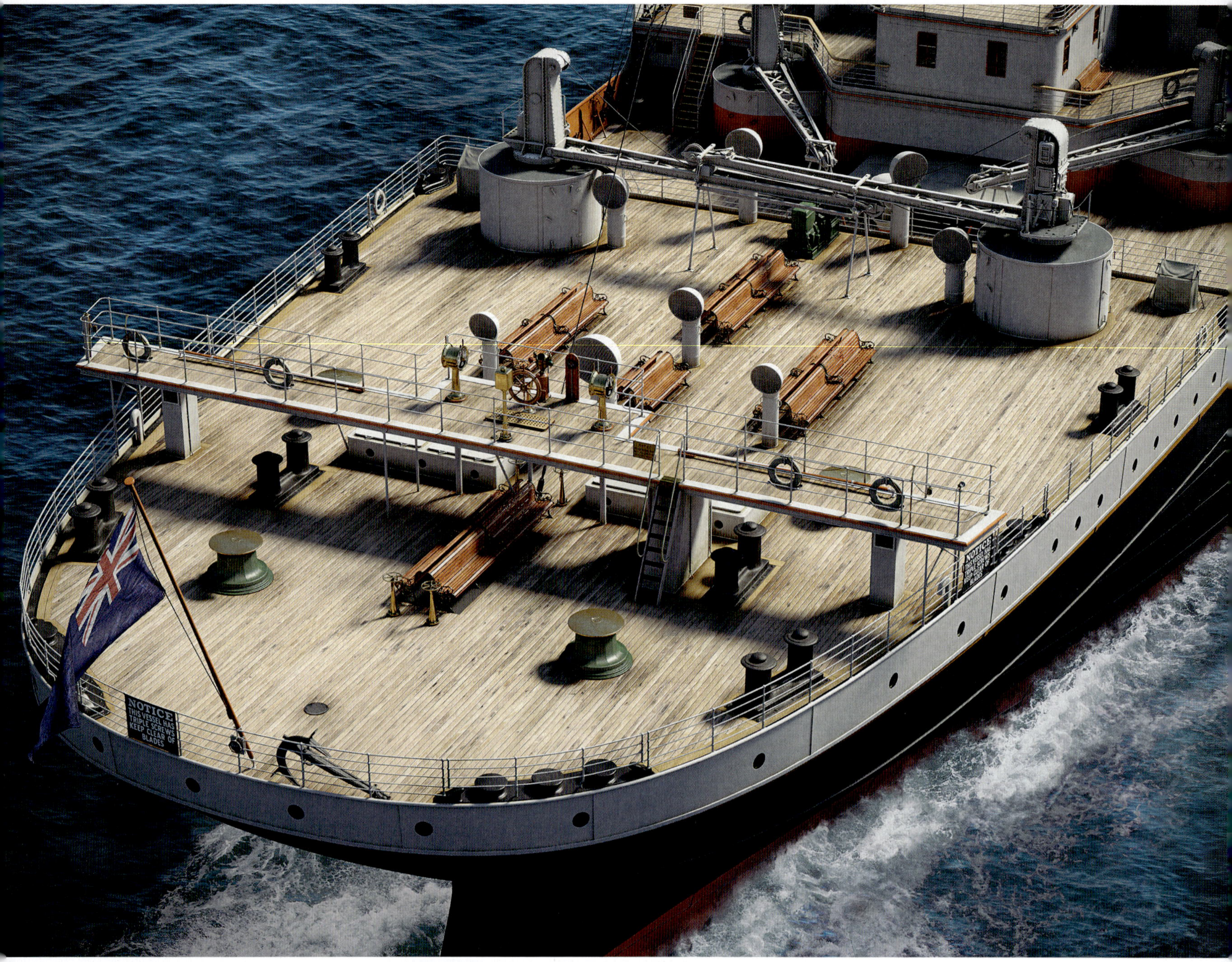


*Above:* An aerial view of the entire Poop Deck and after portion of *Titanic*. The outboard edges of the Aft Docking Bridge extended over the edge of the Poop Deck by several feet, a change from *Olympic*'s original design. (Vasilije Ristović)

*Opposite: Titanic* steams away from the coast of Ireland in the afternoon of 11 April. (Vasilije Ristović)

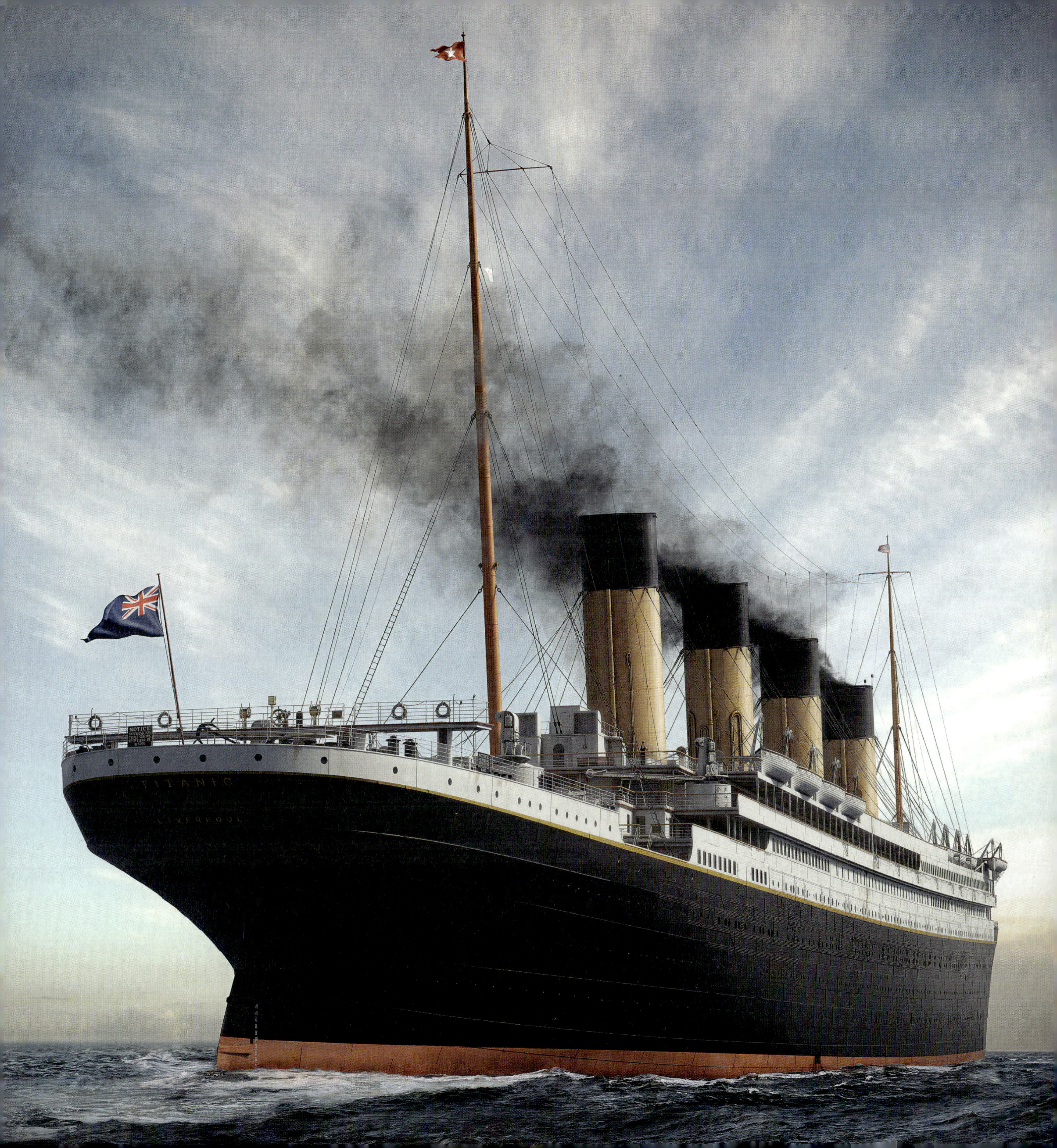
NOTICE
TITANIC
LIVERPOOL

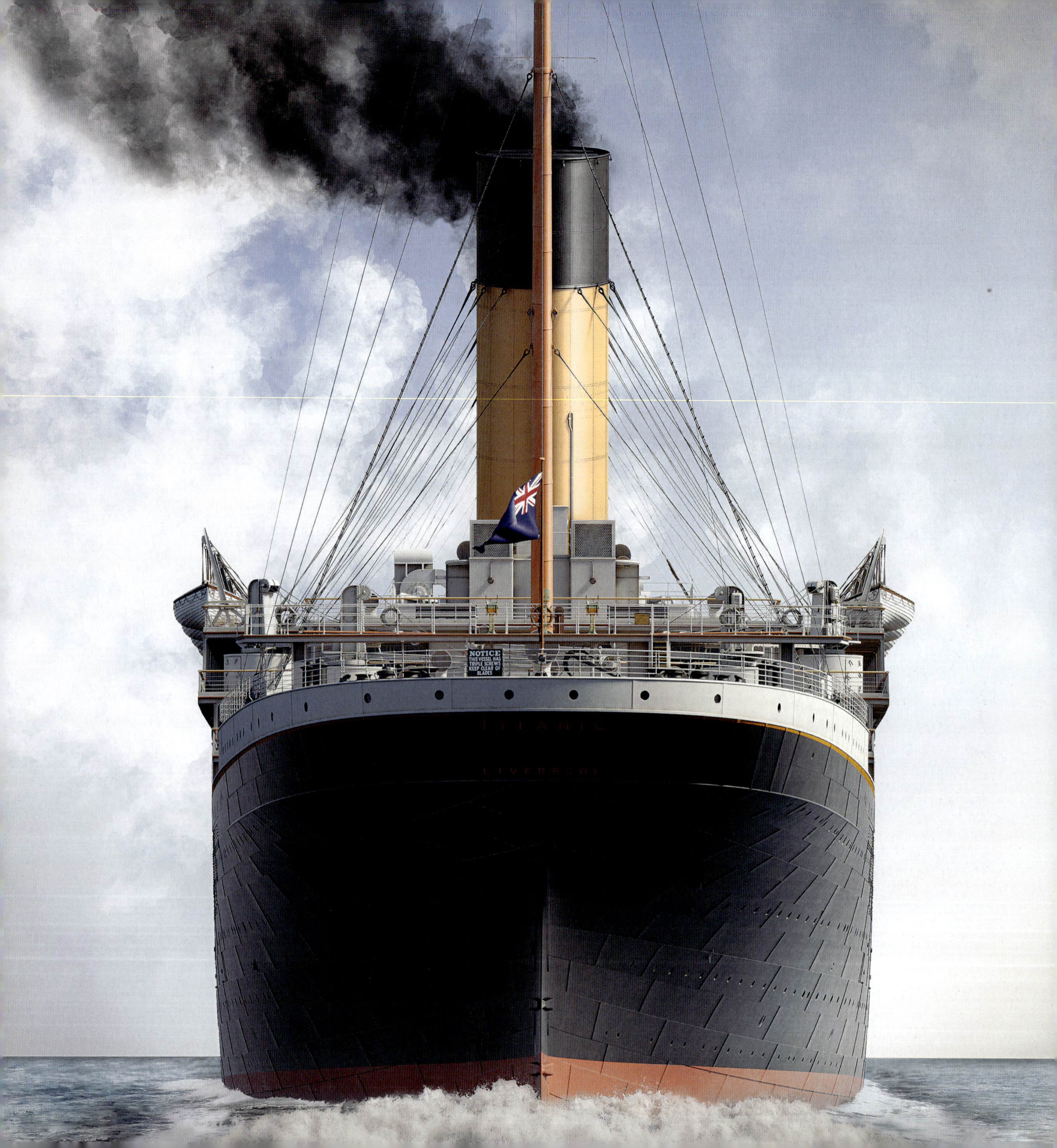
NOTICE
THIS VESSEL HAS
TRIPLE SCREWS
KEEP CLEAR OF
BLADES
TITANIC
LIVERPOOL

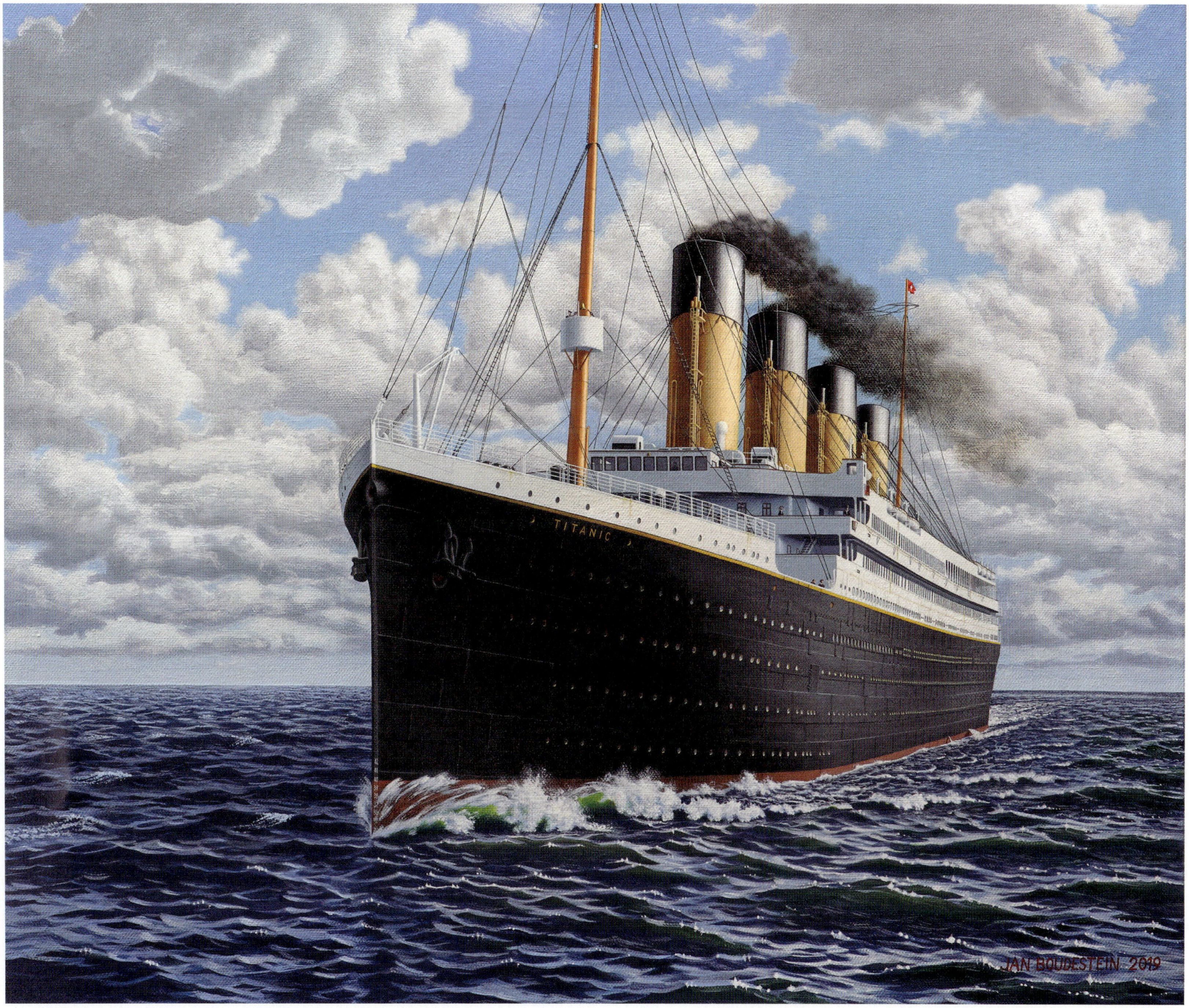


*Opposite:* This fine stern-on view shows *Titanic* sailing away into the North Atlantic Ocean and toward her final rendezvous with destiny. (Vasilije Ristović)

*Above:* As morning rain showers disperse around dawn, the weather on Sunday morning, 14 April, becomes clear and fine with a strong wind that gradually dies off during the evening hours. (Jan Boudestein)

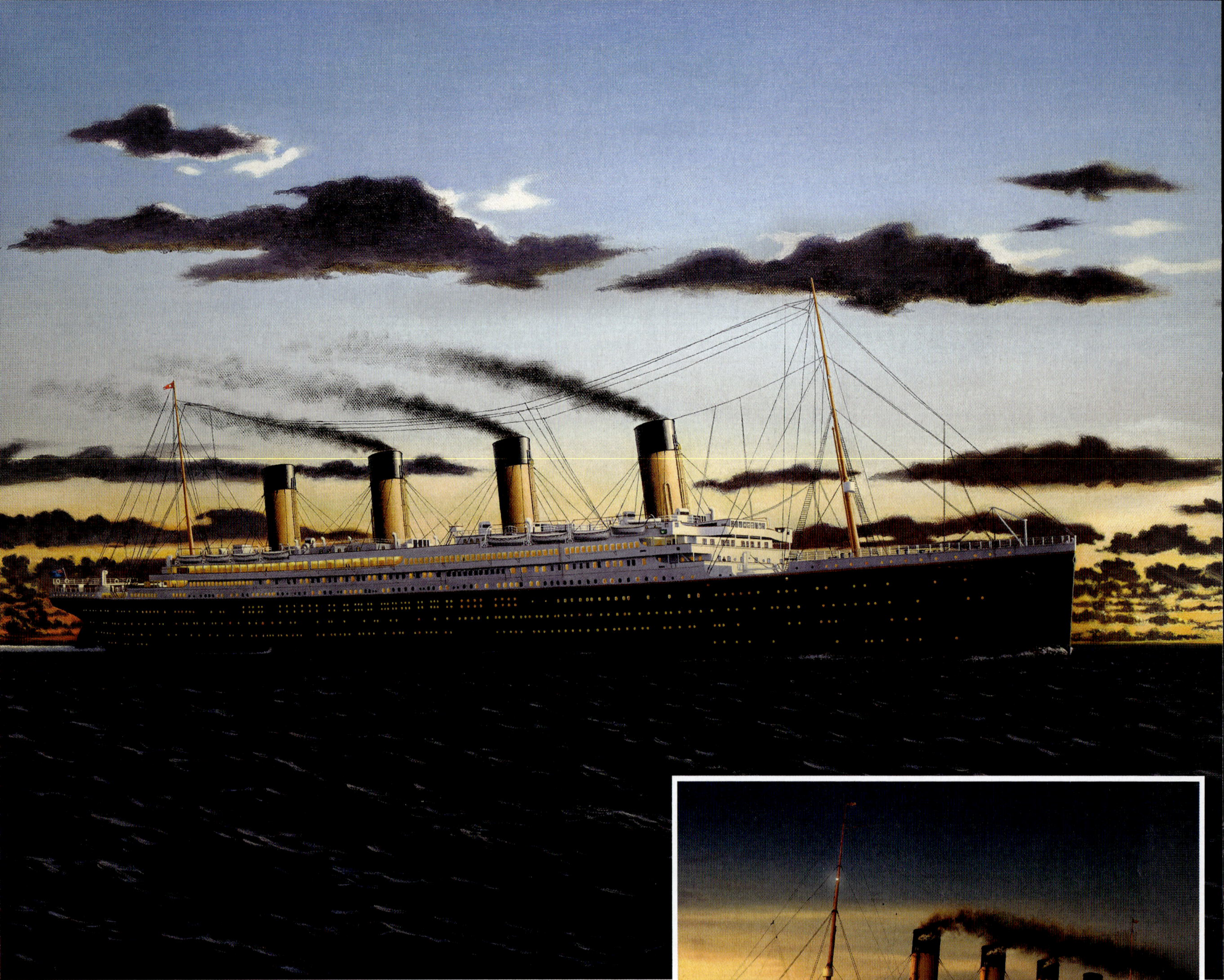

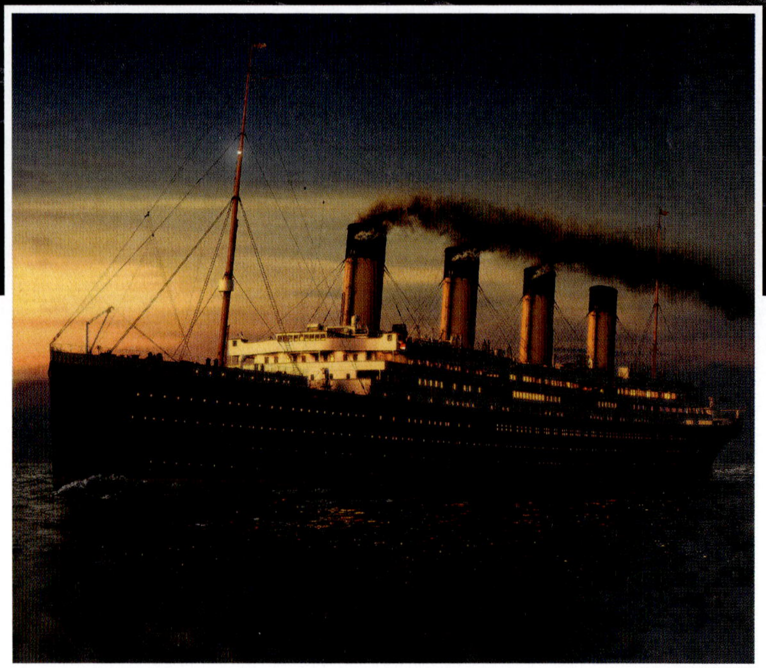

*Above:* At about 6.52 p.m. on Sunday, 14 April, *Titanic* steams into the gathering darkness. Dinner would soon be served, and passengers would enjoy one last evening aboard. (Jan Boudestein)

*Right: Titanic* sails confidently into the night of Sunday, 14 April. She is making her best speed of the crossing yet, and hopes run high that she will arrive in New York ahead of schedule. (Vasilije Ristović)

THREE

# THE SINKING OF *TITANIC*

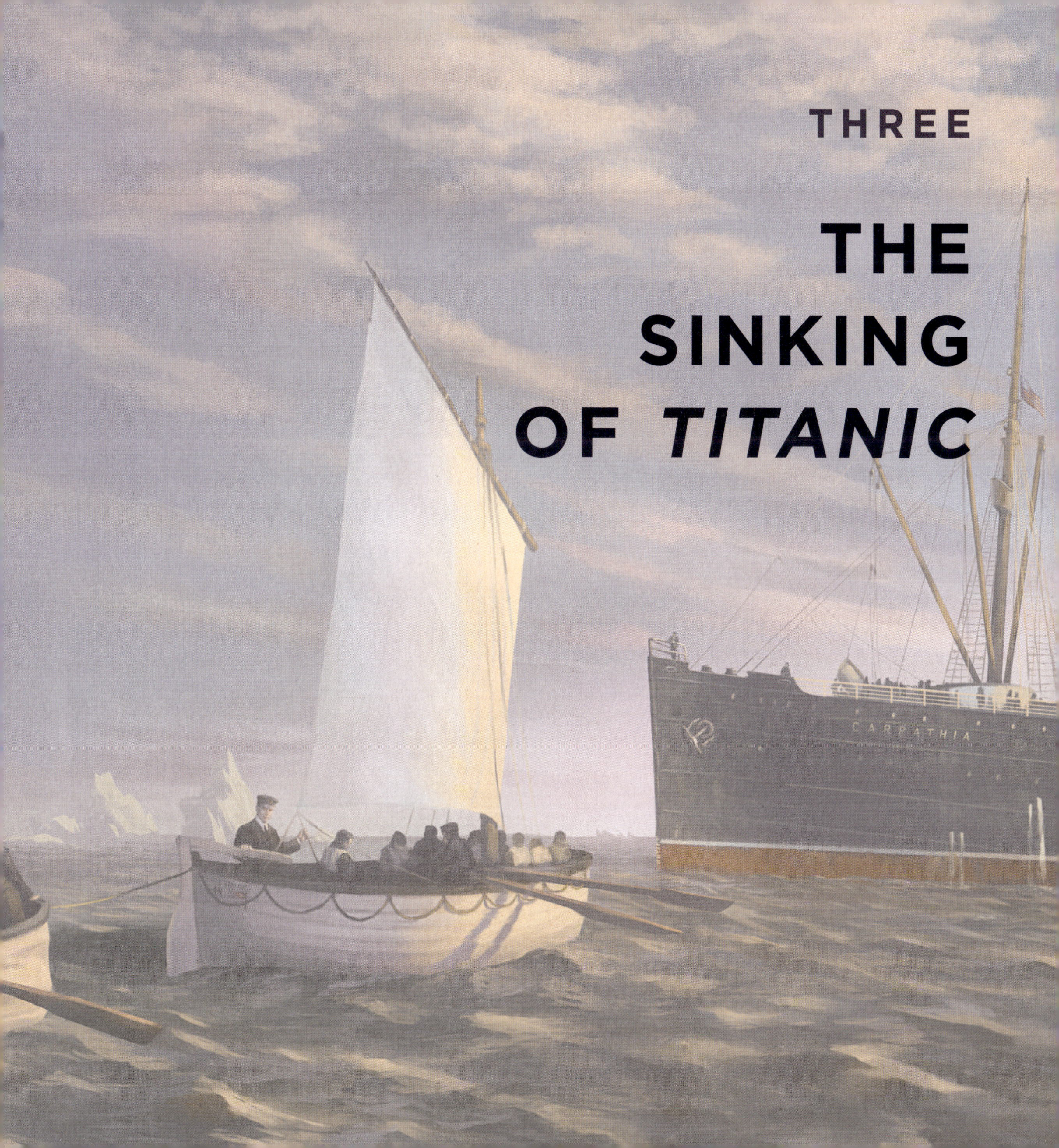

At 11.39 p.m., *Titanic* continued to steam full ahead, having averaged 22.16 knots since noon that day. After the hustle and bustle of preparing *Titanic* for her maiden voyage, things had settled down to a mostly quiet routine after leaving Queenstown, Ireland. The trip had been entirely free of incident and the ship was making good time – so much so, in fact, that word was beginning to spread aboard that she would reach New York late on Tuesday evening, rather than Wednesday morning, as scheduled.

Thomas Andrews' 'Guarantee Group' had been busy sorting out the inevitable bugs with a new ship: some of the lavatory taps in Second Class had not yet been installed and there seemed to be a rather irksome issue with the ship's system of heating the air in passenger spaces. Throughout most of the trip passengers found the liner quite chilly – a situation that had grown even more acute as temperatures fell through the day on Sunday until the air temp read only 27°F, and the water temperature was just one degree warmer, at a frigid 28°F. Andrews' stateroom, A-36, on the port side of the A Deck Aft First Class Entrance, was strewn with charts and plans, upon which Andrews had busied himself writing notes for future improvements and fixes.

Many shipboard friendships had been struck up during the crossing, and a camaraderie had begun to develop among many of the passengers, particularly in First Class. That evening, it seemed that everything came together at dinnertime, both in the *à la carte* Restaurant, where a special dinner was held in Captain Smith's honour, and in the Dining Saloon.

The meal itself seemed to be the best of the trip yet, but most remembered that there was something more than that in the air. Men wore their usual white tie and tails, while women donned their finest gowns and jewels. After dinner, the ship's band held their regular 8.00–9.15 p.m. concert in the Reception Room, but even when that ended, many passengers did not seem interested in calling it a night.

Some were clearly feeling the nearness of New York and the end of the crossing, which was only about fifty hours away, so, while some retired early, others went to the Lounge or Smoking Rooms to play cards, enjoy some good conversation, or to have a smoke or a drink, determined to make the most of what little time remained of the trip.

Shipboard gossip was always in full gear. The wealthiest man on the ship, by far, was 47-year-old John Jacob Astor IV, travelling with his 18-year-old wife, Madeleine. Madeleine was Astor's second wife, and the scandal of his 1909 divorce and 1911 remarriage to a much younger woman had rocked New York society circles. The couple were returning from an extended European honeymoon and attracted much interest. Yet perhaps the most noteworthy event of the crossing had come when Mrs Irene Harris, wife of Broadway theatrical producer Henry Harris, had slipped on the Grand Staircase and broken her arm in the resulting fall.

Meanwhile, no fewer than eight ice warnings had been received via Marconi wireless. The first arrived on Friday, and seven more had come in throughout the day on Sunday. Although not every one of these messages had made its way to the Bridge, Captain Smith and the senior officers were well aware of the threat of ice ahead and knew what time they might encounter it.

At about 12.45 that afternoon, Captain Smith entered the Bridge carrying a Marconigram and bumped into Second Officer Lightoller, who was then on duty. He held it out to Lightoller and told him to read it. Lightoller did so and took careful note of the co-ordinates laid out in the warning of ice ahead. The message was from the Cunard liner *Caronia*, and Captain Smith apparently posted it on the noticeboard in the Chart Room for all to see.

When First Officer Murdoch returned from his lunch, Lightoller discussed the message with his friend, and then went down to lunch himself. Just over an hour later, at 1.54 p.m., the White Star liner *Baltic* sent another ice warning. When Captain Smith received it, he gave it to Bruce Ismay when the two men crossed paths. Ismay glanced at it casually, tucked it in his pocket, and went down to lunch. That evening, before dinner, he showed it to two ladies from First Class who he stopped to speak with, before giving it back to Captain Smith shortly thereafter.

Fifth Officer Lowe had noticed, shortly after 6.00 p.m., a slip of paper stuck in the frame of the Chart Room table with a set of co-ordinates and the word 'ice' on it. The note had apparently been written down by Fourth Officer Boxhall earlier that afternoon. At about 6.00 p.m, Third Officer Pitman had seen co-ordinates denoting the position of ice marked on the chart showing their course.

Captain Smith came onto the Bridge at about 8.55 p.m. and had a detailed conversation with Second Officer Lightoller about the cold, the flat calm, and the ice they knew lay ahead. Lightoller even remarked that it was a pity that the breeze they had throughout much of the crossing had not kept up, because if it had, they would have been able to see water breaking around the base of a berg, making it easier to spot in the dark. The two men agreed that if things became in the least bit hazy, or there were any other doubts, that they would have to slow the ship down.

When First Officer Murdoch took over as Officer of the Watch from Lightoller at 10.00 p.m., the two men discussed it further and Murdoch certainly had the subject in the forefront of his mind. Earlier that evening, just after sunset, he had even ordered that the fore scuttle hatch be sealed up tight to make sure that the night vision of both the Lookouts and the Officers of the Watch was not hindered by the glow that had been coming from it.

Yet, in spite of all the warnings, up to 11.30 p.m., nothing had yet materialised from the darkness. The ship was beginning to quiet down. The last passengers were leaving the frigid Café Parisien about that time and the last few stragglers were clearing out of the Lounge on A Deck. Up in the Smoking Room, however, a number of gentlemen were still up chatting, having a game or a drink and enjoying the last of a thoroughly enjoyable evening. Everything seemed perfectly normal.

Lookouts Frederick Fleet and Reginald Lee were on watch in the freezing cold Crow's Nest. Suddenly, Fleet spotted a 'black mass' in the water, right in the ship's path. After taking a moment to discern the object, Fleet reached across the Crow's Nest and rang the warning bell three times, signalling that something was in the ship's path. Fleet then telephoned the Bridge.

*Titanic* approaches a rendezvous with destiny at 11.39 p.m. on 14 April. (HFX Studios)

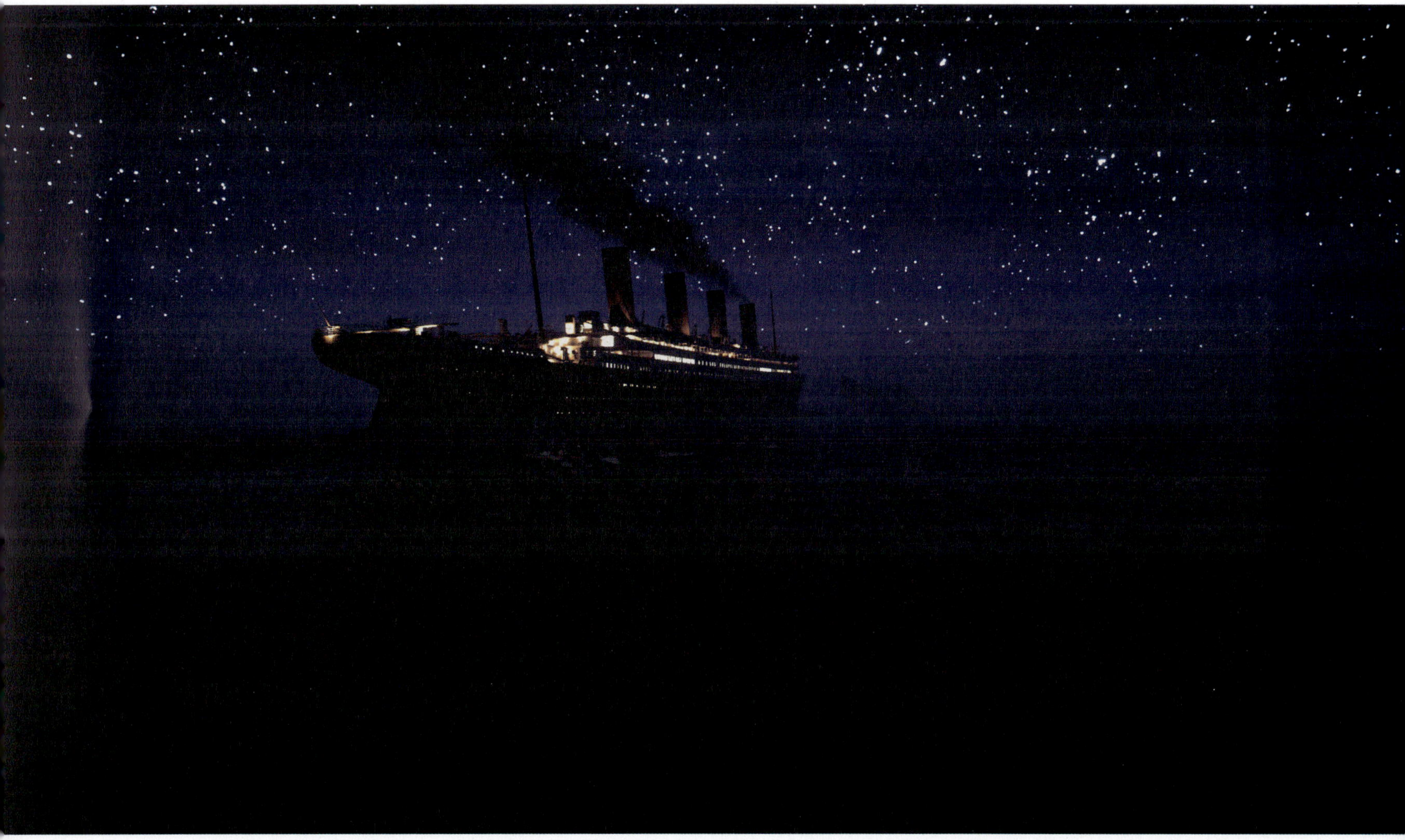

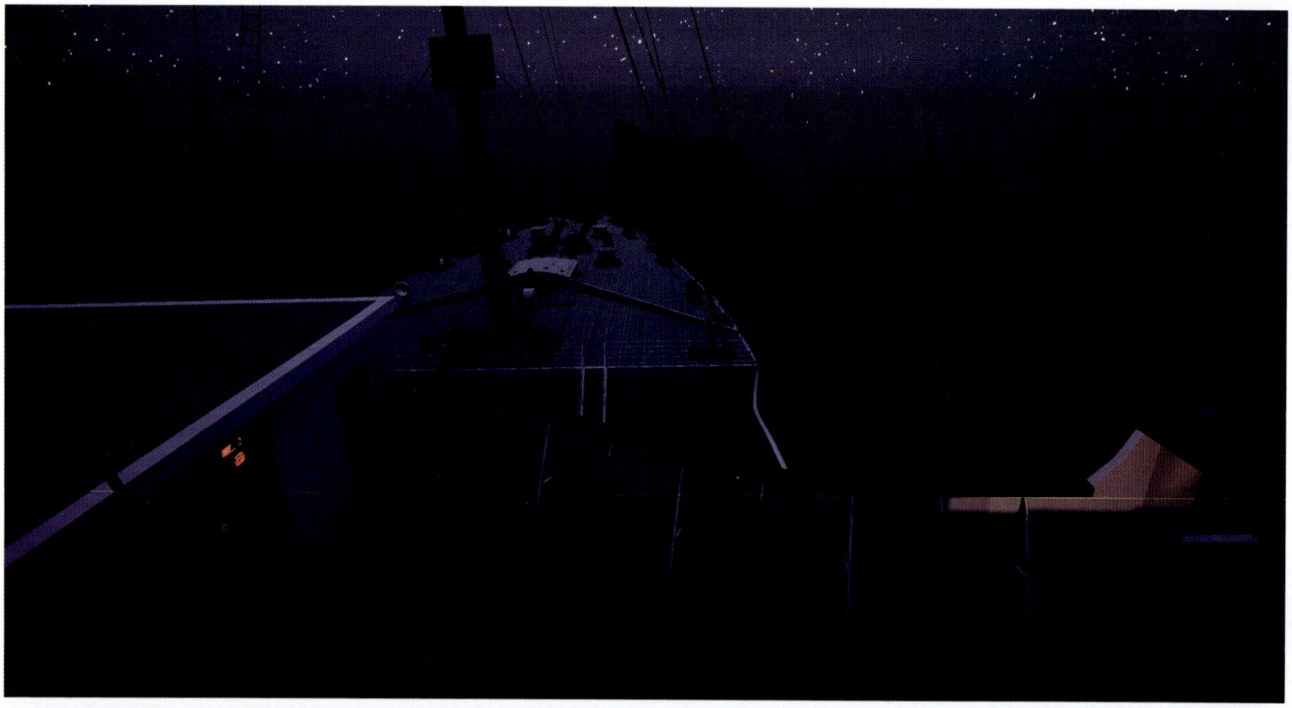

*Right:* From the starboard Bridge wing, First Officer Murdoch assesses a nearly impossible situation and tries to decide how best to react. (HFX Studios)

*Below:* As Murdoch watches from the starboard Bridge wing, the iceberg passes astern, just feet from the ship and very close to Emergency Boat No. 1. (HFX Studios)

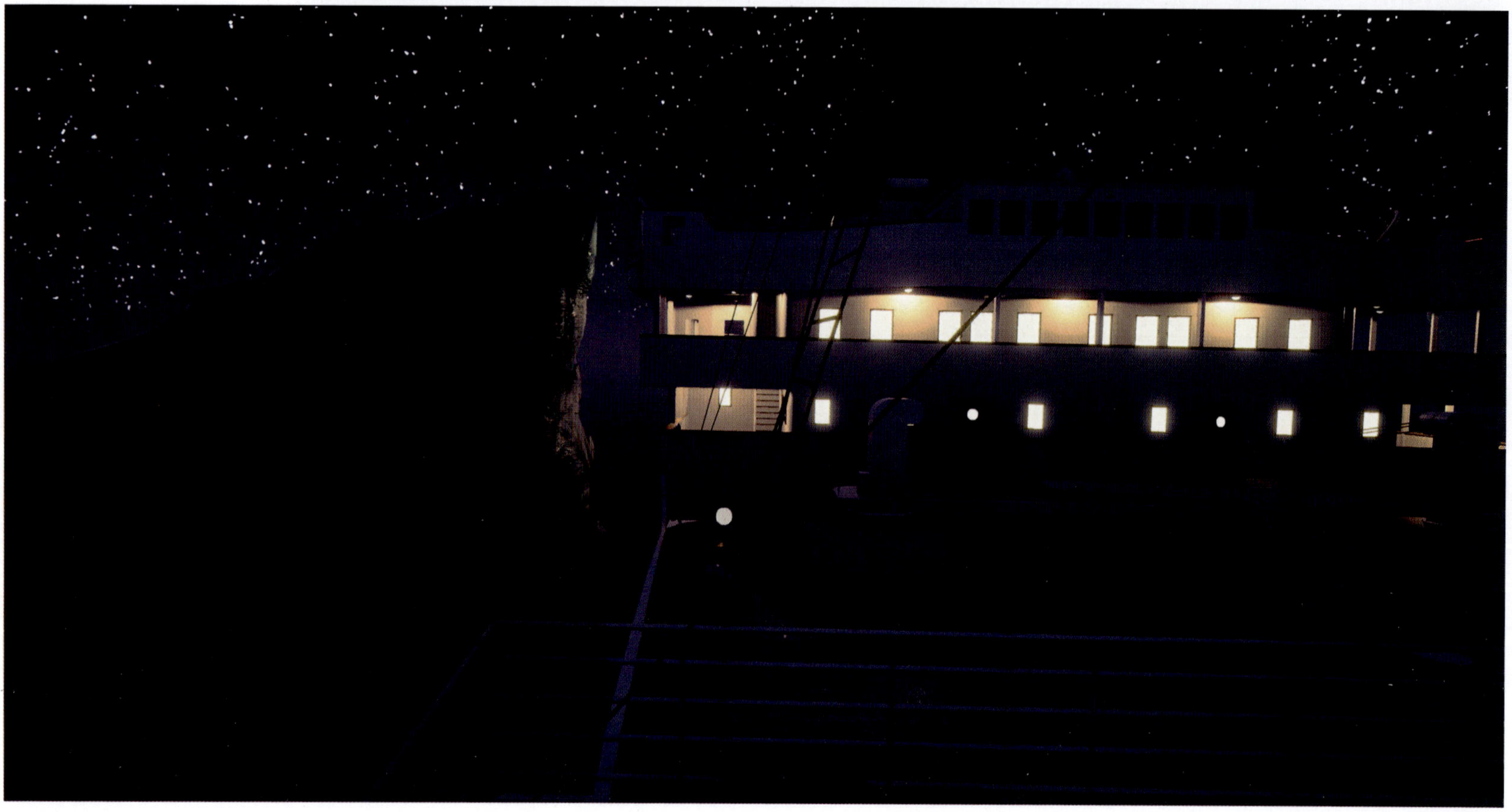

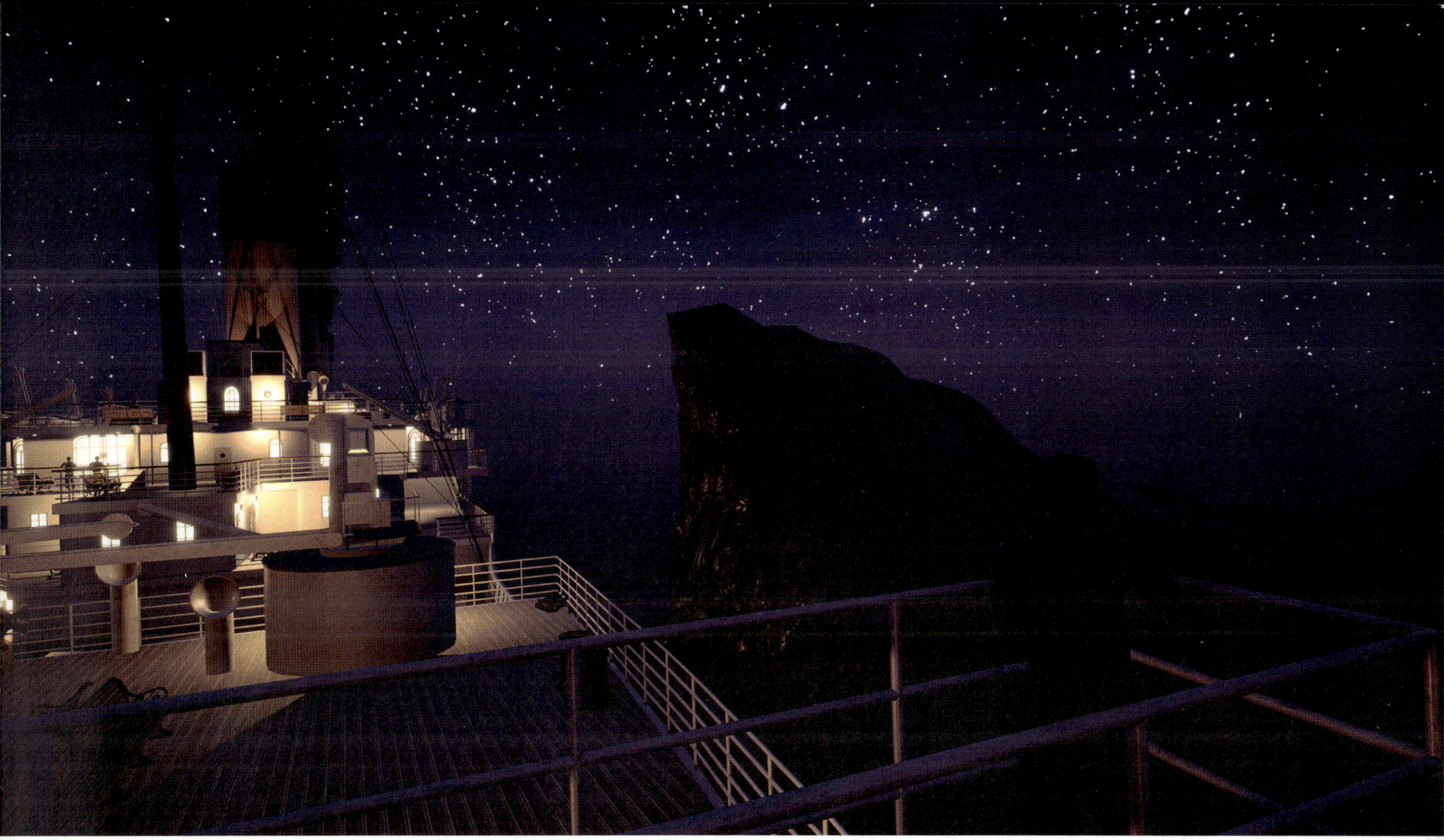

Sixth Officer Moody answered the call, 'Yes, what did you see?' When Fleet responded, 'Iceberg right ahead!' Moody calmly thanked Fleet and relayed the warning to First Officer Murdoch. Murdoch had likely already seen the iceberg by that point. Once he heard the warning bells, he would have been scrutinising the water ahead, to see what had been spotted. Murdoch took a few moments before giving an order, considering the best course of action in order to avoid a collision, rather than making one more likely with a rash decision.

The decision, when it came, was clear. 'Hard a starboard!' Murdoch called, apparently as he was rushing in from the starboard Bridge wing. Quartermaster Robert Hichens, at the ship's wheel, carried out the order, as Sixth Officer Moody watched to ensure it was carried out properly. It had been approximately thirty seconds since the Crow's Nest warning bell had been rung. At the same time, Murdoch gave the helm order, he signalled 'Stop' on both of the Engine Room telegraphs.

The First Officer intended to port around the berg. As the seconds ticked by, the iceberg loomed closer and closer. Finally, *Titanic*'s bow began to turn to port and, at first, with the liner's stem shooting into the clear, it appeared as if a collision would be narrowly avoided. However, the iceberg had simply been spotted too late.

As the berg began to glide past the ship's starboard side, a strange scraping and grinding sound emanated from below. As the starboard hull contacted the berg, the dense ice dented plates and popped riveted seams. The damage extended intermittently from the bow to just past the forward bulkhead of Boiler Room No. 5. Chunks of ice broke off and slid down the peak of the iceberg as it passed, cascading down onto the Forward Well Deck.

Feeling the impact, First Officer Murdoch raced toward the watertight door controls on the open Bridge. As the peak of the iceberg passed the Bridge, he ordered, 'Hard a port!' This action swung the stern and propellers clear of the iceberg, preventing further damage. Murdoch then turned over the lever to close the watertight doors below deck, while signalling the alarm.

Fourth Officer Boxhall, who had just arrived on the Bridge, later claimed that Murdoch had ordered the engines full

Quartermaster Rowe on the Aft Docking Bridge steps back as he sees the iceberg passing by, afraid it would strike the ship – it looks to him like a 'windjammer'. Passengers come out on deck to see what is happening. (HFX Studios)

astern prior to the collision, but this is contradicted by the other two surviving witnesses on the Bridge, Quartermasters Hichens and Olliver, as well as by every surviving witness from below deck in the Boiler Rooms or Engine Room, all of whom mentioned a 'Stop' order, not 'Full Astern'.

As the ship cleared the ice and continued to move ahead, with its bow now starting to turn in a north-westerly direction, Murdoch asked Quartermaster Olliver to note the time and told one of the junior officers to record the event in the ship's log. It was 11.40 p.m. on the Wheelhouse clock.

While the collision did not feel severe to those on deck, it had made a more startling impression on those working below in the Boiler Rooms. This was particularly true in the forward-most Boiler Room, No. 6. Leading Fireman Barrett had just heard the telegraph ring and had seen the red warning light, indicating 'Stop', flash on when he actually witnessed the iceberg damage occur. As the iceberg caved in the hull plates with a crash, Barrett saw water pouring in 2ft above the stokehold floor plates. Immediately heading aft, Barrett and Second Engineer John Hesketh barely made it through the closing watertight door to Boiler Room No. 5.

Many passengers also felt the collision, and some even saw the iceberg as it glided past the ship. A few even had the alarming experience of having ice from the berg actually come into their cabins through open portholes, but most did not recall it as startling. Many who were on upper decks remembered the collision as slight, often comparing it to a bumping or grinding feeling, a series of slight shocks – nothing that seemed serious. More were actually alerted to something being wrong when they felt the soothing vibrations of the ship's engines cease. Others slept through the whole incident.

A group of First Class passengers in the Smoking Room actually ran aft, through the Verandah Café, and out of the sliding doors onto the aft Promenade Deck in time for some to see the iceberg fading astern. Despite this, many of the men, unalarmed, simply returned to the Smoking Room as the iceberg began to disappear astern.

Down below, many Third Class passengers had a more alarming experience. Housed both forward and aft, most were in the lowest passenger accommodation on the ship. Many of them felt the collision more violently, with some being jarred awake or even thrown from their bunks. When they stepped out into the corridor, many of them asked what had happened, but were reassured by stewards that everything was all right and they should go back to bed. In the forward-most cabins, some even got out of bed or stepped into the corridor outside their cabin, only to find water on the floor. To them, the seriousness of the situation was immediately apparent. Some began gathering their belongings and suitcases and headed aft or up to the Forward or Aft Well Decks.

Moments after the collision, Captain Smith appeared on the Bridge. 'What have we struck?' he asked his First Officer. Murdoch explained that it was an iceberg and he had tried to port around it, but it was too close. 'Close the watertight doors,' Smith ordered instinctively. Murdoch confirmed that the doors were closed. Following this, Smith, Murdoch and Boxhall went out on the starboard Bridge wing to look aft and try to spot the iceberg. Boxhall then left to inspect for damage below deck.

Meanwhile, passengers, having felt the collision or the engines stop, began rousing themselves out of bed and asking stewards and crew members what had happened. They were reassured that everything was okay and the ship would soon be underway again. J. Bruce Ismay, not being content with general statements such as these, returned to his cabin and put on a dressing gown over his pyjamas, intent on heading to the Bridge to see what had happened.

On the Bridge, at around 11.43 p.m., Captain Smith ordered *Titanic* 'Slow Ahead'. He then directed Quartermaster Olliver to go below and find the Carpenter, and to instruct him to inspect the ship. The ship began moving forward again, but still proceeding north-north-west, rather than being put back on a westerly heading.

Meanwhile, Fourth Officer Boxhall proceeded on his inspection below deck, moving down to the Well Deck, then under the Forecastle. He next moved through the lower passenger sections in the bow, all the way down to F Deck. Finding no damage, he started to head back up to the Bridge.

At the time of the collision, Second Officer Lightoller and Third Officer Pitman were both off duty and in bed. Feeling the collision, Lightoller got up and went out on deck for a look around. Pitman, who had also just turned out, conferred with his fellow officer. With nothing appearing seriously wrong, both men decided to go back inside, where it was warm.

Quartermaster Hichens was still at *Titanic*'s helm when Captain Smith came back into the enclosed Wheelhouse and looked at the inclinometer in front of the compass. It was approximately five minutes following the collision and the ship already had a 5-degree list to starboard. Clearly there had been significant damage. Following this, Smith ordered *Titanic*'s engines stopped for the last time, and the ship began slowing its forward momentum.

Around this time, Bruce Ismay arrived on the Bridge and asked the Captain what had happened. Smith told him that they had struck ice. White Star's Chairman asked whether the Captain believed the damage was serious. Although he had not yet received a full damage report, Smith had seen enough to reply that 'he thought it was'. Ismay turned to leave the Bridge and head back below to his room.

With *Titanic*'s engines stopped, steam was no longer feeding into the engines and the pressure in the ship's boilers began to quickly build up. Soon the automatic safety valves began to trip in order to prevent a violent rupture. This excess steam was released through pipes on the forward and aft ends of the funnels, where it began to 'blow off' with a thunderous roar that could be heard throughout the ship. For those who had slept through the collision, this was an immediate sign that something unusual had happened.

Meanwhile, heading back towards his suite, Bruce Ismay met Chief Engineer Joseph Bell at the top of the staircase and asked him if he believed the ship was seriously damaged. Bell told him 'he thought the damage was serious, but that he hoped the pumps would be able to control the water'. Hearing this, Ismay briefly returned to his room.

Forward inside the ship there were ominous signs. Boatswain's Mate Albert Haines had discovered that the Forepeak Tank was flooding and went aft and down to look at the No. 1 Hold. He found the tarpaulin 'bellying up', pushed outward by air escaping from the hold. This was also noted by Seaman Edward Buley. Haines headed for the Bridge to report this disturbing find.

Elsewhere, Fourth Officer Boxhall emerged onto the Forward Well Deck after failing to see any damage during his initial inspection below. Stepping out into the cold night air, he saw a man holding a piece of ice 'about as large as a small basin'. Asking where he had got it, the man led him across to the starboard side, where there were small chunks of ice covering the deck, 3–4ft from the bulwarks.

Around 11.50 p.m., Boxhall returned to the Bridge, reporting that he had not seen any damage below deck. It was the last piece of good news that Captain Smith would receive about the condition of the ship. A few minutes later, seemingly unconvinced by his Fourth Officer's preliminary findings, he ordered Boxhall back below deck to 'find the carpenter and get him to sound the ship'. Apparently, Quartermaster Olliver had not yet returned from his own search for the Carpenter.

Below deck, things were looking alarming. In Boiler Room No. 6, Fireman George Beauchamp was working on drawing fires in his area. Due to the starboard list that had developed, Beauchamp's workspace on the port side was only shallowly flooded. He stayed put as the water continued to rise until, around 11.55 p.m., he was told 'that will do', and promptly exited the stokehold.

Meanwhile, after fleeing aft to Boiler Room No. 5 to escape the flood of seawater, Leading Fireman Barrett heard Second Engineer Hesketh give the order for all hands to stand by their stations. For Barrett, this meant returning to Boiler Room No. 6. Arriving there by way of 'Scotland Road', he went down the ladder and found that there was now 8ft of water on the starboard side of the compartment. In the ten minutes since he had left, the water level had risen 6ft at that location.

Returning to Boiler Room No. 5, Barrett, along with Junior Assistant Second Engineer Jonathan Shepherd, found Senior Assistant Second Engineer Bert Wilson and Junior Assistant Second Engineer Herbert Harvey attending to the pumps. Water was pouring into the forward coal bunker on the starboard side, 2ft above the plates. However, the stokehold itself remained dry. For now, the pumps were keeping up with the flooding.

In the First Class Dining Saloon, Saloon Steward James Johnstone – whose name is often misspelled – was serving as the night watchman. At around 11.50 p.m., he walked forward towards the Grand Staircase in the Reception Room. He arrived just in time to see Thomas Andrews racing down the working stairs to E Deck, then turn aft, apparently heading for the Engine Room for an inspection of the damage. A small group of First Class ladies had spotted and approached Andrews, and Johnstone heard him reassure them that everything would be all right, before heading below. This also encouraged Johnstone.

Further below, in the lowest crew areas in the ship – even in those that did not immediately begin to flood, like Boiler Room No. 6 had – damage was beginning to become apparent. Water was inundating the mail hold on the Orlop Deck, and the ship's five postal clerks were working desperately to lug the sacks of mail up to the next level to escape the rising water. They managed to complete this task, only to find that the water continued to climb the staircase behind them, towards G Deck.

In the minutes immediately following the collision, in First, Second and Third Class alike, passengers who had awakened and come out of their rooms – whether into the hallways or going up on deck – were all being reassured that everything was okay. They were encouraged to go back to their rooms and were even told to go back to sleep. To some, particularly to those in Third Class, this delay may have lessened their chances of survival. However, at this stage, none of the crew knew the exact condition of the ship, so they can hardly be blamed.

Meanwhile, before Fourth Officer Boxhall could get very far on his second inspection trip below deck, he crossed paths with Carpenter Hutchinson, who reported that the ship was quickly taking on water. Boxhall told Hutchinson to report it to the Bridge, while Boxhall proceeded further below. Next, Boxhall ran into British Postal Clerk John Smith, who reported that the Mail Room was flooding. Boxhall sent Smith to the Bridge to report this as well. Boxhall then proceeded to the Mail Room. Arriving around 11.52 p.m., Boxhall could actually hear the water rushing in, and saw bags of mail floating in the water. The water had risen to within 2ft of G Deck.

In the First Class Reception Room, Steward Johnstone and Saloon Steward Mackay saw Captain Smith going down the same working staircase to E Deck that Johnstone had seen Thomas Andrews rush down a few minutes prior. Carpenter Hutchinson had just earlier reached the Bridge and told Smith of the flooding in Cargo Hold Nos. 1, 2 and 3, and the Captain was likely heading below to check in with Chief Engineer Bell, Thomas Andrews, or both.

At that same time, an order came down for the firemen to be sent up on deck. However, the electric lights then went out in the stokeholds, impeding the damage-control efforts of those trying to keep ahead of the flooding. Second Engineer Hesketh then ordered Leading Fireman Barrett to go get lamps for Boiler Room No. 5, so that the crew working there could see.

Just before 11.50 p.m., Bruce Ismay had a brief conversation with Chief Engineer Bell at the top of the Grand Staircase. Bell told Ismay he thought the damage was serious, but that he remained hopeful. Ismay returned to his stateroom, staying just long enough to throw a suit on over his nightclothes, before heading back up on deck to the Bridge.

Around this time, Boatswain's Mate Haines arrived on the Bridge and found Chief Officer Wilde, whom he had seen a few minutes prior on the Forecastle Head, near the exhaust vent for the Forepeak Tank. He reported seeing evidence that the No. 1 Hold was flooding. Wilde, as a precaution, told Haines to go wake his men up and then to get the lifeboats out. Even though Captain Smith, who was on his inspection below, had not yet given an official order for all hands to stand by, Chief Officer Wilde told Haines to do so as experienced seamen would be needed to get the boats uncovered, swung out and ready for loading, should this become necessary. Haines hurried back below deck to start the task. He was overheard passing the order to the men in the Forecastle Head that all hands should stand by, since they 'may be wanted at any moment', at about 11.55 p.m.

At 11.55 p.m. Steward Johnstone, still in the Reception Room, watched Thomas Andrews come back up to D Deck, apparently having already been to the Engine Room. Andrews proceeded forward, before heading down another set of stairs to the Mail Room. After a minute, curiosity got the best of Johnstone and he followed Andrews down. From forward on E Deck, they could look down the stairs straight through to G Deck. There, Johnstone could see that the First Class Baggage Room on G Deck was flooding. Johnstone headed back to his berth and got dressed in his shore suit.

Around this time, First Class passenger Alfred Nourney, already having seen the iceberg fragments covering the Forward Well Deck, decided to go below deck and look for damage. He headed down to the Squash Racquet Court. The Squash Court consisted of two levels, with a gallery for spectators on F Deck and the actual court down below on G Deck. The Squash Racquet Court was the lowest and furthest forward First Class passenger space on the *Titanic*, and was directly inboard from the Post Office Sorting Room. Nourney was stunned to find between 3 and 6ft of water on the court. He turned around and headed back up the stairs.

By this time, as more and more passengers began to wake up, it was becoming clear that something was not right. Besides the silent engines, some passengers could perceive that the ship was slightly down by the bow and listing to starboard. Still, there were no signs of panic or confusion on deck, so most who noticed this were not too alarmed, and waited for word from crew members about what to do.

Around midnight, Fourth Officer Boxhall had finished his second inspection trip below and had returned to the Bridge, confirming to Captain Smith that the Mail Room was taking on water. Smith had just returned from his own inspection below, making it as far as E Deck aft. Immediately after sharing this news, Smith said 'all right' and left to head below for his own personal inspection of the flooding in the forward areas.

Prior to leaving the Bridge, Captain Smith had officially given the order for all hands to stand by. Boxhall heard this order come down, and the crew members began uncovering and clearing the lifeboats. Bruce Ismay had just returned to the Boat Deck and heard Captain Smith 'give the order to clear the boats'. Smith did this out of an abundance of caution, since he did not yet know the full extent of the damage to the ship, nor that the ship was sinking. This is one of many points that shows that – contrary to the myth that he was in a daze following the collision – Captain Smith was actually hard at work and being proactive in case an evacuation did become necessary.

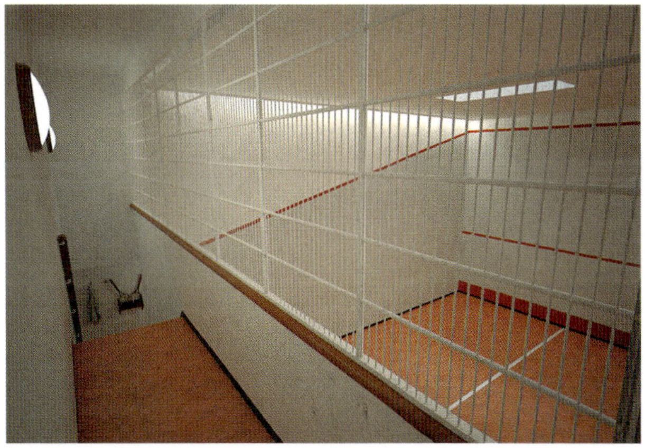

*Left:* By 11.55 p.m., a mere fifteen minutes after the collision, Alfred Nourney was stunned to look down from the spectator's gallery on F Deck and see several feet of water covering the Squash Racquet Court floor below. (Chris Walker)

*Below:* By 12.05 a.m., twenty-five minutes after the collision, the ship is noticeably down by the head, with a 5-degree list to starboard. Here, the forward ports along the port side of F Deck are about to slip beneath the surface. (HFX Studios)

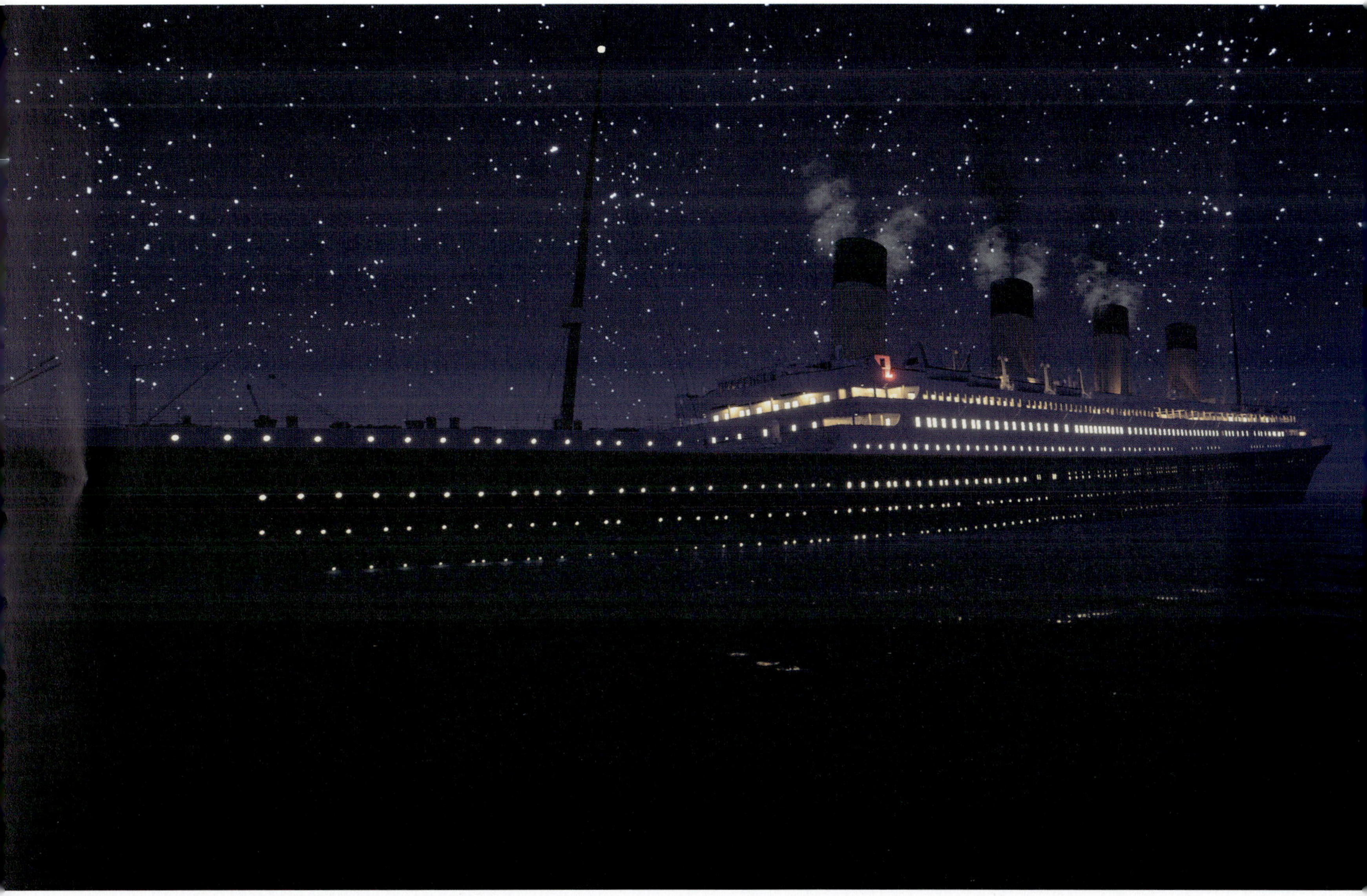

Also around midnight, Sixth Officer Moody told Quartermaster Olliver to get the muster list for the lifeboats. At the same time, Fourth Officer Boxhall received an order, possibly from Chief Officer Wilde, to wake up Second Officer Lightoller, Third Officer Pitman and Fifth Officer Lowe. Lightoller and Pitman had come out on deck following the collision but had since returned to their rooms. Sixth Officer Moody was then tasked with heading to the aft port boats, to begin the prep work there.

Back in the stokeholds, the lights finally came back on, just as Leading Fireman Barrett was returning to Boiler Room No. 5 with lamps. Since the firemen had been ordered up on deck, Barrett was sent to find some of them, return and draw the fires from the boilers in No. 5. With the ship's engines stopped, the high-pressure steam from the boilers was continuing to be vented from the funnels on deck. Barrett returned with fifteen men and quickly got to work.

Elsewhere, Assistant Second Steward Joseph Wheat, having witnessed water coming onto G Deck near the Mail Room, decided that he should check whether anyone had closed the watertight doors near the Turkish Baths. Descending to F Deck, Wheat closed the door near the bottom of the staircase. He noted that there was not any water there at the time. However, the second watertight door, on the starboard side, between the Turkish Bath and Swimming Bath, had to be closed from the deck above. Wheat headed up to E Deck, where Second Steward George Dodd and Turkish Bath Attendant James Crosby used the key that operated the watertight door control and managed to close it.

Back on deck, at midnight, Lookouts Hogg and Evans proceeded up to the Crow's Nest to relieve Fleet and Lee, who were still on duty, despite the ship having stopped. Heading below, Lee saw firemen and greasers coming up to C Deck with their kits, because their quarters had begun to flood. Lee also looked down Hatch No. 1 and could see water coming into the hold. Just after this, the order for 'all hands' was given, and Fleet and Lee followed the other men up to the Boat Deck.

Below deck, Saloon Steward William Ward saw steerage passengers heading aft from the bow, along 'Scotland Road'. Some were carrying lifebelts and belongings with them, and some of them appeared to be wet. Ward had initially thought that the ship had thrown a propeller. Despite this, Ward did not believe that the situation was serious, and decided to stay put for a while longer.

Up on deck, Fourth Officer Boxhall had arrived at Third Officer Pitman's cabin and found him awake and unhurriedly getting dressed. After informing him that they had struck an iceberg and that the Mail Room was flooding, Pitman quickened his pace. Boxhall also attempted to wake up Fifth Officer Lowe, but the exhausted Lowe did not completely wake up, and fell back to sleep.

Boxhall then arrived at Second Officer Lightoller's cabin, only to find him awake. He exclaimed, 'You know we have struck an iceberg!' Lightoller responded that he knew they had struck something. After learning that the Mail Room was flooding, Lightoller knew there must have been significant damage, but he still concluded that there was not any danger of actually sinking. Lightoller got dressed as quickly as he could and went out on deck, leaving his watch behind.

By this time, Third Officer Pitman had gone out onto the port side of the Boat Deck aft, where he found Sixth Officer Moody and some crewmen uncovering the lifeboats there. Moody told Pitman that he had not seen the iceberg pass, but that there was ice on the Forward Well Deck. Moody was inside the enclosed wheelhouse next to Quartermaster Hichens at the time of the collision, and with the shutters closed, had been unable to see outside. Pitman decided to go and see the ice for himself.

By this time, Alfred Nourney had returned from his trip below deck and had informed his friends about the flooding and that the ship was in danger. None of them believed him, and one told him that it 'doesn't matter because the *Titanic* cannot sink'.

Shortly after midnight, First Class passengers Norman and Bertha Chambers were heading back to their cabin; they reached E Deck forward, and Mr Chambers looked down the stairs at the end of their starboard-side passage, where it led down to the First Class Baggage Room and Post Office Sorting Room. He found a couple of the mail clerks wet to their knees, hauling up mail bags. Chambers could see far enough down to observe that water was within '18 inches or 2 feet of the deck above' F Deck.

Despite seeing this, Chambers did not believe that there was any danger, since the water was 'forward of the bulkhead'. While he and his wife stood there, three of the ship's officers descended the stairs and looked into the Baggage Room, immediately coming back up and remarking, 'We were not making any more water.' Right after this, the Chambers encountered their Bedroom Steward, who told them that they could go back to bed. Mr Chambers believed that they could, but regardless decided to change into heavier clothing.

On deck, Third Officer Pitman headed forward to see if he could spot the ice on deck that Sixth Officer Moody had mentioned. He descended to the Well Deck and saw the ice, before heading all the way forward underneath the Forecastle. He did not see any signs of damage.

Back on the Boat Deck, work was progressing on uncovering the lifeboats, as the officers supervised. First Officer Murdoch took charge of the boats on the starboard side, while Chief Officer Wilde oversaw efforts forward on the port side, with Second Officer Lightoller assisting. Meanwhile, with Smith, Pitman and Boxhall otherwise occupied and Lowe still asleep in his cabin, Sixth Officer Moody did not have much assistance in preparing the aft port boats.

When Second Officer Lightoller came out of his cabin onto the deck, he found that the steam venting from the funnels was extremely loud and made it difficult to communicate. Meeting Chief Officer Wilde just outside the door into the Officers' Quarters, he asked whether the order for 'all hands' had been given, and Wilde confirmed that it had. Lightoller noted that none of the covers had been stripped off of the lifeboats near him at that point, so there was much work to be done.

Meanwhile, below deck, Captain Smith was still completing his own inspection of the damage. Stewardess Annie Robinson spotted him on E Deck, heading towards the Mail Room with one of the mail clerks and Purser McElroy. Apparently, Smith met Thomas Andrews while there since, at 12.10 a.m., Robinson saw both men coming back from the Mail Room. As they passed by, Robinson heard Andrews state, 'Well, three have gone already, Captain', surely a reference to the first three watertight compartments. Curious at what Smith, Andrews and the others had been looking at, Robinson went to see for herself, and saw water within six steps of coming onto E Deck, near the stairs leading to the Mail Room.

After this sighting, Captain Smith and Andrews separated, with Smith heading back up to the Bridge, while Andrews continued his inspection below. At this point, Smith was well aware that there was significant damage. However, what he had seen on this inspection, coupled with the statement made previously by 'the officers', and overheard by Norman Chambers, that 'we were not making any more water', makes it seem as if when he separated from Andrews, the situation appeared very serious, but did not indicate that the ship was actually in danger of sinking.

Captain Smith arrived back on the Bridge around 12.12 a.m. Even though he did not yet know that the ship was sinking, he proactively gave the order to swing out the lifeboats and to begin getting the passengers up on deck with their lifebelts on. Around this same time, Smith worked out the ship's approximate position using dead reckoning. Not taking any chances, he then stopped by the Marconi Room to inform Wireless Operators Jack Phillips and Harold Bride that the ship had struck an iceberg, they were carrying out an inspection and they had 'better get ready to send out a call for assistance'. He then told them not to send it until he gave the order.

The Captain's order to get the passengers up on deck was immediately acted upon, and as the word spread, crew members began pitching into the effort to help First, Second and Third Class passengers get dressed and get their lifebelts on. Around the same time that passengers were being roused, an order came down to begin provisioning the lifeboats. Even though the lifeboats already had biscuits and water stored in them, Chief Baker Joughin sent thirteen men up on deck with four surplus loaves of bread each, to supplement those supplies. Joughin then went below to his room to get a drink.

At 12.20 a.m., First Officer Murdoch was joined by Third Officer Pitman on the Boat Deck, and Murdoch instructed him to start working on Boat No. 5. At that point, the crewmen on hand had just started to remove the canvas cover from the boat. Pitman immediately went to work. Along both sides of the ship, the officers and crewmen were hard at work in preparing the boats for loading. Besides stripping the covers, the gripes holding the boats in their positions on deck had to be removed, the lifeboat davits had to be swung out, and then the boats had to be lowered down level with the edge of the deck. All of this took time, and even though there had not yet been an order to begin loading passengers into the boats, the officers wanted to make sure that they were ready in the event that it became necessary to do so.

Below deck, Thomas Andrews was continuing his inspection. A disturbing picture was beginning to emerge as he investigated and observed more regarding the extent of the iceberg damage. Meanwhile, music began to drift through the ship's upper decks, emanating from the First Class Entrance forward on the Boat Deck. The ship's band began playing to calm the passengers who were coming up from below. The music was upbeat, mostly popular ragtime and waltzes at this point. Although some passengers were nervous, the atmosphere in First Class was jovial, perhaps even party-like.

Up in the Crow's Nest, Lookouts Hogg and Evans were still on duty. While they had only been there for twenty minutes, it had been a strange start to their watch. There was the collision prior to when they went on duty, the ship was still stopped, there was ice on the deck and they could hear the steam blowing off from the boilers. Curious about what was going on and unable to hear anything due to the roar of the steam, the two men turned around and lifted the canvas weather screen that was blocking their view of the Boat Deck. They saw people 'running about on deck with life belts on'.

Realising that something serious was going on, and that they had apparently been forgotten, the two men called the Bridge, but nobody answered. This is perhaps not surprising, since nearly all of the officers were then engaged in the efforts to get the lifeboats ready. Not wanting to take any chances, Hogg and Evans decided to leave their posts and head to the Boat Deck.

Around this time, further observation of the flooding in the forward compartments had given Thomas Andrews the full picture of the extent of the damage to *Titanic*. Flooding in Boiler Room No. 6 was catastrophic, rendering that compartment a lost cause. Based on his earlier observations before Captain Smith and he had separated, Andrews already knew that Cargo Hold Nos. 1, 2 and 3 were also flooding beyond control. The Forepeak Tank was also flooded.

As Andrews did the mental math, he reached an inescapable and unbelievable conclusion: *Titanic* was doomed. Andrews turned and headed back up on deck immediately to share this grim news, no doubt in a hurry to find Captain Smith. By this time, curious passengers and crew had gathered in the ship's public spaces, in varying stages of dress and undress.

Around 12.22 a.m., First Class passenger Anna Warren saw Andrews rushing up the Grand Staircase on D Deck, taking three steps at a time and with a 'look of terror' on his face. Just after this, First Class passenger William Sloper and some fellow passengers saw Andrews at A Deck, hurrying towards the Bridge. Despite having stopped to reassure passengers earlier in the night, Andrews now ran past them without saying anything. One look at his face had convinced Sloper that Andrews was 'worried'.

By 12.25 a.m., forty-five minutes after the collision, *Titanic*'s bow was noticeably down, and the ship was still listing to starboard. The E Deck portholes, which were normally 20ft above the waterline in the bow, were now nearing the sea. At this time, the Seamen's Quarters on E Deck flooded. Normally 48ft above the waterline, this gives an idea of how quickly water was inundating the ship. Seaman John Poingdestre was in these quarters at the time when a wooden bulkhead gave way, flooding the compartment with 3ft of water. Poingdestre waded through icy seawater up to his waist in order to escape, clearing out of the room in about thirty seconds, and headed up to the Boat Deck.

Also around 12.25 a.m., Captain Smith and Thomas Andrews met on the Bridge. Andrews had dire news to report. Water was intruding into the six forward watertight compartments. Although the flooding was currently under control in the aft-most of these, Boiler Room No. 5, the forward five compartments were flooding uncontrollably. *Titanic* was designed to stay afloat with its first four watertight compartments completely flooded. However, with the weight of water in a fifth, the ship's bow would be pulled down until the water flooded over the top of the next watertight bulkhead. Once that watertight compartment flooded, the ship would settle far enough that the next compartment aft would also flood. This process would repeat until the ship sank.

Unlike the portrayals in movies, there is no evidence that Captain Smith or Thomas Andrews called a conference of all of the ship's officers in order to deliver this grim news. While word would spread to some of the officers and crew who needed to know, no formal announcement was ever made to the officers or crew as a whole that the ship was sinking. Because of this, some crew members refused to believe that the ship would sink, even relatively close to the end. It is less surprising that no formal announcement was ever made to the passengers, since this was surely in an effort to avoid a panic.

Smith and Andrews' conversation must have been truly sobering. Both men knew that there were nowhere near enough lifeboats aboard to rescue all 2,208 souls aboard *Titanic*. Realising this, Smith asked Andrews how long they had. After doing some calculations, Andrews estimated an hour, perhaps an hour and a half, but not much longer. It is hard to imagine what was going through Captain Smith's head when he was given that information. He must have known that unless another ship was close enough to come to their rescue, that there was bound to be a massive loss of life.

As the enormity of the situation began to set in, the conversation must have been a short one. Smith quickly issued the order to begin getting the women and children into the boats. While some have questioned why it took forty-five minutes after the collision for Captain Smith to give the order for passengers to be put in the boats, there were a number of good reasons for this. First, the extent of the damage and the seriousness of the situation in general took some time to discover following the collision. Inspections had to be made and then all of the pieces of the puzzle put together from different observers. Smith had taken proactive measures even prior to knowing the ship was sinking, including having all hands called, ordering the lifeboats uncovered and having passengers assembled on deck with their lifebelts on. He also had prompted the wireless operators that they may need to send a signal, and to standby.

Secondly, the crewmen had only had twenty-five minutes to prepare the boats for loading prior to when the order to begin loading women and children aboard had been given. It would have been completely foolhardy and unsafe for Captain Smith to have ordered passengers into the boats without even knowing for sure just how dire the situation was. Only after Andrews had delivered the update from his inspection did Smith know that they would have to risk the danger of a mid-ocean evacuation in the middle of the night.

After giving the order to begin lowering the boats, Captain Smith immediately headed aft to the Marconi Room. Barely popping his head in the door, he handed the wireless men *Titanic*'s estimated position and gave the order to call for assistance. When Jack Phillips asked what call to send, Smith told him, 'The regulation call for help. Just that.' Then he left again. Phillips, as the Senior Operator, took the wireless key from Harold Bride, and at 10.25 p.m. New York Time, or 12.27 a.m. *Titanic* time, began sending CQD.

By the time the first distress signal was sent, things were proceeding more quickly on *Titanic*'s Boat Deck. All of the officers, including Lowe, were now up and assisting, and passengers continued to arrive on deck. Hindering the situation was the fact that many of the crew members, even though fully aware that there had been an accident, were not told that the ship was sinking and were thus not aware of the urgency of the situation.

After having awakened the off-duty officers, Fourth Officer Boxhall was helping unlace some of the canvas covers on the lifeboats. Suddenly, he heard someone report that there was a light of a ship ahead. Boxhall was heading back to the Bridge to investigate when he ran into Captain Smith. After conversing, Boxhall asked whether he should have a distress signal sent. Smith stated that he had already requested a distress signal be sent. Boxhall, concerned, then asked, 'What position did you send it from?' Smith replied that it was from the 8.00 p.m. dead-reckoning fix.

Boxhall replied that the celestial fix had shown that they were about 20 miles ahead of the dead-reckoning position. Smith then suggested that Boxhall double-check the position that he had worked out. Before parting, Smith asked his Fourth Officer how work was progressing on the boats. Boxhall said, 'They are carrying on all right,' then asked Smith if the situation was really serious. Smith responded directly, stating, 'Mr Andrews tells me he gives her from an hour to an hour and a half.' With this news on his mind, Boxhall proceeded directly to the Navigating Room aft of the Bridge and began using the stellar observation taken around 7.30 p.m. to work out their location.

Even as Boxhall engaged in this work, *Titanic*'s 12.27 a.m. distress call was immediately heard by the Cape Race, Newfoundland, station, as well as by the ships *La Provence* and *Mount Temple*. The *Ypiranga* also heard the CQD call at around 12.30 a.m. However, nobody yet fully grasped the enormity of what was happening.

Meanwhile, after his difficult conversation with Captain Smith, Thomas Andrews had proceeded to A Deck, where he assisted the crew members in waking up passengers and encouraging them to put their lifebelts on. Crossing paths with Stewardess Annie Robinson, he told her, 'Put your lifebelt on and walk about and let passengers see you.' When she was reluctant to do so, he responded firmly and told her, 'If you value your life, put your belt on.' Around this same time, Stewardess Mary Sloan recalled seeing Andrews 'here, there and everywhere, looking after everybody, telling the women to put on lifebelts, telling the stewardesses to hurry the women up to the boats, all about everywhere, thinking of everyone but himself'.

Things were not nearly as organised in Third Class. Many male passengers in the bow section of the ship had literally been chased out of their cabins by rising water. They fled aft or up onto the Forward Well Deck, many dragging their belongings with them. Others in the further aft sections of

Third Class were awaiting instructions. Steward John Hart recalled how word had been passed down at 'about 12.25 or 12.30' for the women and children to be sent up to the Boat Deck. Instead of leaving the passengers to fend for themselves and having to find their path through unfamiliar sections of the ship, Hart gathered a group of around thirty women and children, and personally began leading them to the lifeboats.

Below deck, there is some evidence that the impact with the iceberg may have exerted enough pressure to slightly warp or deform the hull frame near the forward funnel casing on D Deck. Alternatively, it is possible that as the weight of water accumulated in the bow, it began to warp the structure of the ship. Based on eyewitness evidence, there appear to have been at least two passengers who got stuck behind jammed cabin doors on D Deck. The most famous instance of this was when First Class passenger Richard Norris Williams freed a trapped passenger by breaking down the door, at which point a steward threatened to report him for damaging White Star Line property.

Additionally, there was evidence presented in the inquiries after the sinking and from some eyewitnesses that some of the manually operated watertight doors had jammed in the open position. One example of this was the watertight door on the starboard side of F Deck, near the casing for the second funnel. First Class passenger George Harder saw crew members attempting to lower the door, which had to be done from E Deck above, and complaining that it wouldn't work. It is impossible to say whether this door was stuck from an issue with installation, or whether this is indicative of additional stresses that *Titanic*'s hull was subjected to that night.

On the Boat Deck, Lookout George Hogg had been helping uncover the lifeboats when the Boatswain sent him to retrieve a Jacob's ladder. After he returned, First Officer Murdoch ordered him into Boat No. 7 to begin preparing it for passengers.

Fifth Officer Lowe had by this point joined Murdoch, assisting at Boat No. 7. Lowe had fallen asleep after Boxhall's attempts to wake him and, in fact, didn't even recall Boxhall entering his cabin. He was eventually awakened when he heard commotion outside of his room and saw ladies in the Officers' Quarters with lifebelts on. He got up and dressed quickly; coming on deck, he found more passengers wearing lifebelts. Lowe was then told that the ship had struck an iceberg, and he could feel the ship listing under his feet.

Lowe returned to his cabin and retrieved his personal revolver, a Browning automatic, because as he later said, 'you never know when you will need it'. Crossing over to starboard, he found First Officer Murdoch working at Boat No. 7 and began assisting there.

On the port side, Second Officer Lightoller – after having checked on the starboard side and seeing that Murdoch had things well in hand there – asked Chief Officer Wilde whether they could swing out the boats, but Wilde was reluctant to do so. Lightoller then ran across Captain Smith and asked him the same question. Smith gave permission to begin swinging out the boats. After this task had been accomplished, Lightoller returned to Smith. Cupping his hands over the Captain's ear, he shouted over the escaping steam, 'Hadn't we better get the women and children into the boats, sir?' Smith responded affirmatively.

On the Bridge, Fourth Officer Boxhall had just finished his own calculation of the ship's position, which he promptly gave to Captain Smith. Smith ordered him to take the revised position to the Marconi Room. Arriving there, and finding the roar of escaping steam too loud to communicate over, Boxhall wrote the new position on a slip of paper, rather than verbally telling the wireless men what to transmit. Jack Phillips then began sending out the CQD for the position of 41° 46' N, 50° 14' W.

Within a minute, Phillips had connected with the Cunard liner *Carpathia*, which was around 58 miles away. *Carpathia*'s Marconi Operator, Harold Cottam, immediately informed his Captain, Arthur Rostron. Rostron instantly ordered *Carpathia* brought about and started for *Titanic*'s reported position at full speed. Phillips had Harold Bride find and report this news to Captain Smith, who was in the Wheelhouse at the time. Despite the initial good news, Smith must have been horrified once he realised that the *Carpathia* was around four hours away, meaning that there was no possibility of them arriving prior to *Titanic* sinking.

Now, even though it had been nearly an hour since the collision, the noise from the escaping steam continued to interfere with the evacuation efforts. In the Marconi Room, it was causing Jack Phillips to have extreme difficulty in even hearing responses from other ships. At 12.38 a.m., the SS *Ypiranga* heard Phillips sending CQD, and also saying, 'Require immediate assistance. We have collision with iceberg. Sinking. Can hear nothing for noise of steam.'

On the port side of the Boat Deck, Second Officer Lightoller had Boat No. 4 lowered to the Promenade,

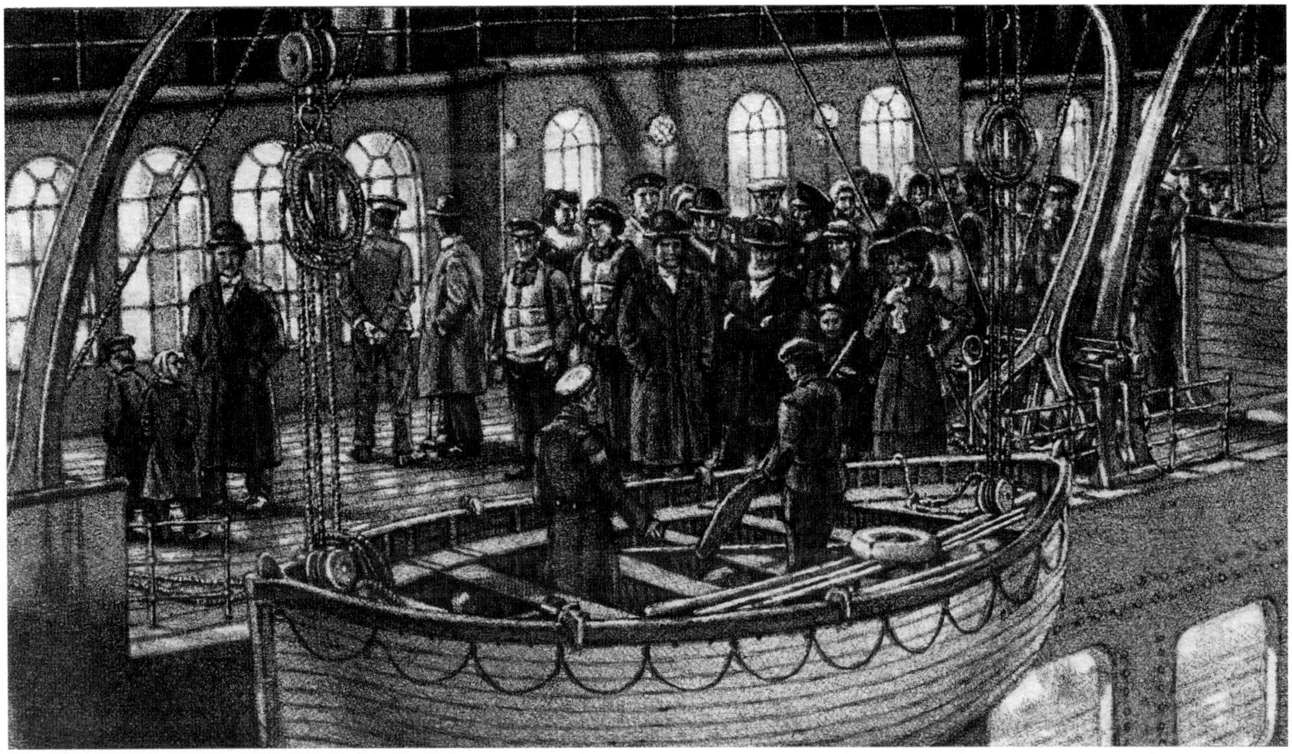

Boat No. 4 being lowered to A Deck. (Ballpoint pen on paper by Eugene Nesmeyanov)

or A Deck. It seems likely that he did so at Captain Smith's suggestion, since First Class passenger Hugh Woolner overheard him discussing his intent for passengers to go down to A Deck and board from there. The thought process was that it would be easier for passengers to get into the boat from that location. There was a coaling wire that ran underneath A Deck, and Lightoller tied off No. 4 to that, just in case the ship developed a list. The process of lowering the boat to that deck and tying it off took some time, at which point Lightoller ordered crew members who were on the scene down to the Promenade to help with the loading.

This quickly led to a dilemma, since this portion of the Promenade had enclosed windows, which had to be cranked open. After this was pointed out to him, Smith exclaimed, 'By God, you are right. Call those people back!' The exasperated passengers, who had just trudged their way down below, then had to turn around and head back up to the Boat Deck. Lightoller then sent some stewards to find the handles and crank open the windows. However, all of this was going to take some time.

Around 12.40 or 12.45 a.m., rather than wait around, Lightoller moved on to Boat Nos. 6 and 8 and began working there. He quickly gave another order, sending Boatswain Alfred Nichols and six other crewmen below with the intention of opening the gangway doors on D Deck. He thought that the lifeboats could take aboard additional passengers from that location, once afloat.

However, this plan never materialised. While it has often been claimed that the men drowned, since they were never seen by Lightoller again, this is very unlikely considering how slowly the flooding was progressing near that location at the time. It is more plausible that the men simply returned on deck and ended up employed in some task elsewhere on the ship. Regardless, this exacerbated an already critical shortage in seamen available to assist with the loading of the boats.

The plan of loading additional passengers into the boats from the gangway doors may have come far closer to working than many have acknowledged. One of the two gangway doors on the port side of D Deck is open on the wreck today; it is unlikely that this is because it opened upon impact with the sea floor since its interior gates are also open. It seems possible that the crew members Lightoller sent below did manage to get the doors open, but it took too long and, by the time they did so, the lifeboats in that vicinity had already lowered and rowed away from the ship.

Meanwhile, on the forward starboard side, First Officer Murdoch was still hard at work at Boat No. 7. Lookout George Hogg put the plugs into the boat and had been completing the prep work. He then jumped out of the boat to assist at the falls, when Murdoch ordered him to man the boat. The crewmen on the scene then began passing in all of the women and children who were standing nearby. They were still having difficulties in getting the passengers to come out on deck, with the cold and noise. *Titanic* appeared safer than the tiny boats – at least at that point.

Steward Henry Etches lent a hand at No. 7, after seeing that Boat No. 5 was still being prepared for loading. He noticed Murdoch, Pitman, Ismay and other crew members all assisting. Etches helped to clear the falls, which passengers were tripping over. Sometime after arriving on deck, Etches saw First Class passenger Benjamin Guggenheim and his valet Victor Giglio. Their appearance stunned the steward as they were dressed in formal evening clothes. When he had sent them out of their cabin earlier in the night they were wearing nightclothes, sweaters and their lifebelts.

'What's that for?' Etches asked. 'We've dressed up in our best, and are prepared to go down like gentlemen,' Guggenheim replied. Apparently, the gentleman had somehow been told that the ship was sinking, when so many were still unaware of this. Guggenheim then told Etches, 'If anything should happen to me, tell my wife in New York that I've done my best in doing my duty.' Guggenheim and Giglio then went to assist the crew at Boat Nos. 7 and 5.

As Boat No. 7 finished loading, Murdoch instructed the crewmen to row away a short distance, but to stand by the gangway to take on more passengers. Again, this is an indication that the officers were thinking that the boats could be more fully loaded once afloat. Fifth Officer Lowe was handy and helped lower the boat. As No. 7 touched down in the abnormally still ocean, it had around twenty-eight aboard, despite having room for sixty-five.

By 12.38 a.m. *Titanic*'s bow has settled well into the sea and the forward portholes on the port side of D Deck are about to vanish below the surface. (HFX Studios)

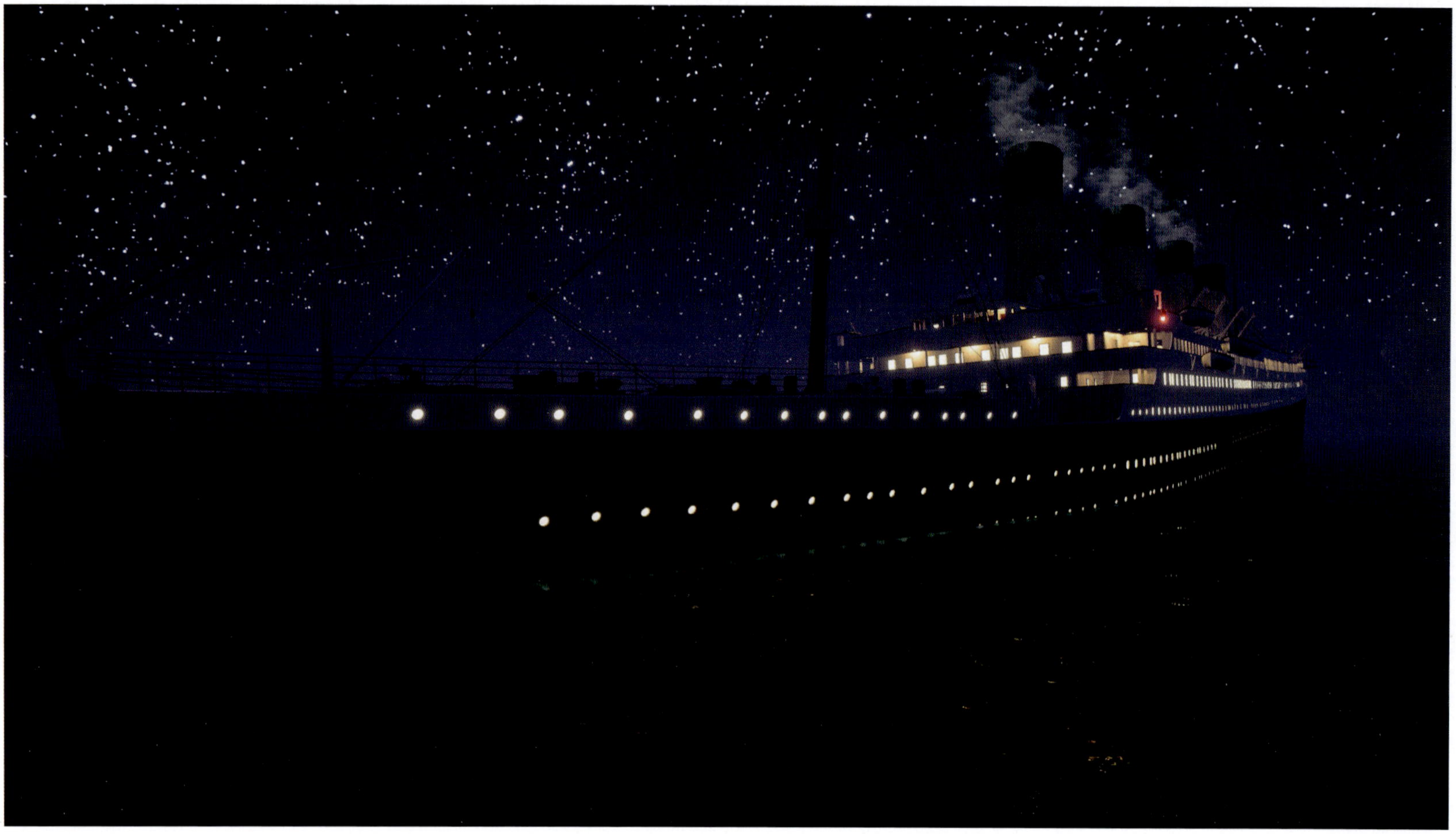

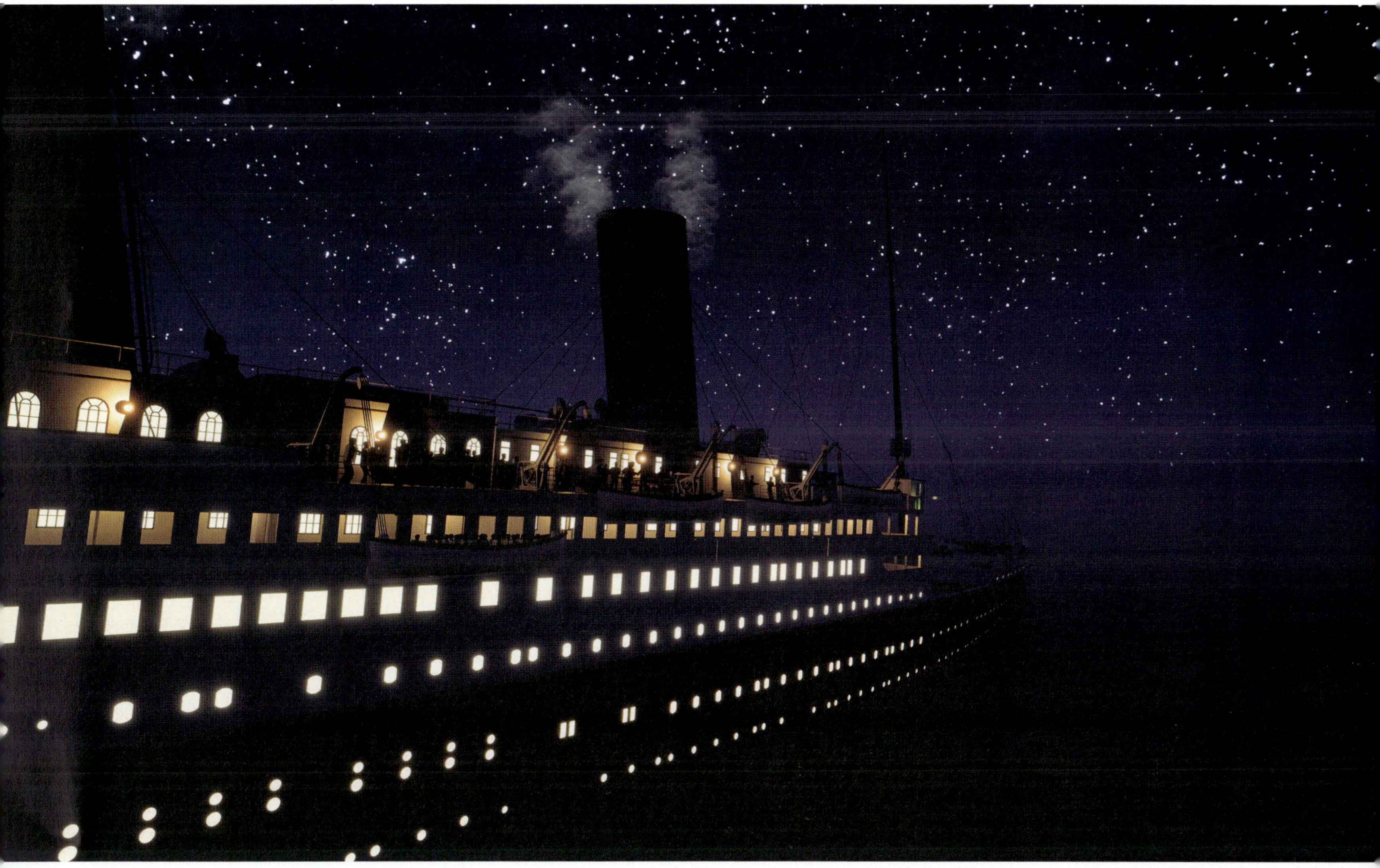

This view shows the scene at the forward-starboard lifeboats at 12.41 a.m. Boat No. 7 is being lowered away, passing A Deck, and just forward, No. 5 is being loaded. Work would thereafter progress to No. 3 and then to No. 1. (HFX Studios)

## LIFEBOAT LAUNCH TIMES AND OCCUPANCY

The British inquiry of 1912 attempted to compile a lifeboat launch table as part of their Final Report. However, in the years since, research by a number of historians has found a number of flaws in their version.

The revised table was originally published online in 2000, and other than a minor tweak or two in the years since, the times have remained stable in spite of ongoing research. The current authors have also added columns with the approximate number of people in each of the lifeboats.

All times are approximate within a five-minute timeframe, as exact times from the survivors are impossible to determine. The number of survivors in each lifeboat is also approximate. Only in one case do we know the exact number aboard: No. 1, with twelve people. Because of these approximations in the number of survivors per boat, the total survivors comes to 724, which is more than the known number of survivors as determined by a number of researchers: 712. Unless each known survivor can be assigned to a lifeboat, this is likely the closest approximation that can be determined.

## LIFEBOAT LAUNCH TIMES AND ESTIMATED OCCUPANCY

(From '*Titanic*: The Lifeboat Launching Sequence Re-Examined' (at www.wormstedt.com/Titanic/lifeboats/lifeboats.htm) and '*Titanic* Lifeboat Occupancy Totals' (at www.wormstedt.com/Titanic/lifeboats/occupancy.pdf)

| PORT SIDE | | | | | STARBOARD SIDE | | | | |
|---|---|---|---|---|---|---|---|---|---|
| **Time** | **Boat** | **In Charge of Loading** | **As Lowered** | **At *Carpathia*** | **Time** | **Boat** | **In Charge of Loading** | **As Lowered** | **At *Carpathia*** |
| 12.40 | | | | | 12.40 | 7 | Murdoch, Lowe | 28 | 34 |
| 12.45 | | | | | 12.45 | 5 | Murdoch, Lowe, Pitman (at 12.43) | 36 | 30 |
| 12.47 | Rockets first fired | | | | | | | | |
| 12.50 | | | | | 12.50 | | | | |
| 12.55 | | | | | 12.55 | 3 | Murdoch, Lowe | 32 | 32 |
| 1.00 | 8 | Lightoller, Wilde, Smith | 27 | 27 | 1.00 | | | | |
| 1.05 | | | | | 1.05 | 1 | Murdoch, Lowe | 12 | 12 |
| 1.10 | 6 | Lightoller, Smith | 23 | 24 | 1.10 | | | | |
| 1.15 | | | | | 1.15 | | | | |
| 1.20 | 16 | Moody | 53 | 52 | 1.20 | | | | |
| 1.25 | 14 | Lowe, Wilde, Lightoller | 40 | 25 | 1.25 | | | | |
| 1.30 | 12 | Wilde, Lightoller | 42 | 69 | 1.30 | 9 | Murdoch, Moody | 40 | 40 |
| 1.35 | | | | | 1.35 | 11 | Murdoch | 50 | 50 |
| 1.40 | | | | | 1.40 | 13<br>15 | Murdoch, Moody<br>Murdoch, Moody (at 1.41) | 55<br>68 | 55<br>68 |
| 1.45 | 2 | Wilde, Smith | 17 | 17 | 1.45 | | | | |
| 1.50 | 10<br>4 | Murdoch<br>Lightoller | 57<br>30 | 55<br>60 | 1.50 | | | | |
| 1.52 | Rockets cease firing | | | | | | | | |
| 1.55 | | | | | 1.55 | | | | |
| 2.00 | | | | | 2.00 | C | Murdoch, Wilde | 43 | 43 |
| 2.05 | D | Lightoller, Wilde | 20 | 35 | 2.05 | | | | |
| 2.10 | | | | | 2.10 | | | | |
| 2.15 | (B) | Lightoller<br>(Bridge goes under) | 28 | to 4 & 12 | 2.15 | (A) | Murdoch, Moody<br>(Bridge goes under) | 13 | to 14 |
| 2.20 | | | | | 2.20 | | | | |
| Totals | | | | 364 | | | | | 364 |
| | | | | | | | FINAL TOTALS | | 728 |
| | | | | | | | Buried at sea from *Carpathia* | | 4 |
| | | | | | | | Survivors | | 724 |
| | | | | | | | Actual Survivors | | 712 |

It had taken around an hour from the collision until the first lifeboat was away, even with Captain Smith having proactively had them uncovered before he knew that the ship was sinking. While this seems like an extremely long time, one has to keep in mind that it took fully forty to forty-five minutes before Thomas Andrews discovered that the ship was sinking. Even once this was known, the officers and crew had extreme difficulty in assembling the passengers to load into the boats. Many were already asleep and had to be woken up and dressed before turning out.

As discussed previously, many passengers simply refused to believe that there was any danger of the ship sinking, and the noise and cold outside were further deterrents. The physical task of the crew uncovering the boats, swinging them out and lowering them to the deck's edge would conservatively take a combined twenty to thirty minutes. In some ways, it is miraculous that the crew managed to get the first lifeboat away as soon as 12.40 a.m. Once the first boats began to leave the ship, the pace of the loading and lowering process accelerated.

Third Officer Pitman moved on and had been working on Boat No. 5. While there, a man in slippers and a dressing gown quietly said to him, 'There is no time to waste.' Pitman ignored him and helped swing out No. 5. Once this was done, the man spoke to Pitman once again, saying that they had better get the women and children into the boat. Pitman replied, 'I await the commander's orders.' Suddenly realising that the man was likely Bruce Ismay, Pitman went to the Bridge and told Captain Smith that Ismay wanted him to get the boat away with women and children. Smith gave Pitman the okay. Pitman noted that Ismay assisted with getting passengers into the boat from that point on, and proved himself helpful.

First Officer Murdoch, having seen Boat No. 7 away, came along and called out for more ladies, which Ismay repeated. A woman nearby hesitated, stating that she was 'only a stewardess'. Ismay said, 'Never mind, you are a woman, take your place.' When no more women responded to the call, the crew members let a few men board it.

Quartermaster Olliver, who had been standing in the boat, stepped back onto the Boat Deck, apparently having no intention of leaving the ship. Third Officer Pitman, upon learning that there was no sailor in No. 5, ordered Olliver to get back into the boat. First Officer Murdoch then told Pitman, 'You go in charge of this boat, old man', and ordered him to 'Keep handy to come to the after gangway when hailed'. Pitman did not like having to leave, as he felt he was better off remaining on the ship. Murdoch stuck out his hand and the Third Officer shook it. Murdoch said, 'Goodbye. Good luck to you.'

Pitman fully believed that he would see Murdoch again and the lifeboats would return to the ship once more in 'two or three hours'. He later came to believe that Murdoch had not been expecting to see him again. Boat No. 5 was lowered away, and as it began dropping down, First Class passengers Henry and Isaac Frauenthal jumped down into it, landing on top of fellow First Class passenger Annie Stengel, injuring her ribs.

The crew were furious at the Frauenthal brothers, and someone yelled, 'Throw that man out of the boat!', but it had already started to descend. Murdoch stated, 'I will stop that. I will go down and get my gun.' Just thirty-six out of the sixty-five seats in Boat No. 5 were filled as it lowered away.

As No. 5 lowered along the ship's side, Bruce Ismay leaned over the edge of the Boat Deck, hanging from a davit. Fifth Officer Lowe stated that he was 'overanxious and was getting a trifle excited', waving his arms and yelling, 'Lower away! Lower Away!' over and over again. Lowe, not recognising Ismay, and annoyed at him getting in the way, snapped. He yelled, 'If you will get to hell out of that I shall be able to do something!' Lowe then continued, 'Do you want me to lower away quickly? You will have me drown the whole lot of them!' Ismay did not respond, and instead walked forward, towards Boat No. 3, to assist there.

As Boat No. 5 descended towards the water, the men manning one of the davits accidentally let the falls play out faster than those manning the other davit. The boat tipped precariously for a few moments and the occupants in the boat feared that they were about to be hurled into the ocean. Finally, the boat was evened out and continued to lower. There was a bit of difficulty freeing No. 5 from the falls, but once that was accomplished, the lifeboat began to row away from *Titanic*. As it did, passengers in the boat noticed that the ship's bow was significantly down by the head. First Class passenger Helen Østby noted how the lights of some of the forwardmost portholes were beginning to go underwater.

Back on the port side of the ship, Second Officer Lightoller was hard at work when Chief Officer Wilde, possibly prompted by First Officer Murdoch, came over from the starboard side and asked if he knew where the firearms were. Murdoch had not been able to find them, owing to the last-minute shake-up in the officer ranks before the maiden voyage. Prior to Wilde being assigned to the ship, Lightoller was serving as First Officer, and the firearms and navigation instruments would have been his responsibility. Murdoch, who was supposed to have been Chief Officer, did not know where they were located.

Since Lightoller knew where they were, he led Captain Smith, Wilde and Murdoch to his original cabin, now Murdoch's cabin, and showed them where he had stored the firearms. As he turned to leave, Wilde handed Lightoller one of the Webley revolvers and a handful of ammunition, saying, 'Here, you may need it.' As they were exiting the cabin, Wilde told Lightoller that he was going to put on his lifebelt. Lightoller, not even having thought of his own lifebelt at that point, followed suit.

Around this time, Captain Smith ordered Lamp Trimmer Hemming to make sure the boats had lights. Hemming complied right away, going below deck and bringing up the lighted lamps four at a time. He put lamps in the boats that were not yet lowered away, and asked the crewmen to pass lights down to the boats that had already left.

Below deck, damage control efforts were still taking place. At 12.45 a.m., Greaser Scott and other crewmen were told to open all of the watertight doors aft of the Engine Room. There was not yet any flooding in those compartments and opening those doors allowed the engineers to bring a portable suction pipe forward to help fight the flooding in one of the forward stokeholds.

Coming back up the stairs from the Turkish Baths on F Deck at about 12.45 a.m., Steward Wheat notices water coming down the steps from E Deck above. Climbing up to the Entrance on E Deck, he sees water on deck, snaking around the corner from the starboard-side passageway. (Chris Walker)

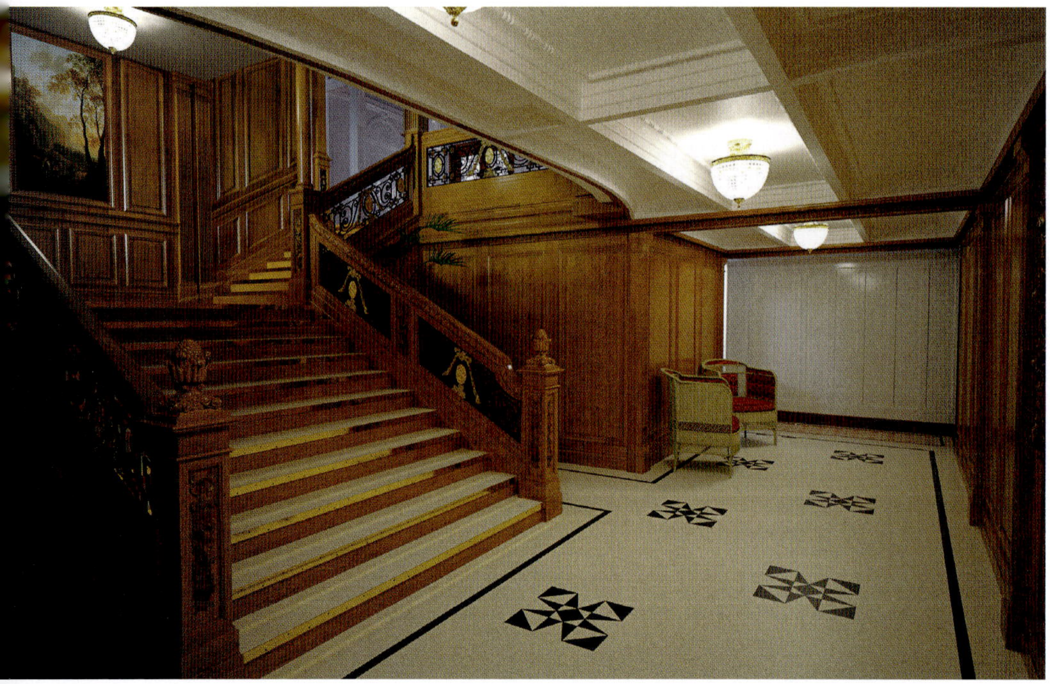

Meanwhile, Steward Joseph Wheat had a startling experience. He had gone below deck to his quarters on F Deck to get some personal items. His quarters were on the port side of the ship, directly behind watertight bulkhead F, at the bottom of the stairs leading to the Turkish Baths on the starboard side.

Earlier, when he had closed the watertight door between the stairs and the Swimming Bath, forward of the bulkhead, there had been no water there. Everything had seemed okay as he went down the stairs. While there, he ensured that all of the Turkish Bath attendants and stewards had turned out. Yet when he came back up the stairs, he noticed that there was water flowing down from E Deck, deep enough to cover his heel.

Climbing the wet stairs up to E Deck, he saw water coming from the First Class corridor, forward on the starboard side, then turning the corner, entering the vestibule and flowing down the stairs. Peering through the emergency door onto the port side, he saw that 'Scotland Road' – the corridor there – was dry. The water that Wheat saw coming down from E Deck, aft of watertight bulkhead F, meant that water was collecting on the deck above Boiler Room No. 4. It was also collecting above Boiler Room No. 5.

On the Bridge, Fourth Officer Boxhall finally had a chance to see what the light on the horizon was that had been mentioned previously. He could clearly see the light, and with the aid of binoculars, was able to see that it was actually the two masthead lights of a steamer. Boxhall attempted to contact the vessel, using the Morse lamp.

As he did so, Captain Smith instructed him, 'Tell him to come at once, we are sinking.' Receiving no discernible response, Boxhall went to find the distress rockets that were kept on the Bridge. He placed one in the socket on the starboard Bridge wing and then pulled the cord to fire it off. The missile soared into the air above *Titanic*, leaving a faint white streak behind it as it rose. At 600–800ft, it exploded with a deafening report, throwing out a burst of white stars, which slowly descended. The explosion of the first rocket was the moment when many passengers aboard the ship began to realise that the situation was serious. Everyone knew what rockets at sea meant, and many individuals began to rethink their situation and what they needed to do.

Around this time, aft, Boat No. 9 was being swung out. A group consisting of Steward William Ward, Boatswain's Mate Haines and about sixteen other men was working to swing

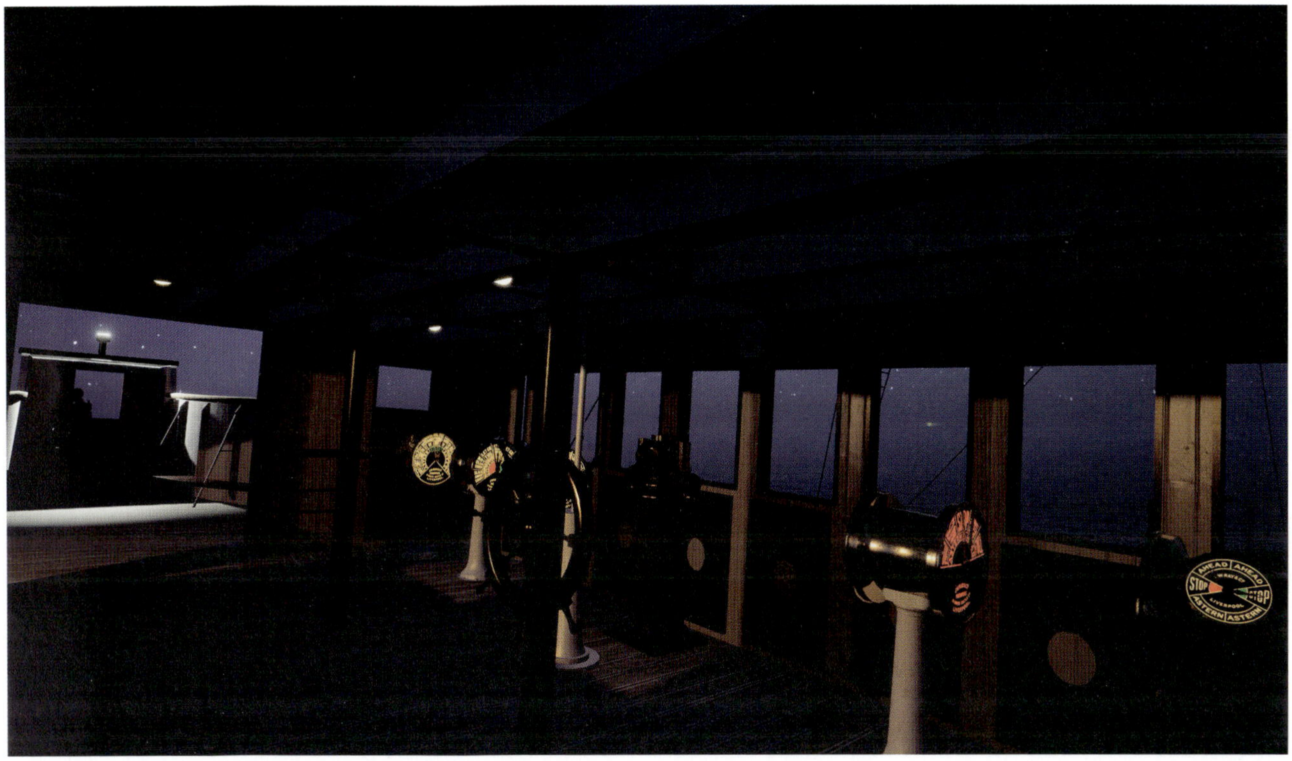

In the minutes before 1.00 a.m., Fourth Officer Boxhall is hard at work in the port Bridge wing, signalling by Morse lamp and attempting to get the attention of the ship on the horizon. Starting at 12.47 a.m., he also fires a series of white distress rockets in the effort. (HFX Studios)

it out. First Officer Murdoch and Chief Purser McElroy were spotted nearby. At this point, there were only 'one or two' passengers standing in the vicinity, and it would take some time to get the boat ready for loading.

On the Aft Docking Bridge, Quartermaster Bright showed up to relieve Quartermaster Rowe. Rowe was the forgotten man of the night. Having seen the iceberg brush past and initially thinking that it was a windjammer with its sail set, Rowe was well aware that there had been an incident, but not the specifics. In the chaos following the collision, nobody on the Bridge thought to call him and inform him regarding the 'all hands' order, so Rowe had dutifully continued to stand watch on the Docking Bridge. Bright was supposed to show up and take over the watch at midnight, or 12.23 a.m., but time passed and nobody showed up.

As it turned out, Quartermaster Bright had slept right through the accident. He was awakened by Quartermaster Walter Wynn with both the news that the ship was going down by the head, and the realisation that he had not relieved Rowe at the proper time. Once Bright arrived, Rowe and he stood together on the Docking Bridge, talking and unsure what to do. When the first rocket exploded, Rowe spotted Boat No. 7 in the water, possibly illuminated by the flash of the detonator. He pointed this out to Bright and decided to call up the Bridge.

On the Bridge, Boxhall was just putting the firing lanyard inside the Wheelhouse when the phone rang. The man at the other end of the line informed him that one of the starboard boats had left the ship. This caught Boxhall off guard, as he had not heard the order to lower any of the boats. He asked if the man was Third Officer Pitman, to which Rowe responded in the negative. Once he realised it was Rowe, he instructed the Quartermaster to bring the extra box of detonators from the stern to the Bridge. Rowe and Bright each fetched a box of rockets and headed forward towards the Bridge.

When Rowe and Bright arrived, Boxhall worked with them to send up more distress rockets, at intervals of roughly five minutes. This was in agreement with international maritime regulations, which indicated that rockets fired 'at short intervals' signalled distress. As the group prepared to fire each rocket, Boxhall would chase people away from the area around the firing tube by Boat No. 1, so that no one nearby would be injured if something went wrong with the launching. Throughout this process, Boxhall and Rowe also attempted to signal the ship on the horizon with the Morse lamps. They received no definite response from it, despite the ongoing attempts.

Meanwhile, nearby, on the Leyland liner *Californian*, all was quiet. The ship had prudently stopped in ice earlier in the night. The officers on watch and at least one off-duty crewman all saw white rockets fired from a ship to the south-east of them, spaced out at intervals of several minutes. Earlier in the night, the ship on the horizon was seen to approach, then suddenly stop and shut out its lights. The ship's Captain, Stanley Lord, was napping on the Chart Room settee and was alerted about the rockets at least once. He gave orders for the crew on watch to signal the other ship via Morse lamp, and to let him know if there was a response. He never ordered the ship's sole wireless operator to be woken up, and did not even bother to go up to the Bridge to investigate the situation for himself. By a little after two o'clock in the morning, the lights of the steamer on the horizon had disappeared from sight ...

## THE *CALIFORNIAN* AFFAIR

One of the most controversial issues relating to the *Titanic* disaster involves whether the SS *Californian* was the ship seen on the horizon as *Titanic* sank. While survivor accounts vary in the specific details, most agreed that this mysterious ship was visible on the horizon off the port bow of *Titanic* all night, and remained visible even after the ship foundered. According to details provided by Quartermaster Rowe, Fifth Officer Lowe, passenger Lawrence Beesley and other lines of evidence, *Titanic* stopped with her bow facing in a generally north-north-west to northerly direction. In an attempt to contact the other vessel, those on the *Titanic* used Morse lamps and fired a number of distress rockets, which were white according to the crewmen who launched them.

On the night of the sinking, the *Californian* had stopped for the night, rather than try to navigate through ice in the dark. A liner was witnessed approaching from a general easterly direction and then appeared to shut out some of her lights and stop. A while later, crewmen from the midnight watch saw the same ship fire about eight rockets. Captain Lord, who was taking a nap on the Chart Room settee, was informed that the other vessel had fired a number of 'white rockets'. Lord told his men to try to contact the other ship via Morse lamp, but did not ask the wireless operator to be awakened to investigate the situation further, nor did he arise to check the situation himself. The men on the Bridge discussed this mysterious liner. As they watched, the ship took on an odd appearance. Officer of the Watch Stone even remarked to Apprentice Gibson that the ship looked 'queer', and it seemed to 'have a big side out of the water'. Eventually, the other ship seemed to disappear over the horizon. The crew did not find out that the *Titanic* had sunk until the morning, after the ship's wireless operator woke up. At that point, Captain Lord leapt into action, arriving on the scene after the rescue ship *Carpathia* had already picked up all of *Titanic*'s survivors.

After the *Californian* arrived in Boston, the press caught wind of the fact that some aboard had seen rockets the night that *Titanic* sank. Captain Lord denied this. Subsequently, men from the *Californian* were questioned at the American and British inquiries into the disaster. Both inquiries concluded that the Leyland liner saw the distress rockets, saw *Titanic* sink, and could have arrived on the scene in time to rescue everyone, had action been taken. Captain Lord eventually lost his position with the Leyland Line, although he was subsequently able to find employment with other steamship companies.

In the subsequent years, many defenders of Captain Lord have argued that he was wrongly scapegoated for the disaster, and the *Californian* was not the ship sighted from *Titanic* and vice versa. Lord's defenders are correct that he was made a scapegoat and faced a disproportionate amount of blame for the deaths that night. Even the most optimistic estimates indicate that, even if Lord had gotten up and taken immediate action, the *Californian* could only have arrived on the scene of the sinking shortly before *Titanic* sank, too late to save everyone. Yet no serious researcher has argued that the rockets seen by the men of the *Californian* were from any ship other than *Titanic*. Indeed, since the discovery of the wreck of the vessel – and with it, the actual location of *Titanic* when it sank – this has been proven beyond any doubt. Given the position of *Californian* as given by its crew members, even if that estimate was off by a number of miles, the rockets would have still been visible above

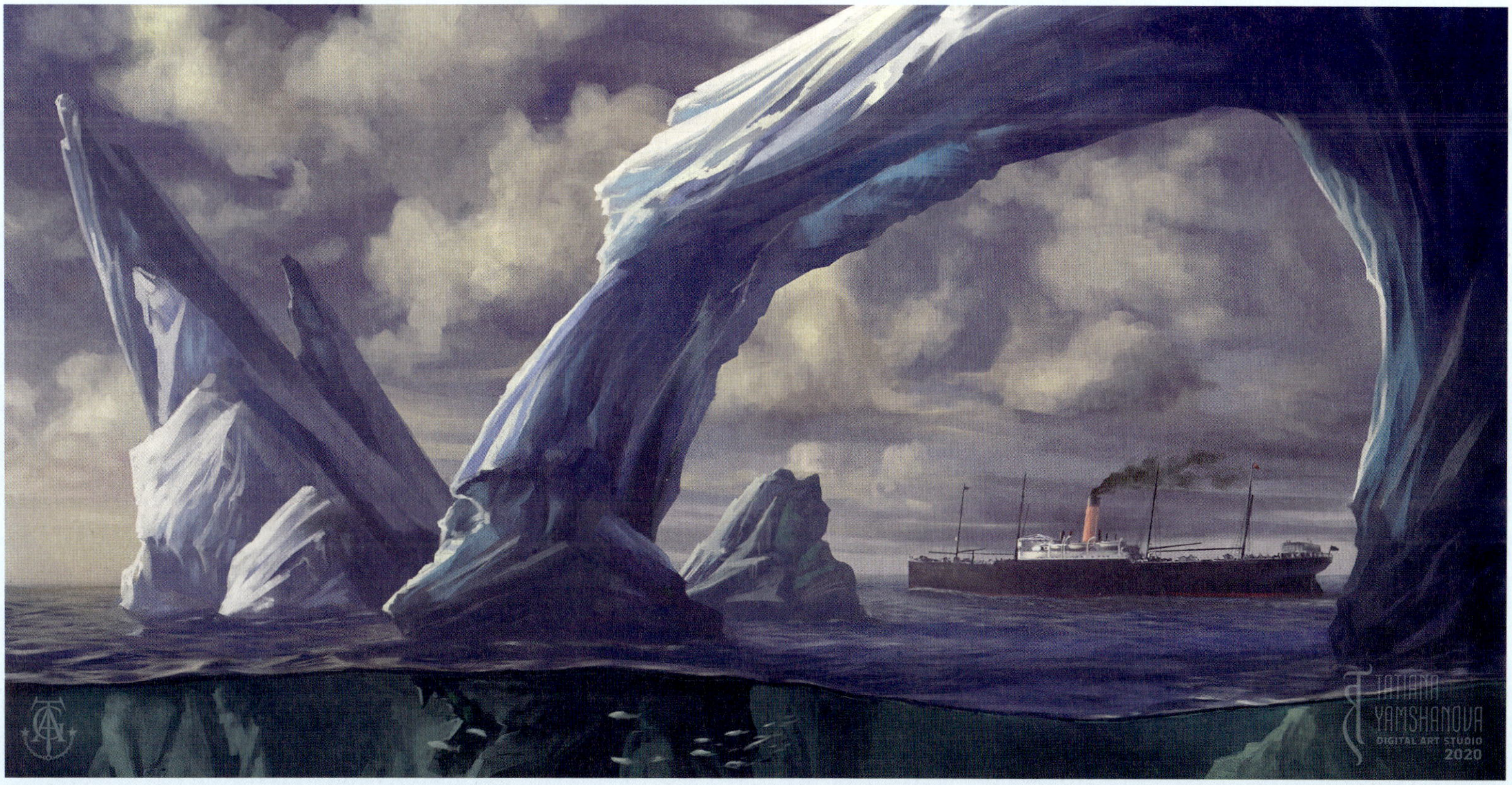


The Leyland liner *Californian* in the ice field that sank *Titanic*. (Tatiana Yamshanova)

the horizon. There would not have been room for another ship in between to be firing rockets, without witnesses on both *Titanic* and *Californian* seeing them.

It is clear that at the time of the collision and throughout the sinking, *Titanic*'s clocks were running two hours and two minutes ahead of New York Time, while *Californian*'s clocks were running one hour and fifty minutes ahead of New York Time. In other words, there was a difference of about twelve minutes between clocks on *Californian* and clocks on *Titanic*, with *Californian*'s being behind those on *Titanic*.

With this time difference in mind, when one examines and compares the accounts of eyewitnesses on both ships that night, the similarity between the times of *Titanic*'s collision and stopping, the time she started firing rockets and the time when she fired her last rocket, and even the time that she sank, all appear to be a close match to the times that those on the *Californian* saw the steamer they were watching shut out her lights, stop, fire rockets and eventually disappear. As a result, the present authors feel that there is good reason to suspect that *Titanic* and *Californian* were watching each other directly, without the presence of any other ships in between them.

While Lord's defenders have passionately argued in favour of there being one or even two 'mystery ships' in between *Titanic* and *Californian* – and a number of other variations – at the end of the day, this is all irrelevant. There is now zero doubt that *Californian* saw *Titanic*'s rockets, if not the ship itself, that she did not take action, and that Captain Lord tried to cover it up after the fact. Lord and the crew of the *Californian* would most likely not have been able to save everyone even if they had acted immediately, yet by trying, they would surely have been the first rescue vessel on the scene, could possibly have saved a small number of additional lives, and doubtless would have been above all reproach.

## *TITANIC*'S 'MYSTERY SHIP' – *MOUNT TEMPLE*?

For over 100 years, there have been disputes over the 'mystery ship' seen off *Titanic*'s port bow on the night of the sinking. Both the American and British inquires came to the conclusion that this ship was the *Californian*, which by her own reckoning, was north-north-west of *Titanic*'s distress position by about 19 miles.

In the years since, certain 'experts' have claimed the *Californian*, and her master Captain Stanley Lord, have been unjustly accused. They have pointed the finger at a number of other ships that happened to be on the North Atlantic that night, and in particular, *Mount Temple*. The latest example of this is the 2020 British TV programme, *Abandoning the Titanic* (alternately called *Titanic: A Dead Reckoning*). This show attempts to prove that *Mount Temple* had to be the ship *Titanic* saw on the night of 14–15 April 1912.

The theory outlined by this programme is that the *Titanic* continued moving roughly westward after striking the iceberg and, upon sighting the light off the port bow, was looking at a ship to the west of them. This leaves the *Californian* out of the equation, as she was to the northward. *Mount Temple*, originally 59 miles south-west of *Titanic*, upon receiving the distress message, headed toward the collision point, 41° 46'N, 50° 14'W, and finally was stopped by the icefield to the west of *Titanic*. This makes her a prime candidate, they say, to be 'the mystery ship'.

There are two major flaws to this theory. The first is that *Titanic* was not headed westward after striking the iceberg. There is the well-known 'hard-a-starboard' order, which would have had the effect of turning *Titanic* slightly to port (southward) from her initial westward direction. What is not so well known is that there was a second order, 'hard-a-port', just as the ship struck the iceberg. This second order had the effect of turning the stern of *Titanic* away from the iceberg, which was critical to keep the aft portions of the ship safe from further damage, and the side of the ship from being ripped open for a longer distance than it already had. This second order turned the ship to starboard. As the stern moved away from the iceberg, the ship gradually kept turning to starboard, and ended up pointing slightly west of north.

The second order, 'hard-a-port', was first brought up by Quartermaster Alfred Olliver in the American inquiry. Though heading for the Bridge, he was not close enough to hear the earlier 'hard-a-starboard' but was very specific that he heard a 'hard-a-port' as he entered the Bridge.

This order was confirmed later by the man at the wheel, Robert Hichens. Hichens told passenger Howard Chapin on *Carpathia* that a 'hard-a-port' order immediately followed the 'hard-a-starboard' order. Other confirmation comes from Seaman Joseph Scarrott, who testified that 'the starboard bow was going as if to make a circle round it', and also Quartermaster George Rowe, who stated that by the time he had reached the Bridge almost an hour later, the stern of the *Titanic* had swung around to 'practically dead south', putting the bow to the north. Finally, the wreck of the bow is pointed to the northward, after dropping 2 miles to the sea floor.

The second flaw with this theory has to do with *Mount Temple* herself. She received the corrected position at around 12.21 a.m. (12.37 a.m. *Titanic* time) and changed course to head to *Titanic* at 12.26 a.m. (12.42 a.m. *Titanic* time). The two ships were then about 61 nautical miles apart, well beyond the distance that either vessel could see the lights of the other. *Titanic* first reported seeing the light off the port bow at about 12.13 a.m. her own time – before *Mount Temple* had received the distress messages, was still headed westward and had not even turned and started her rescue mission!

*Mount Temple* increased her speed to about 11.5 knots, and at this speed would have taken over four hours to reach the distress position that was given. *Mount Temple* continued her course and speed until she reached the west side of the icefield about three hours into her run, when her Captain reduced her speed to avoid the ice. *Mount Temple* finally stopped at 4.46 a.m. her own time, which equates to 5.02 a.m. *Titanic* time – over two and a half hours after *Titanic* had sunk!

The chart below gives the various times on the three ships – *Titanic*, *Mount Temple* and *Californian* – in relation to each other, and how they also relate to events on the *Titanic*.

There is just no way that *Mount Temple* can be the 'mystery ship'. The above-mentioned 'experts' are flat-out wrong.

**Chronology of times 14 - 15 April 1912**
***Titanic*, *Mount Temple* and *Californian***

| NY Time | *Titanic* ATS | *Mount Temple* ATS | *Californian* ATS | Event |
|---|---|---|---|---|
| 8.30 p.m. | 10.32 p.m. | 10.16 p.m. | 10.20 p.m. | *Californian* stops in ice |
| 9.38 p.m. | 11.40 p.m. | 11.24 p.m. | 11.28 p.m. | *Titanic* hits iceberg |
| 10.23 p.m. | 12.25 a.m. | 12.09 a.m. | 12.13 a.m. | Light reported ahead of *Titanic* (approx) |
| 10.25 p.m. | 12.27 a.m. | 12.11 a.m. | 12.15 a.m. | *Titanic* sends out first position |
| 10.35 p.m. | 12.37 a.m. | 12.21 a.m. | 12.25 a.m. | *Titanic* sends out Boxhall's corrected position |
| 10.40 p.m. | 12.42 a.m. | 12.26 a.m. | 12.30 a.m. | *Mount Temple* changes course to northeast |
| 10.45 p.m. | 12.47 a.m. | 12.31 a.m. | 12.35 a.m. | *Titanic* sends up first rocket (approx) |
| 11.50 p.m. | 1.52 a.m. | 1.36 a.m. | 1.40 a.m. | *Titanic* sends up last rocket (approx) |
| 12.18 a.m. | 2.20 a.m. | 2.04 a.m. | 2.08 a.m. | *Titanic* sinks |
| 2.08 a.m. | 4.10 a.m. | 3.54 a.m. | 3.58 a.m. | *Carpathia* reaches first lifeboat (approx) |
| 3.00 a.m. | 5.02 a.m. | 4.46 a.m. | 4.50 a.m. | *Mount Temple* stops in ice |

*Titanic* ATS was 2 hours and 2 minutes ahead of NYT
*Mount Temple* ATS was 1 hour 46 minutes ahead of NYT
*Californian* ATS was 1 hour 50 minutes ahead of NYT

These times were all based on where each vessel was when local apparent noon occurred for them on 14 April 1912.

From the article 'Abandoning the *Titanic*, Abandoning Reality', available online at www.wormstedt.com/titanic/abandoning_the_titanic.pdf.

Back on *Titanic,* Boat No. 8 was loading. First Class passengers Isidor and Ida Straus came alongside, and after some convincing, managed to get Mrs Straus' maid, Ellen Bird, to board the boat. Unlike on the starboard side, where First Officer Murdoch and the crew there were allowing men into the boats after all willing women and children had boarded, the rule of women and children only was being strictly enforced by Second Officer Lightoller. Mrs Straus herself refused to board the boat when prompted by both crew members and her husband, stating, 'We have been together all these years. Where you go, I go.'

On the starboard side, First Officer Murdoch put Able Bodied Seaman George Moore in charge of Boat No. 3 and ordered Able Bodied Seaman James Anderson aboard to assist. The crew members on the scene rounded up as many women and children into the boat as they could and then Murdoch allowed some of the men in the area to climb aboard. No. 3 was ordered to lower away at 12.55 a.m., becoming the third lifeboat to leave the ship – all from the starboard side up to this point. Approximately thirty-two of its sixty-five seats were filled.

In the Marconi Room, Jack Phillips continued to send distress signals. Thus far, the *Carpathia* remained the closest vessel to have responded. Despite the seriousness of the situation, Harold Bride joked that Phillips should begin sending SOS, because 'it's the new call, and it may be your last chance to send it'. Phillips laughed, and at 12.57 a.m., switched from sending CQD to SOS.

Far below, in Boiler Room No. 5, the desperate attempts at damage control were ongoing. Around 12.55 a.m., Junior Assistant Second Engineer Harvey had ordered Leading Fireman Barrett to lift a manhole plate on the floor, so that they could get access to the pump valves located there. Due to the firemen having to throw water on the burning coal as the fires were being drawn, there was a thick blanket of steam in the compartment. With the reduced visibility, Junior Assistant Second Engineer Shepherd was rushing through the area and, unable to see, fell through the manhole, breaking his leg. As he was unable to walk, Harvey and Barrett lifted and dragged Shepherd into the pump room.

Although things appeared stable in Boiler Room No. 5 with the pumps running and it remaining dry, this was extremely deceiving. At that very moment, water was flooding aft along E Deck, making its way as far aft as the rooms above their heads, and the next compartment aft, Boiler Room No. 4. It was inevitable that the water would find its way back down into those boiler rooms. The men below were unaware of the danger lurking overhead and continued to work heroically.

Back on deck, a rumour began to circulate among the men on the starboard side that women were to be loaded into the boats there, but men were to be taken off on the port side. The origin of this rumour is unknown and, in fact, was the complete opposite of what was being allowed at the boats. Despite this, Second Class passenger Lawrence Beesley observed that this report was 'acted on at once by almost all the men', essentially leaving the aft starboard side abandoned. This had the effect of significantly delaying the loading of Boat No. 9 and simultaneously causing a large crowd to swell at the aft port boats, where they began to get unruly.

At 1.00 a.m., the crewmen working at Boat No. 9 had just lowered it level with the deck when a sailor with his kitbag came along and jumped into the boat, saying that he had been placed in charge of it by the Captain. Boatswain's Mate Haines had no patience for this and ordered the man out of the boat. The sailor meekly complied.

Meanwhile, on the forward port side, a group of about '100' stokers with their kitbags emerged on deck, crowding around the lifeboats there. Unlike many of the passengers, these men knew the full extent of the situation, having just escaped from the flooding below decks. Chief Officer Wilde, not wanting the men to crowd out women and children, came along and drove them away from the area.

Second Officer Lightoller remained hard at work at Boat No. 6, when he noticed for the first time how far the boat was down by the head. Mercifully, the deafening noise of the steam venting from the pipes on the funnels had finally ceased after continuing for over an hour. Lightoller and the officers found it much easier to communicate from that point forward. Now that the racket had ceased, the sounds of music from the ship's band began to drift out from inside the Boat Deck level of the First Class Entrance.

Just aft of Lightoller, Captain Smith was working with Chief Officer Wilde and other crewmen to load Boat No. 8. This process was slow going, since many passengers were still reluctant to leave the ship or were simply unwilling to separate from their loved ones. Captain Smith called out, 'Any more ladies?' A small number of additional passengers boarded, at which point the call for more women and children was repeated twice more. At 1.00 a.m., with none forthcoming, the order to lower away No. 8 was given. Before the boat began descending, Smith ordered Able Bodied Seaman Thomas Jones to row towards the ship on the horizon, drop off his passengers there, and then return to the ship. No. 8 had just twenty-seven of its sixty-five seats occupied.

Meanwhile, back on the starboard side, First Officer Murdoch was preparing Emergency Boat No. 1 for loading. Located just aft of the Bridge, this boat had a capacity of only forty persons, rather than the sixty-five for the main lifeboats. The loading of this boat was delayed because Fourth Officer Boxhall was chasing passengers away from the area as he was firing the distress rockets. Now, as the boat was loading, Sir Cosmo and Lady Duff Gordon, along with their maid, Laura Francatelli, were standing nearby. Sir Cosmo had pleaded with his wife to leave in one of the earlier boats, but she had refused to separate from her husband.

At the time, there were very few passengers around Boat No. 1, and seeing this, Sir Cosmo asked whether they might be permitted to board the boat. Murdoch responded, 'Please do,' and the three climbed aboard. Around this same time, First Class passenger Charles Stengel, having seen his wife safely away in Boat No. 5, arrived alongside No. 1. Seeing others in the boat and no one else around, he approached Murdoch and asked if he could get into the boat.

Murdoch told him to 'jump in', but as Stengel climbed over the bulwark rail, he got caught and ended up rolling into the boat. Despite the stress of the night, Murdoch laughed heartily, exclaiming, 'That is the funniest sight I have seen tonight.' Murdoch's humour reassured Stengel that perhaps the situation was not as dire as what he had believed.

Once a total of twelve people, consisting of five passengers and seven of the crew, were aboard, and as there were no other passengers nearby, Murdoch ordered the boat lowered away at 1.05 a.m. As it descended, it got hung up on a 'painter' or 'wire', and the descent to the water was delayed while a crew member was sent to free the boat from the obstruction. This prevented No. 1 from reaching the water until approximately 1.15 a.m. As the boat reached the water, the crew aboard were ordered to stay close to the ship and to be prepared to return when called.

On the port side, Second Officer Lightoller had been struggling to find enough manpower ever since he had sent Boatswain Alfred Nichols and the six other crewmen below to open the gangway doors. This dilemma was part of what contributed to the forward port boats getting away from the ship so much later than the forward starboard boats. Because of the sailors not returning from below deck, Lightoller was 'reduced to sending one seaman away in a boat'. As Lightoller gave the order to lower No. 6 away, he suddenly realised that there were no qualified crewmen manning the aft falls and called out for assistance. When there was no response, Lamp Trimmer Samuel Hemming, despite having been ordered into No. 6, stepped back onto the deck and manned the falls as the boat lowered away.

Boat No. 6 began descending at 1.10 a.m. Quartermaster Robert Hichens, who had been at the ship's helm during the collision, was aboard, as was Lookout Frederick Fleet and First Class passenger Margaret Brown. Hichens sat at the tiller in the stern of the boat and appointed himself in charge. There were about twenty-three of its sixty-five seats occupied. By this point in the night, it appears that the initial 5-degree list to starboard had lessened somewhat, although the edge of No. 6 did rest against the ship's side, causing the occupants to have to push the boat away from the hull on the way down.

Boat No. 6 had lowered halfway to the water when it began to tip precariously, with its bow pointing towards the sea. This was similar to the problem that was encountered during the lowering of No. 5 earlier in the night. Hichens shouted up to Lightoller, and the lowering was halted until the boat was evened out. Hichens shouted that he had no seamen to help row. Lightoller asked the crowd near the boat if there was 'any seaman there'. When none responded, First Class passenger Arthur Peuchen volunteered his services, pointing out that he was a yachtsman.

Lightoller accepted, telling Peuchen, 'If you're seaman enough to get out on those falls, and get down into the boat, then you may go ahead.' Peuchen then grabbed the falls and lowered himself safely into the boat. The boat then continued its descent towards the calm ocean. As the boat was lowering, Margaret Brown noticed how water was entering the ship through an open porthole on D Deck. As soon as No. 6 reached the water, Peuchen helped unhook the falls. As he did so, Quartermaster Hichens told him, 'Hurry up. This boat is going to founder.' Peuchen initially thought he was referring to the lifeboat, before realising that he had meant *Titanic*.

Deep below deck, it had been fifteen minutes since Junior Assistant Second Engineer Jonathan Shepherd had severely injured his leg in Boiler Room No. 5. Things seemed to be stable, with the pumps continuing to work. Even though the coal bunker on the forward side of the compartment was full of seawater, the boiler room itself was remaining dry.

All of this changed at 1.10 a.m. Frederick Barrett saw a large rush of water suddenly coming through the pass in between the boilers. Junior Assistant Engineer Herbert Harvey ordered Barrett up top, and he complied right away, barely escaping the flood of water. Coming up to E Deck, Barrett saw water coming down the alleyway from forward of where he escaped Boiler Room No. 5. He speculated that this might have been due to water reaching open portholes on the lower decks.

Meanwhile, in Boat No. 1, the crew were rowing away from the ship. As they did so, Lookout George Symons noticed that the ship was down by the head to an alarming degree. He watched as the second row of portholes underneath *Titanic*'s nameplate became awash. Back on deck, with all four of the forward starboard boats away, First Officer Murdoch and his men moved aft to begin working at the boats there.

On the forward port side, Boat No. 4 still sat level with the Promenade Deck, neglected, waiting for the crew members to get the windows cranked open. As crowds continued to grow at the aft port boats, the crew turned their attention there. Boat No. 4 would have to wait.

Meanwhile, Fifth Officer Lowe, fresh off the lowering of Boat No. 1, immediately headed to the aft port boats, 'as they seemed to be busy there'. Other crewmen who had assisted at the forward starboard boats, such as Able Bodied Seamen Frank Evans and Edward Buley, also headed to the aft port boats. The need to assist and calm the crowds on the aft port side helps to explain the delay in the loading and lowering between the forward and aft starboard boats.

As it became more and more apparent that *Titanic* was in serious danger, and as there were fewer and fewer boats left, the mood aboard the ship began to change. Passengers were suddenly more willing to board the boats, although there were still some who denied the reality of the situation, refusing to believe that the ship could sink.

However, as more individuals came to the realisation that their lives were in danger, some began to get aggressive. At Boat No. 14, a group of men tried to push their way into the boat, after Chief Officer Wilde had given the order to start loading women and children into it. Able Bodied Seaman Joseph Scarrott had to resort to using the boat's tiller to persuade the men to stay back.

As Fifth Officer Lowe arrived on the scene, he bumped into Sixth Officer Moody. 'What are you doing?' Lowe asked. Moody responded, 'I am getting these boats away.' Lowe mentioned how he had seen a number of boats leave without an officer, and suggested one of them should go in Boat Nos. 16 or 14. Even though, as the ship's junior officer, Moody was entitled to leave, he deferred to Lowe, stating, 'You go, I will get in another boat.' Moody's selfless decision to stay behind and continue to assist in the evacuation resulted in Lowe's survival, a fact that Lowe would never forget.

The two junior officers then moved on to help at the boats, Lowe to No. 14 and Moody at No. 16. Lowe spotted Second Officer Lightoller briefly while working at No. 14, before the latter moved on to oversee the loading at No. 12. Arriving alongside No. 14, the crewmen there informed Lowe of the difficulties they had been having with the crowd. Lowe immediately began loading passengers into the boat.

At Boat No. 16, there was a group of firemen and sailors standing around the boat with their bags, waiting to board it. However, they do not appear to have interfered with the loading process. As First Class Stewardess Violet Jessop boarded No. 16, Sixth Officer Moody, who she described as looking 'weary and tired', handed her something, saying, 'Look after this, will you?' Jessop reached out and was surprised to receive somebody's baby. She held the infant close to her chest, concerned about how little comfort the hard cork lifebelt would provide.

Between 1.20 and 1.25 a.m., a desperate scene is beginning to unfold around the aft port boats. (Eugene Nesmeyanov)

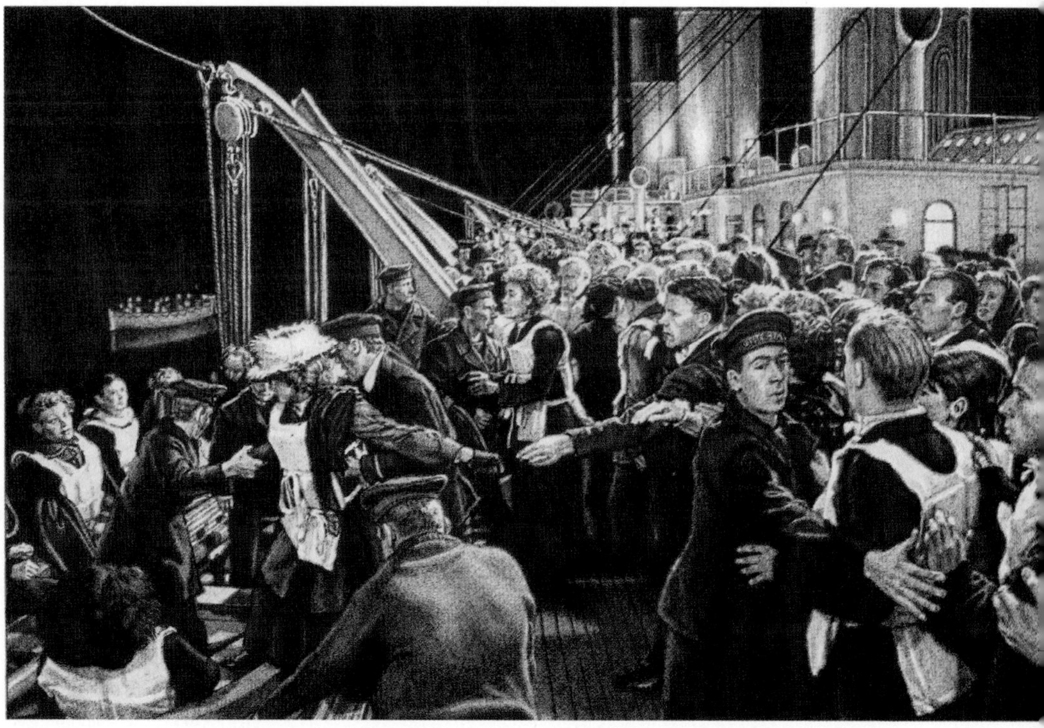

At around 1.20 a.m., the remaining men working below in the Engine Room and stokeholds were released from duty and permitted to go up on deck. Leading Fireman Thomas Threlfall recalled Junior Engineer John Hesketh saying, 'We've done all we can men, get out now.'

Trimmer Thomas Dillon was just headed aft from Boiler Room No. 4 to the Engine Room when he heard the order releasing the men below deck. For him, this could not have come at a better time. He had just seen water starting to come up from under the stokehold plates in Boiler Room No. 4.

Long-standing myths claim that all of the engineers remained below deck, going down with the ship. However, this is contradicted by the recollection of multiple crewmen, many of whom remembered this same order coming down at 1.20 a.m., as well as the fact that Chief Engineer Bell, Senior Second Engineer Farquharson and some of the other engineers were later spotted up on deck. The body of at least one of the engineering crew, and possibly others, was also recovered, which would have been unlikely had they all remained below deck.

Around this same time, loading finally began on Boat No. 9. It had been swung out and ready for quite some time, but had been abandoned due to the difficulties at the aft port boats. Finally, First Officer Murdoch gave the order to begin loading the boat. Chief Purser McElroy assisted in the process.

Back on the port side, Boat No. 16 began lowering away around 1.20 a.m. As the boat was being lowered, Sixth Officer Moody and the crewmen nearby realised that there was only one able bodied seaman and a steward to man the boat. Master-at-Arms Joseph Bailey then climbed down the falls to take charge and rectify this. Boat No. 16 was more heavily loaded than any of the boats lowered up to that point, with fifty-three of its sixty-five seats filled. As the boat lowered away, Sixth Officer Moody remained on board *Titanic*. At just 24 years old, he would be the only junior officer to die in the disaster.

At Boat No. 14, Fifth Officer Lowe was finishing the loading process that Chief Officer Wilde had started. A crowd was still around the boat, and the crew struggled to hold them back and pass the women and children through. At 1.25 a.m., very shortly after evacuating the engineering spaces below, Thomas Threlfall arrived alongside the boat, being allowed to board just before the order to lower away was given. Approximately forty seats out of the sixty-five in No. 14 were filled.

Lowe was nervous as No. 14 lowered down the ship's side. Standing in the boat, he was concerned that he had overloaded it. As the boat passed the open A Deck Promenade, he saw a large group of men who looked poised to jump down into the boat as it passed. Taking no chances, Lowe yelled at them to stand back and fired three warning shots in between the lifeboat and the ship's hull with his Browning automatic. The threat worked and the men retreated.

As No. 14 neared the water, one of the falls became tangled, and the boat dangled 5ft over the water. Lowe attempted to clear the entanglement, but when this proved impossible, he released the falls, letting the boat drop down into the water. This sharp plunge of several feet frightened many of those on board.

Meanwhile, back on the aft starboard side, First Officer Murdoch, assisted by Chief Purser McElroy, was continuing to load Boat No. 9. However, this was slow going because there were still not many passengers on the scene after the earlier rumour of men being taken off on the port side. As things calmed a bit on the port side, a number of crew members who had gone to assist at the aft port boats began to cross back over to the starboard side. Some of these included Sixth Officer Moody and Able Bodied Seaman George McGough, who had stood by the aft falls of No. 14 as it lowered.

Bruce Ismay, who was seen assisting a female crew member at Boat No. 16, also assisted at No. 9, and lent a hand in the loading process. As the loading resumed, a number of stewards and other members of the victualling department started passing women and children over from the port side in order to fill up Boat No. 9.

On the port side, Boat No. 12 had also been loading. While not as chaotic as what had happened just aft at Boat No. 14, Second Officer Lightoller had to contend with some disorder at this boat as well. Men had tried to rush the boat and were on the falls, and had to be warned to stay back. Lightoller and Able Bodied Seaman Frederick Clench stood on the gunwales of the boat, helping to pass the women and children into it.

Lightoller continued to be a stickler about who was and who was not allowed into lifeboats at which he supervised the loading – in the end, this may have needlessly cost lives. Second Class passenger Emily Rugg recalled how, when No. 12 was loading, a 'stewardess, one that was very popular with the passengers', stepped into the boat with them.

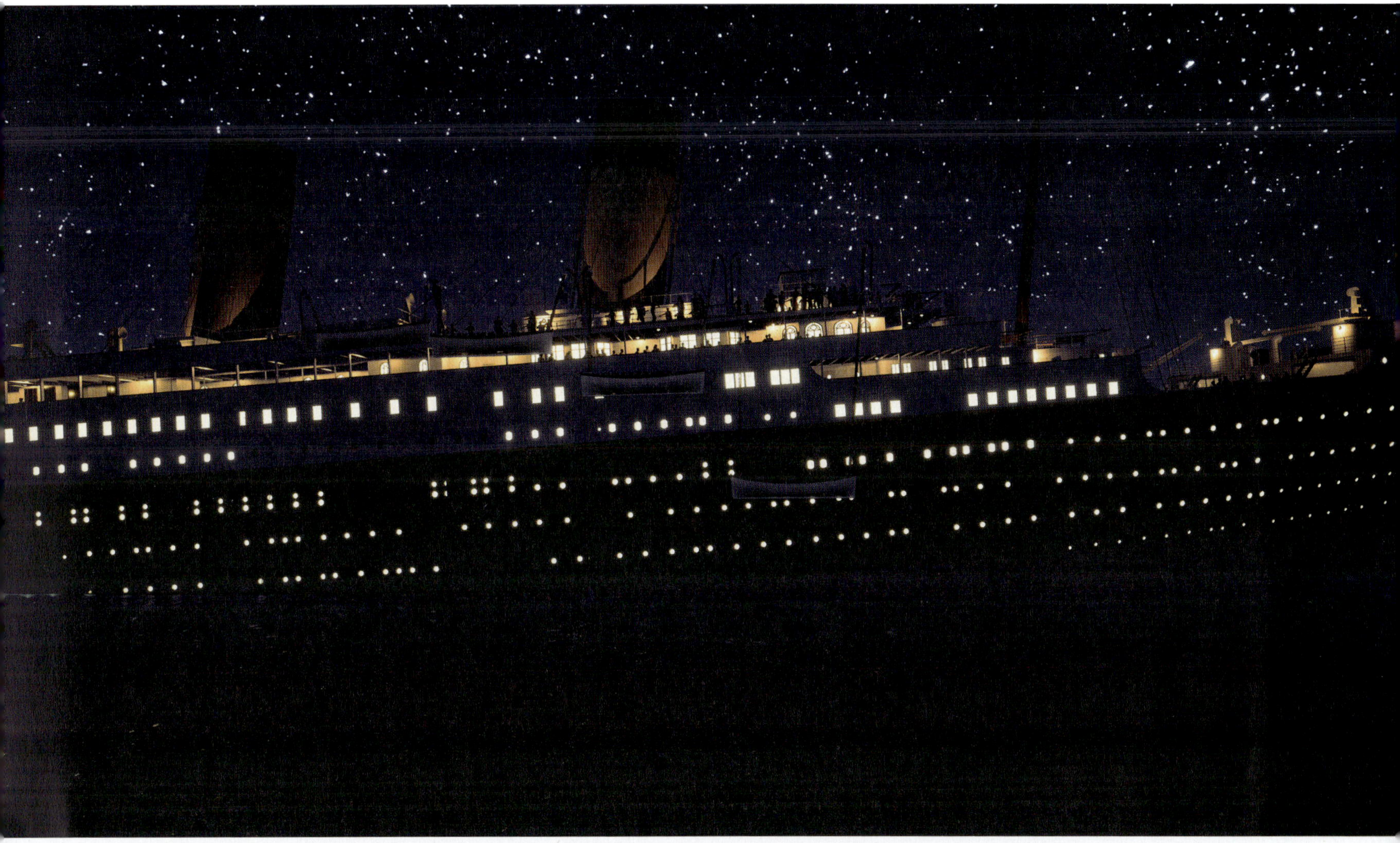

This view at 1.28 a.m. shows the aft port boats. Boat No. 16 is lowering away between D and E Decks, No. 14 has just passed A Deck, where Fifth Officer Lowe fired his revolver, and Boat No. 12 is filling and will be lowered away shortly. (HFX Studios)

She was wearing her uniform at the time. The officer, likely Lightoller, ordered her back on deck, saying, 'Passengers first.' This provides an interesting contrast with the actions of the other officers, as well as Thomas Andrews and Bruce Ismay, who encouraged stewardesses to board boats during the sinking.

Around this time, Chief Officer Wilde arrived back at No. 12 and, seeing that there was only one crewman in it, ordered Clench aboard. He also instructed the able bodied seaman to keep a close eye on No. 14 and to try staying close to it once afloat. At 1.30 a.m., Wilde gave the order to lower away. As No. 12 began descending along the hull, a male passenger took a running leap and jumped down into the boat, injuring at least one woman. There were forty-two of the sixty-five seats occupied.

At this exact same time, on the aft starboard side, the loading at Boat No. 9 was finishing, but it was not going without difficulties. During the process, a 'French lady', possibly First Class passenger Madame Aubart, fell and slightly injured herself while boarding the boat. Subsequently, Chief Purser McElroy instructed two additional crewmen to climb into the boat to assist.

Sixth Officer Moody, fresh from the lowering of Boat No. 16, instructed Quartermaster Walter Wynn to board and take charge of No. 9. However, as the boat began lowering at 1.30 a.m., First Officer Murdoch overruled this order and instructed Albert Haines to take command of the boat. Wynn relinquished control and would instead assist with rowing during the night. The boat had forty of its sixty-five seats occupied.

Once No. 9 was safely away, Murdoch ordered Moody down to A Deck, since Boat Nos. 11, 13 and 15 were to be lowered down to the open Promenade and loaded from that location. Prior to descending to the next deck, a few crewmen were ordered into each of the boats to man them, before they were lowered down to be filled with passengers. Murdoch would oversee things on the Boat Deck, including the crews working on the davits. Crowds began to gather along A Deck, as more and more steerage passengers had reached the deck, only to find many of the lifeboats already gone.

During the loading of Boat No. 11, Able Bodied Seaman Brice climbed down the falls into the boat and shipped the rudder. The crowd gathering along the rail appeared poised to get into the boat, and Murdoch shouted, 'Women and children first!' The loading of the boat was relatively orderly, however, and at 1.35 a.m. it was lowered away with fifty of its sixty-five seats filled. Once afloat, the crew had difficulty freeing the boat from the falls because the after block got jammed, and the ropes had to be cut. A heavy stream of water from the ship's condenser was being discharged just aft of where the boat was lowering, and the occupants struggled to keep the boat from being swamped.

Meanwhile, on the aft end of the starboard Boat Deck, Assistant Saloon Steward Walter Nichols had reported to his assigned lifeboat, No. 15. When he arrived, he saw the officer there, likely First Officer Murdoch, acting calmly but with a revolver in his hand. Unbelievably, even at this late stage, Nichols later recalled, 'We didn't know even then that anything was seriously the matter.' Nichols and a few others were ordered into the boat. Nichols heard the sounds of *Titanic*'s band playing as he rode No. 15 down to A Deck.

Even though nearly two hours had passed since the collision with the iceberg, some steerage passengers continued to struggle to find their way up on deck. Besides the natural obstacles of having to navigate unfamiliar corridors and passages in the ship, language barriers and the like, there is evidence that crew members did, in some cases, hold the Third Class passengers back from coming up on deck. Even

At 1.30 a.m., the aft starboard boats, Nos. 9, 11, 13 and 15, are all being filled. No. 9 is about to be lowered away, and the rest will follow sequentially over the next ten minutes or so. (HFX Studios)

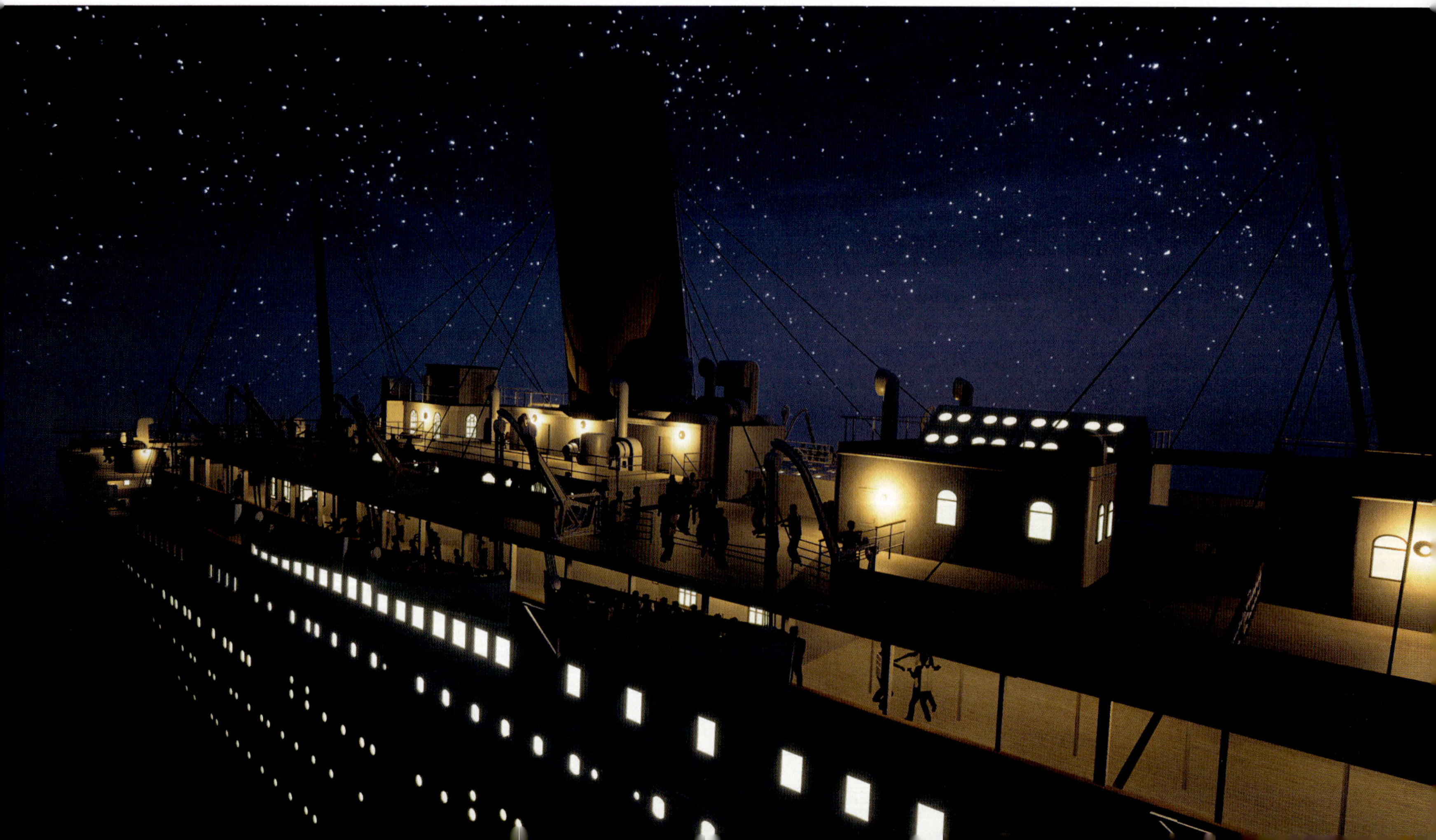

when these passengers reached the Well Decks, they were blocked from proceeding further by locked gates at the top of the stairways, and had to break the locks or climb over these to get through. This appears to have been a case of many crewmen, not knowing that the ship was sinking, enforcing the immigration laws of the day that required the separation of classes, rather than an overt effort to prevent them from getting to the boats. Regardless of the motive, many Third Class passengers did not reach the deck until few boats remained, and they would perish at a far higher rate than either First or Second Class passengers. On the other hand, there were a number of cases, such as Third Class passenger Leah Aks, in which passengers were lifted up the sides of the deckhouses by both crew and other passengers, so they could access the lifeboats higher up.

Meanwhile, by the time Boat No. 15 reached A Deck, *Titanic*'s list had begun to shift 'slightly' to port. It was more perceivable to the crewmen standing in the lifeboat than to those still on deck. The developing port list was not particularly noticeable during the loading of boats that had left prior to this time, or at least not to the point where it significantly interfered with the loading or lowering at those boats.

As the loading began in earnest at No. 15, the crowd on A Deck had swelled even further and Sixth Officer Moody and the crewmen working there struggled to keep control. Some men pushed their way into the boat ahead of women and children and were hauled out. On the Boat Deck, First Officer Murdoch yelled out, 'Stand back! Women first!' The crew quickly got the situation under control, and loading continued.

Around this time, Leading Fireman Frederick Barrett finally arrived on deck, soaked to the bone from his ordeal below. Walking aft along the starboard Promenade, he saw Boat Nos. 13 and 15 loading. Boat No. 13 was nearly full, and he did not see an officer nearby. Knowing the serious danger that they were all in, Barrett climbed into the boat. The call went out for more women, but none stepped forward. At that point, Sixth Officer Moody came along and ordered two crewmen into the boat to help row. Second Class passenger Lawrence Beesley and Kitchen Clerk Paul Maugé both dropped into the boat from the Boat Deck, right before the order to lower away was given at 1.40 a.m. Someone on A Deck tried to pull Maugé back aboard the ship, but No. 13 had already started to descend. It had fifty-five of its sixty-five seats occupied.

First Officer Murdoch and Sixth Officer Moody consistently filled the aft starboard boats more heavily, on average, than the previous boats that had been lowered away. The reason for this was obvious: *Titanic* was increasingly down by the head, and there seemed to be little chance of a rescue ship arriving prior to the ship foundering. Time was clearly running out.

As a result, by the time No. 15 had finished loading, the officers had packed it beyond its sixty-five-person capacity. As the boat lowered away, about 'thirty seconds' after No. 13 had started down, a total of sixty-eight individuals were uncomfortably packed aboard; No. 15 was by far the most heavily loaded boat when lowered.

As soon as No. 15 was ordered away, First Officer Murdoch, realising that there was no time to waste, crossed over to the aft port side to see whether he could be of assistance there. By this point, to better split their efforts, Chief Officer Wilde had headed forward on the port side to work at Boat No. 2.

As No. 15 began to drop down from A Deck, the ship's port list had increased, and the edge of the boat was 'pretty well up against' *Titanic*'s hull. However, this did not significantly interfere with the lowering. As the boat continued down, a man jumped down from the Boat Deck, landing on top of Third Class passenger Bertha Mulvihill, injuring her ribs. Meanwhile, down below, No. 13 was having issues with the same condenser discharge that had nearly swamped Boat No. 11. The outflow of water pushed No. 13 aft, until it was directly underneath the path of the descending No. 15. To make matters worse, this pulled the falls taut, making it impossible for them to be released by the tripper.

The occupants in No. 13 began shouting out to stop lowering No. 15. Hearing these shouts and looking down, occupants of Boat No. 15 realised what was happening and also started shouting up to the crew members on deck, all to no avail. With Sixth Officer Moody on A Deck and First Officer Murdoch having departed for the port side, there had been a momentary power vacuum where apparently no officer was supervising the crews working the davits on the Boat Deck above.

Boat No. 15 was, by this point, so close to No. 13 that the occupants of the latter reached up and touched the bottom of the other boat, trying to push themselves free. At the last moment, a 'sailor' cut the forward set of falls. Leading Fireman Barrett leapt up and over passengers, and after unsuccessfully trying to push the boat away from *Titanic*'s side with oars, began hacking at the other set of falls with a knife. Miraculously, No. 13 swung clear of No. 15, just as the latter boat dropped down into the ocean. They had very narrowly avoided being crushed.

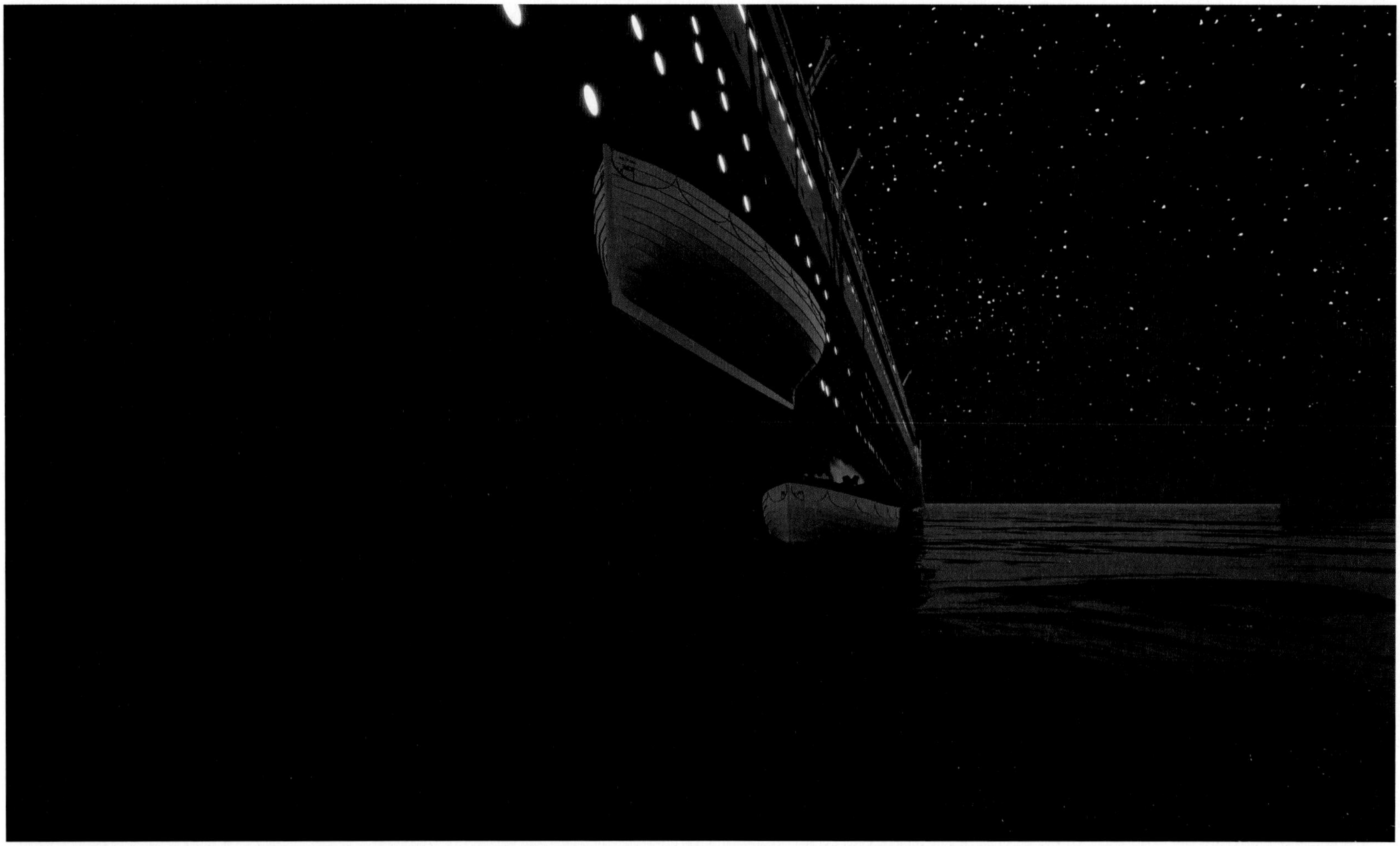

At about 1.41 a.m., a near calamity occurs. Boat No. 15 is being lowered directly down onto Boat No. 13, and the latter boat is pinned by its falls. Just before the oncoming calamity, No. 13 is freed. (HFX Studios)

On the aft port side of the Boat Deck, First Officer Murdoch had just arrived to assist. He found Boat Nos. 16, 14 and 12 already gone, but that Boat No. 10 still had not been swung out over the sea. Murdoch quickly organised two sailors left in the vicinity, Able Bodied Seamen Evans and Buley, and got the boat swung out and lowered level with the deck. He knew that they needed to get passengers loaded into the boat, and fast. There were also several forward boats left to get away.

Meanwhile, on the Bridge, Fourth Officer Boxhall and Quartermasters Rowe and Bright had continued trying to raise the ship on the horizon with the Morse lamp, while also firing distress rockets. Just aft of the port Bridge wing, Chief Officer Wilde started to load Boat No. 2. While this work was commencing, a group of approximately sixteen stokers got into the boat behind Wilde's back, as he was rounding up women and children nearby. Captain Smith, who was standing close by, witnessed this and was furious. He shouted, 'How many of the crew are in that boat? Get out of there, every man of you.' The men sheepishly complied, and the women and children were helped aboard.

Captain Smith kept a close eye on the loading of No. 2 from that point on. When the process was nearly completed and the boat was ready to lower away, Smith saw Boxhall standing on the Bridge. He ordered the Fourth Officer to get into the boat and to take charge of it. Boxhall complied, and then at 1.45 a.m., the order to lower away was given. The boat was ordered to stop at A Deck and take on more occupants there, but when the boat reached that location, nobody was present. Remarkably, given the dire situation, the emergency boat had just seventeen of its forty seats occupied. Unlike the controversy that later surrounded Boat No. 1, which had just twelve aboard, it is remarkable that there was never anywhere near the level of controversy levelled at the individuals who left in No. 2.

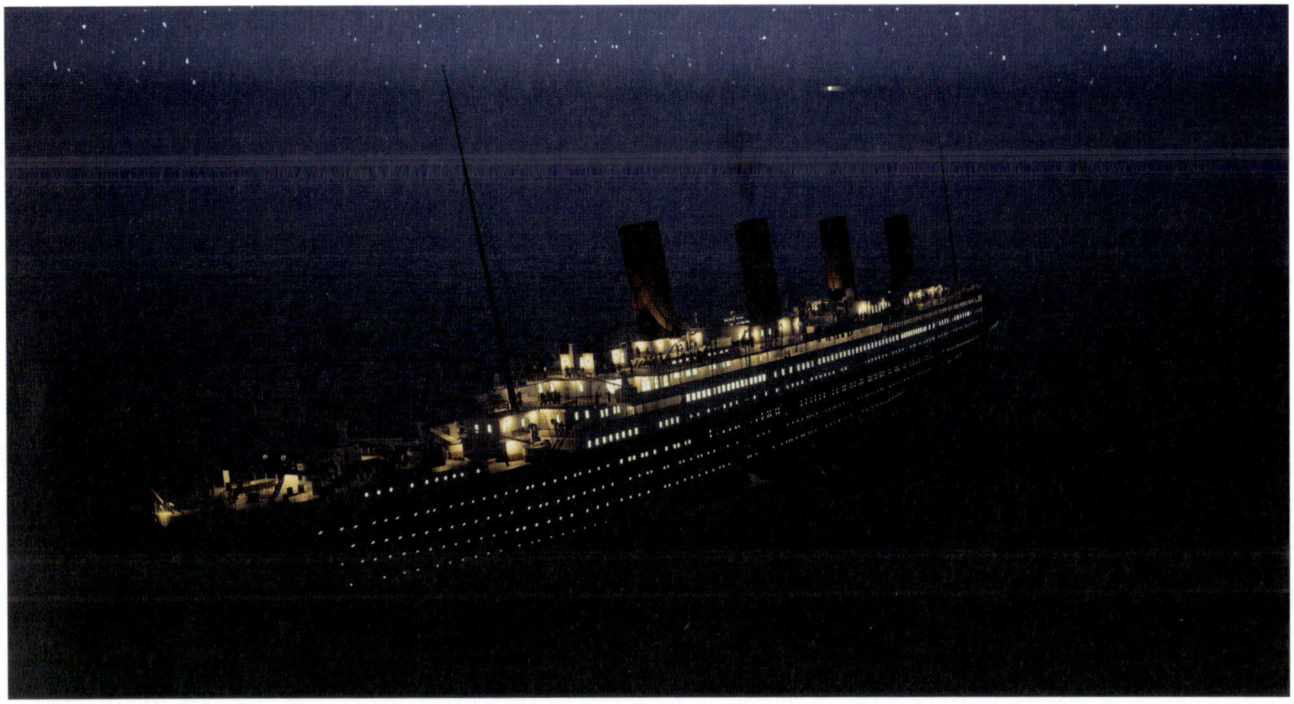

*Left:* From this high angle, Boat Nos. 13 and 15 can be seen in the water, having just escaped catastrophe. *Titanic*'s bow is deeply sunken into the sea and her list is shifting to port. The lights of the *Californian* offer a tantalising, but futile, ray of hope on the horizon. (HFX Studios)

*Below:* Around 1.43 a.m., the open D Deck gangway door begins to be submerged. Prior to 1.00 a.m., Lightoller had ordered crewmen to open it. They may have succeeded, because both the door and its inner gates remain open on the wreck, yet they only opened them after Boat No. 6 was lowered away. (HFX Studios)

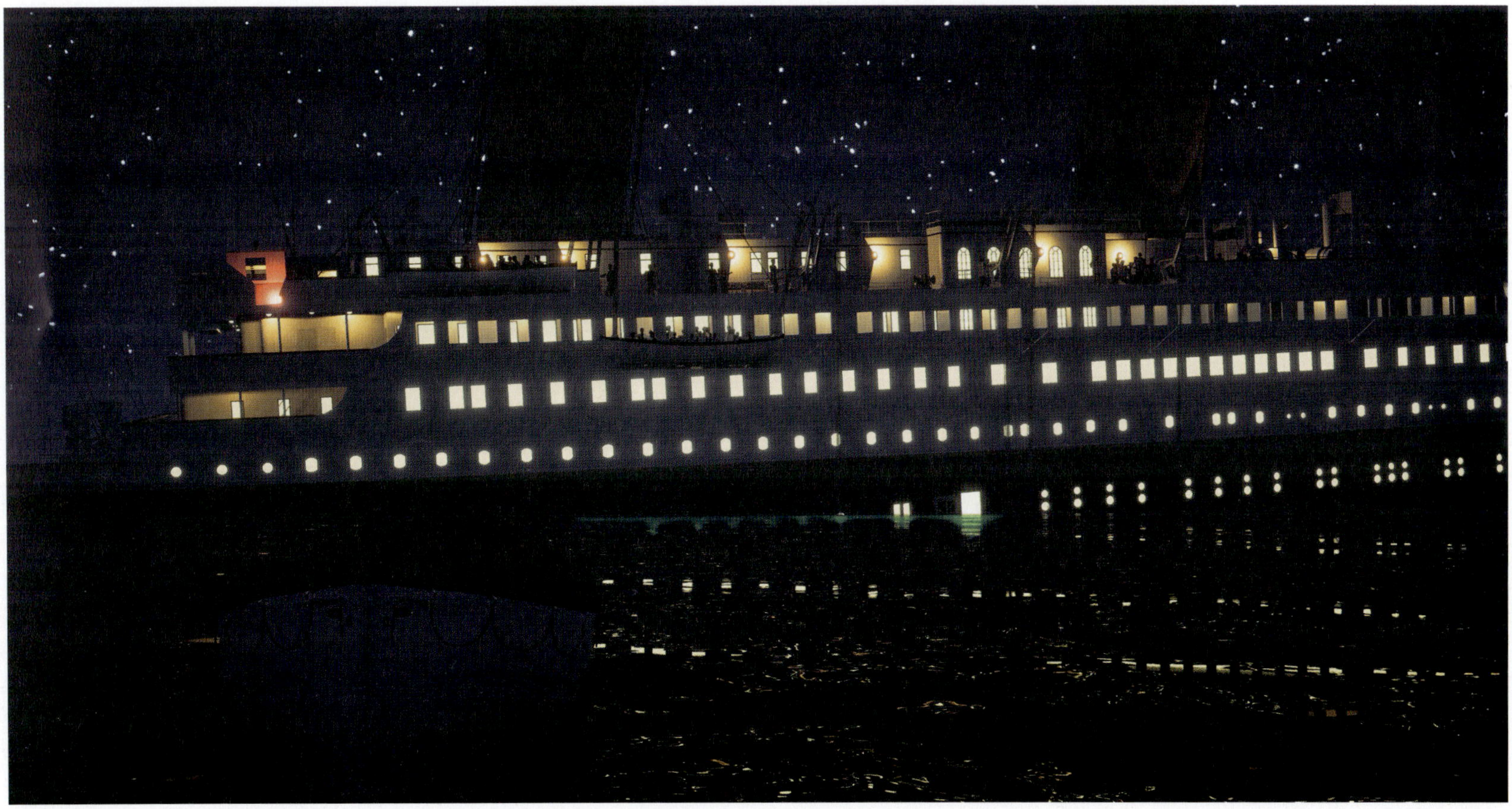

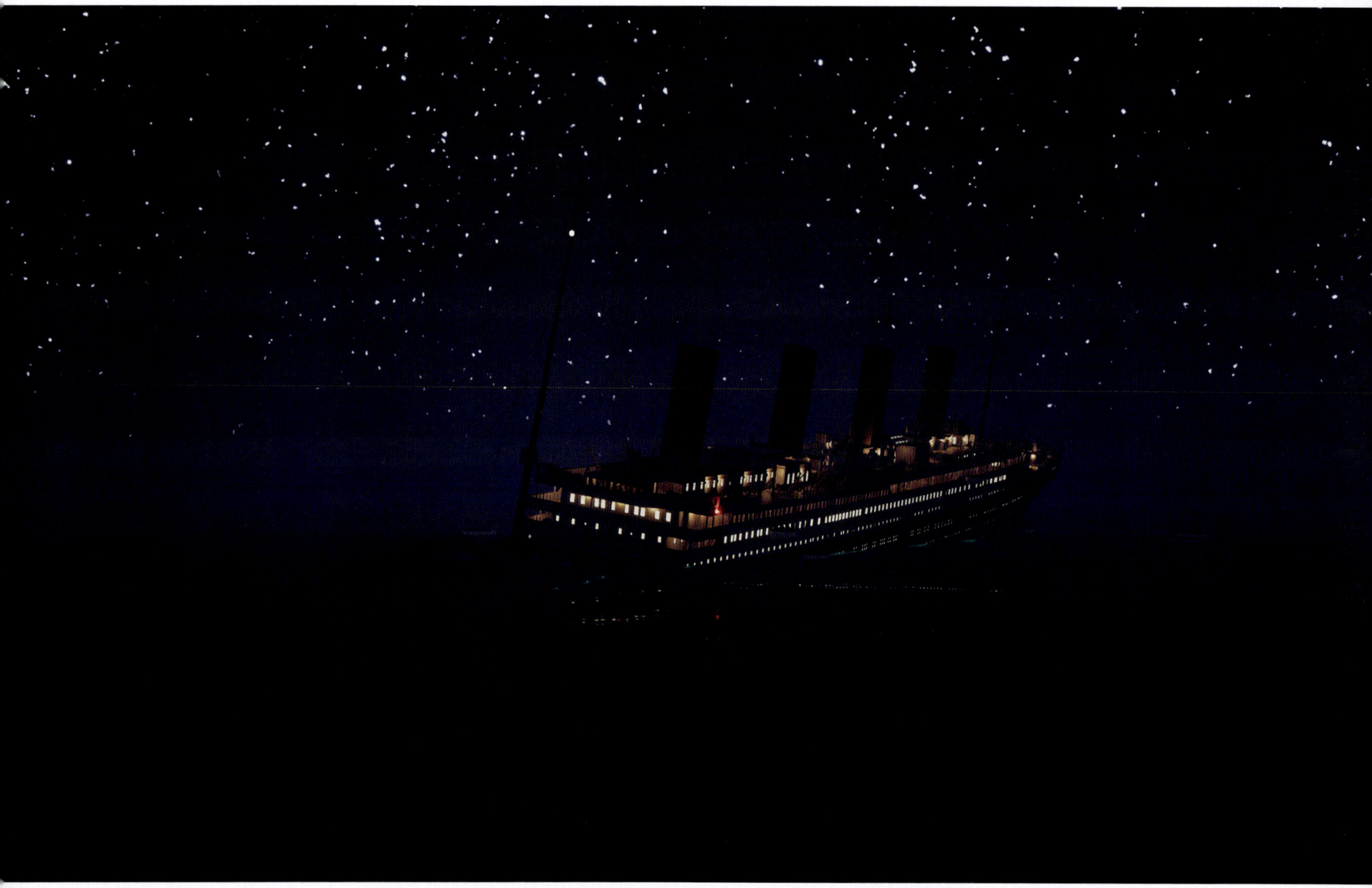

*Titanic* at 1.50 a.m. The ship has a strong port list, and the Well Deck is becoming awash. Boat No. 2 is in the water, while No. 4 is being lowered past B Deck. Toward the stern, No. 10 is also being lowered. (HFX Studios)

After some issues with freeing the boat from its falls, Boxhall began to get the occupants in the boat organised, in order to row away from the sinking ship. He found that he had just one sailor aboard, Able Bodied Seaman Frank Osman, and had to have First Class passenger Mahala Douglas man the tiller to steer the boat. After finally reaching some semblance of organisation, Boxhall had the boat pull away from the stricken ship.

Before No. 2 had gone very far, Captain Smith looked over the side of *Titanic* and, seeing several boats in the water nearby with room for many more passengers, decided to call them back to the ship, with the intention of having them take on more people from one of the gangways. With a megaphone in his hand and an officer – either Wilde or Lightoller – standing next to him blowing on a whistle, he commanded, 'Bring those boats back, they are only half filled!'

Hearing this, Boxhall stopped and listened. Smith ordered him to 'come round to the starboard side' to the gangway doors there. The Fourth Officer complied, although it was going to be a struggle because his boat was so short on experienced sailors. Boat No. 2 immediately began rowing aft towards *Titanic*'s stern.

Meanwhile, in Boat No. 6, Quartermaster Hichens, who was perched at the boat's tiller, reacted far differently to Smith's hails. Passengers described him as 'swearing a great deal' and being 'very disagreeable'. When he heard the whistle, he ordered everyone to stop rowing. After listening to Smith's order, he stated, 'No, we are not going back to the boat. It is our lives now, not theirs.' Despite protests and anger from some aboard, Hichens maintained his stance, and ordered No. 6 to be rowed further away from *Titanic*. At this point, *Titanic* was in its final half an hour. Smith's order had been one desperate, final attempt to save more lives.

Back on the aft port side of the ship, First Officer Murdoch and Able Bodied Seamen Buley and Evans continued to labour at getting No. 10 away. The barely noticeable port list of the ship had now increased to a sharp list of nearly 10 degrees. This caused No. 10 to swing away from *Titanic*'s hull, leaving a dangerous gap of somewhere between 'two feet and a half' and a 'yard and a half'. The crew working at the boat, which now included Chief Baker Joughin, were literally taking children and passengers and throwing them across the gap.

Initially, Murdoch and the crew on hand had difficulty finding passengers to load into the boat. With the situation getting more and more desperate with each passing minute, many passengers, seeing that No. 10 had still been sitting on the deck and hadn't even been swung out, abandoned the aft part of the deck, heading towards one of the few remaining forward boats that were loading. On account of this, Joughin and several other crew members went down to A Deck and rounded up passengers to board from the Boat Deck.

Captain Smith, megaphone in hand, and another officer signal for lifeboats already in the water to come back. Boat No. 6 ignores the order, but Boxhall in No. 2 rows aft on a perilous journey to find an open gangway door on the starboard side. (HFX Studios)

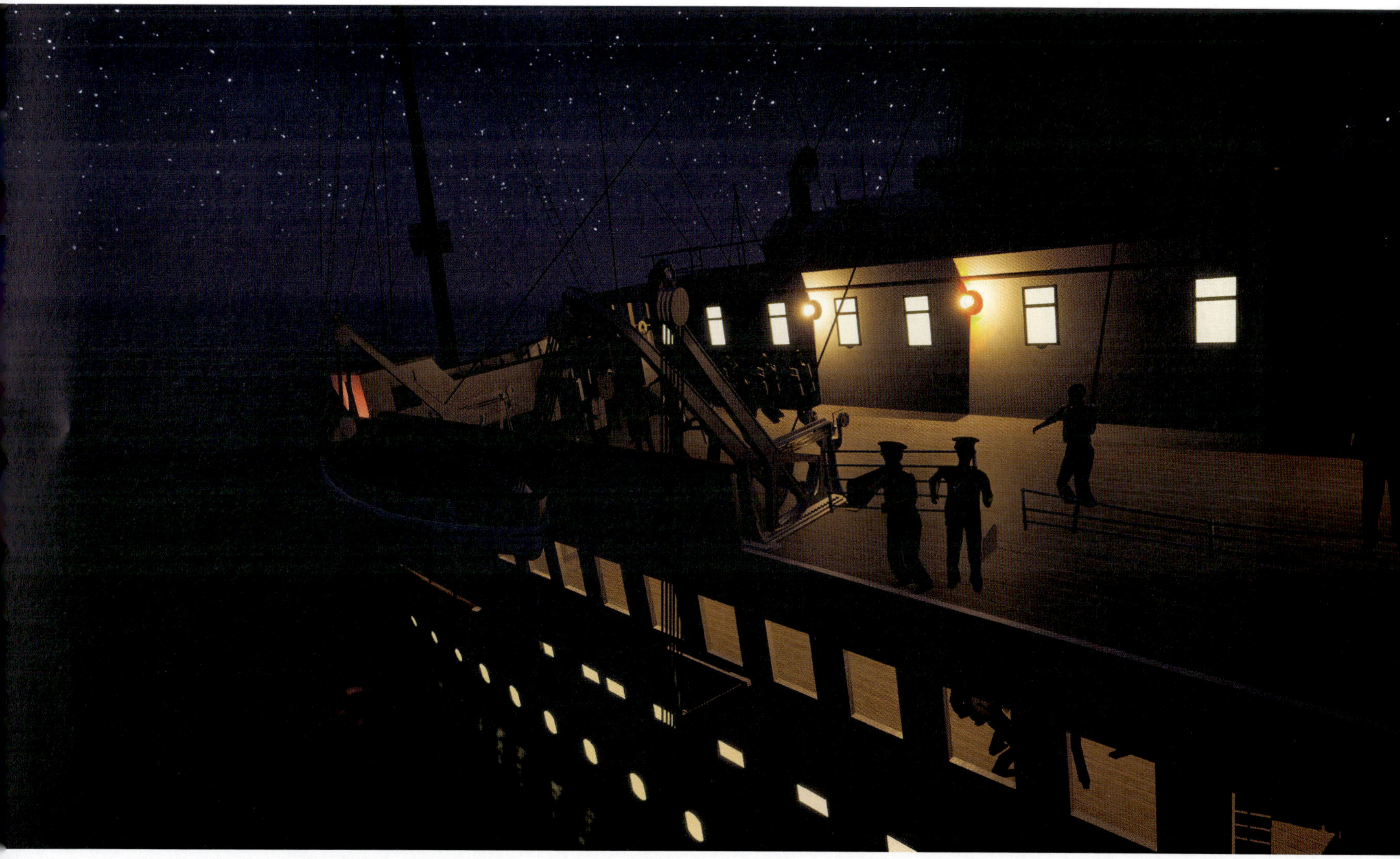

As the loading progressed, a man ran and jumped down into No. 10, injuring the leg of a female passenger. A final call went out, 'Any more women there?' Nearby, Trimmer Thomas Dillon heard this call, and attempted to convince some women who were standing on the Aft Well Deck to head up and board the boat, but they refused, instead moving aft to the Poop Deck. While he stood on the Poop, Dillon heard someone say, 'Go to the first cabin barroom.' Dillon and several others went there and found that a steward was handing out free whisky, saying, 'Go on, lads, drink up. She is going down.' Dillon obliged him, recalling, 'We had our share.'

As the loading of No. 10 was wrapping up, a woman in a black dress caught her heel as she stepped towards the boat, slipping and falling into the large gap between the boat and ship. Someone on A Deck caught her and pulled her back aboard *Titanic*. Thankfully, she was able to proceed back up to the Boat Deck in time to try boarding No. 10 again; this time, she did so safely.

Boat No. 10 began lowering away around 1.50 a.m., with fifty-seven of its sixty-five seats occupied. Murdoch had ordered both Able Bodied Seamen Buley and Evans to man the boat. As it touched down in the water, there was difficulty in unhooking the falls because Evans could not get past the passengers and underneath the thwart to get to the tripper. Eventually, he was able to release them by hand, as someone shouted melodramatically, 'Hurry up, the boilers may explode at any moment!'

While No. 10 was being supervised by First Officer Murdoch, Second Officer Lightoller had proceeded forward on the port side and had finally begun to make progress with No. 4. The enclosed Promenade windows had finally been cranked down, and Lightoller ordered women and children back down to A Deck. Passengers – who had been ordered down to A Deck earlier in the night, then back up to the Boat Deck, and now were being ordered back down – were exasperated. First Class passenger Marian Thayer exclaimed, 'Tell us where to go and we will follow. You ordered us up here and now you are taking us back.'

Soon, Lightoller and the other crew on A Deck began passing the women and children through the windows into the boat. Lightoller's earlier decision to tie No. 4 to the coaling wire had paid off, for unlike what had happened at No. 10, in spite of the growing list Boat No. 4 was not hanging far away from the hull at this point. Still, it was difficult for passengers to climb up and through the Promenade windows and out to the lifeboat, so deck chairs were used as makeshift steps. As passengers were being passed through the windows, Lightoller assisted, standing with one foot in the boat and one on the A Deck rail.

Unfortunately, even at this late hour, Lightoller continued to enforce the rule of women and children only. As First Class passenger Emily Ryerson and her 13-year-old son John were boarding, Lightoller objected, saying, 'That boy can't go.' When Arthur Ryerson objected, pointing out his son's young age, the Second Officer relented, then stated, 'No more boys.'

Nearby, First Class passengers Colonel John Jacob Astor and Colonel Archibald Gracie helped Astor's pregnant wife, Madeleine, aboard. Lightoller assisted her into the boat. After seeing Madeleine's maid, Rosalie Bidois, safely into the boat, Colonel Astor then asked permission to board, given his wife's 'delicate condition'. The Second Officer refused to make an exception, reiterating that no man could board until all of the women were loaded first.

A few more passengers were loaded into No. 4, and at 1.50 a.m., very near to the time when No. 10 began lowering away, No. 4 also began its descent. As it neared the water, one of the sets of falls got caught, tipping the boat at an angle. The falls had to be cut to free them. Suddenly, a sailor in the boat shouted up, 'We need another hand down here!' After informing Lightoller that there was just one seaman aboard, the Second Officer sent Quartermaster Perkis and another sailor down the falls into the boat.

As Boat No. 4 reached the water, it had just thirty of its sixty-five seats occupied, and a number of passengers had been turned away prior to its lowering. Once the boat began rowing away, Emily Ryerson was shocked to see that A Deck, where they had just boarded the boat from, was only 20ft above the water. She could see water rushing into open portholes along the waterline, with the lights still on in the rooms.

Similarly, in the boats that had already rowed away from the sinking ship, any doubts about the seriousness of the situation now vanished. In Boat No. 13, Saloon Steward Littlejohn and the other occupants lay on their oars and watched; they could see the forward portholes on E Deck were under water, and eventually noticed lights further aft going out as the ship continued to settle. In Boat No. 15,

Assistant Saloon Steward Nichols noted how far down the head of *Titanic* was, and watched as the starboard propeller rose halfway out of the water. Any doubts he had about the ship sinking quickly began to vanish.

Around this time, Captain Smith returned to the Marconi Room and informed Jack Phillips and Harold Bride that the ship did not have long to live. He told them that the 'engine rooms were taking water and that the dynamos might not last much longer'. Phillips had just remarked that the power to the wireless was growing weaker. Smith did not release the men from duty before he left.

In response to what Smith had said, Phillips went out on deck to investigate the situation, while Harold Bride worked the wireless key. While he was gone, Bride established contact with the *Baltic*, telling them, 'Engine room getting flooded.' The *Baltic* assured them, 'We are rushing to you.' They were too far away to arrive in time, however. This exchange took place from 11.45–11.50 p.m. New York Time, or 1.47–1.52 a.m. *Titanic* time.

Phillips returned just after this, saying that things looked 'very queer', and adding that the 'Forward Well Deck was awash'. He also indicated that there was a heavy port list, which Bride had already noticed. Phillips decided that they should get their lifebelts and prepare to leave the cabin, but then returned to sending distress calls without following through on his own suggestion. Bride went to his cabin and dressed, then grabbed Phillips' overcoat, which he slipped over his colleague as he continued to work.

The electrical power had decreased to the point where Phillips was unsure of whether he was still transmitting. The lights outside on the deck and along the Promenade had even begun to take on a dull, reddish hue. However, in short order, Phillips discovered that he was still transmitting, and both of the wireless men continued working for another ten minutes.

By the time Boat No. 2 had lowered away at 1.45 a.m., crewmen had already been prepping Collapsible C on the starboard side and Collapsible D on the port side. Both of these boats were stored on the deck, immediately inboard of Emergency Boat Nos. 1 and 2, respectively. Fortunately, some of the prep work had commenced earlier in the night. After being relieved at the ship's wheel at 12.23 a.m., Quartermaster Hichens had got the cover and gripes off D, and Able Bodied Seaman Jones had done further prep work on the boat, prior to leaving in No. 8. It is likely that similar preparations had been made at Collapsible C as well, although this is uncertain.

Regardless, the crewmen on both sides still had to crank back in the same davits that had lowered Nos. 1 and 2, and finish prepping the collapsibles, which included raising their canvas sides, then they still had to ready the falls to receive the boats, raise them up, swing them out and lower them level with the deck, and all of this before any passengers could board the boats.

After finishing at Boat No. 2, Chief Officer Wilde went to the starboard side to work at Collapsible C, helping to load passengers. He then called out for someone to look after the boat, at which point Captain Smith ordered Quartermaster Rowe to help at the collapsible and to take charge of it. Rowe fired one last distress rocket around 1.52 a.m., then abandoned that effort to follow Smith's instructions.

By this time, First Officer Murdoch had arrived at Collapsible C and found the loading progressing. Chief Officer Wilde departed for the port side, to supervise the loading of Collapsible D. A crowd had begun to gather around Collapsible C, which is not surprising since it was the last lifeboat but one left on the starboard side.

At some point during the loading, men clambered into the boat and refused to budge. Having had enough, and knowing there was no time for delay, First Officer Murdoch fired warning shots into the air and yelled at the men to get out. First Class passengers Hugh Woolner, Mauritz Håkan Björnström-Steffansson and others helped to drag the men out of the boat, quelling the situation. Although the details are not entirely clear, it appears that the men who had been thrown out realised there was no chance of them getting into C and likely crossed over to the port side to look for better prospects, leaving the forward starboard side of the Boat Deck largely abandoned.

The pace of loading at both collapsibles now became frenzied, as *Titanic*'s bow settled further and further. At Collapsible D, things became chaotic as a crowd suddenly appeared around the boat, interfering with the loading. Chief Officer Wilde and Second Officer Lightoller, both with revolvers in hand, threatened the men, telling them to stand back. Then they had crewmen around the boat link arms to prevent a rush on it. Only women and children were to be let through the cordon. Shots may have been fired, but the exact specifics around the loading of Collapsible D are a bit murky.

## SHOTS IN THE DARK

In the aftermath of the *Titanic* disaster, there were claims of disorder and shootings during the sinking, as well as allegations that Captain Smith, First Officer Murdoch or others had killed themselves. In the subsequent disaster inquiries, no accounts of shootings or suicides emerged, although at least three separate incidents of warning shots were documented. Because of this, historians largely wrote off the claims as an invention of the yellow press.

What emerged instead over the decades was the myth that the sinking was perfectly orderly, with all of the men chivalrously standing aside and letting women and children board the lifeboats. Now, with the benefit of over a century of study and with modern researchers having access to dozens of additional eyewitness accounts, the true story can be told.

Early in the disaster, many passengers were reluctant to come on deck due to the cold, the noise of the steam venting from the funnels, and so forth. It must be remembered that neither Captain Smith nor any of the officers ever made a formal announcement that the ship was sinking. This led to a lack of urgency to board the boats, since many passengers believed that the ship would stay afloat and it was safer to remain aboard. Because of this, the early lifeboats left less than half full, and the scene on deck was calm.

Eyewitnesses suggest that this began to change around 1.05 a.m. when Boat No. 1 was lowered. Lawrence Beesley stated that a rumour circulated on the starboard side that men were going to be taken off on the port side. With the ship down by the head and fewer and fewer boats remaining, a large crowd began to congregate at the aft port boats. As unrest there began to grow, Fifth Officer Lowe and some of the other crewmen who had been loading No. 1 headed to that location.

As Boat No. 14 was loading, Able Bodied Seaman Joseph Scarrott stopped a rush by hitting men with the tiller. As No. 14 lowered past A Deck, Lowe – fearing that jumpers would buckle the boat – fired three warning shots along the ship's side with his personal firearm. There was also a rush on Boat No. 12, which lowered right after No. 14 at 1.30 a.m., but this appears to have been quelled without gunfire.

After this, a group of men rushed Boat No. 15 as it was being loaded and there is evidence that First Officer Murdoch had his revolver out and yelled at everyone to stand back – again, without shots being fired. A group of officers, including Captain Smith, Chief Officer Wilde, First Officer Murdoch and Second Officer Lightoller, had taken Webley revolvers out of the arms locker earlier in the night. It is known that Wilde requested the revolvers, although some evidence suggests that Murdoch may have asked for them previously.

As it became impossible to deny that the ship was sinking, panic grew. At Collapsible C, multiple eyewitnesses, including First Class passengers Hugh Woolner and Mauritz Håkan Björnström-Steffansson, stated that men rushed the boat, causing First Officer Murdoch to fire warning shots in the air. The men were then hauled out of the lifeboat. There is controversy, however.

Collapsible C is the lifeboat that White Star Chairman Bruce Ismay boarded right before it lowered away at 2.00 a.m., and numerous witnesses described there being no crowd around it at the time. It seems likely that the men were dispersed earlier in the loading, after Murdoch's warning shots.

There was more disorder at Collapsible D, which was the last lifeboat lowered away, at around 2.05 a.m. Crew members linked arms to prevent men from rushing the boat. It appears that both Chief Officer Wilde and Second Officer Lightoller threatened men with their revolvers, and additional shots may have been fired.

There are also multiple claims that one of the ship's officers actually shot passengers and then committed suicide. Some of the eyewitness statements recorded in the press are questionable because many of the witnesses were not in a position to see what they are quoted as seeing, were reporting second-hand information or the details they gave were just too sensational to be believed.

However, some witnesses – such as First Class passenger George Rheims and Third Class passenger Eugene Daly – were on the ship until the last, and both claimed that an unnamed officer fired on men and then subsequently killed himself. Both mentioned this in multiple accounts, including private letters. Daly even mentioned it in his 1915 US District Court testimony during the Limitation of Liability hearings. The description of a boat lowering away and the timing shortly before the end would suggest that this happened at either Collapsible C or D, or during the aborted attempt to lower Collapsible A as the Boat Deck submerged. Indeed, First Class passenger Richard Norris Williams mentioned a gunshot near the Bridge at this exact time, although he did not see what happened.

It has often been suggested that First Officer Murdoch killed himself, since many press headlines claimed this. However, Second Officer Lightoller wrote a letter to Murdoch's widow specifically denying that this was true. On the other hand, Lightoller allegedly told James McGiffin, the Marine Superintendent at Queenstown and Murdoch's friend, that Murdoch shot a crewman through the jaw for leading a rush on a lifeboat. In his later years in Hertfordshire, Lightoller was rumoured to have told others that he knew someone who committed suicide that night.

In the end, there is no way to know for certain whether anyone was shot or if an officer committed suicide, although the evidence suggests that it may have happened. Murdoch is one suspect, as is Wilde – who isn't accounted for after the lowering of Collapsible D – but there are other possibilities as well. In the final analysis, all that can be stated with certainty is that the sinking of *Titanic* was not as orderly as the mythology suggests, and there were multiple incidents of gunfire throughout the later stages of the sinking.

Near Collapsible D, First Class passenger Jane Hoyt reluctantly agreed to get in the boat and leave her husband Frederick, but only after 'much persuasion and after we had bid each other goodbye'. She recalled that the situation around the launch was chaotic, and she feared that this would be the last time she would ever see her husband.

Meanwhile, in the lifeboats that were already afloat, the occupants looked on in horror. It was now apparent that *Titanic* had little time left. In Boat No. 5, Third Officer Pitman watched as water engulfed the last line of portholes below the Forecastle. Even knowing all that he did, Pitman had still been holding out hope that the ship would somehow manage to stay afloat. This observation effectively killed that delusion.

In Boat No. 2, Boxhall and his ragtag group of rowers struggled aft along *Titanic*'s port side, trying to round the stern and reach the starboard gangway as Captain Smith had instructed. Boxhall had seen lines of portholes going under as they went. It was slow going, but now, as No. 2 rounded the stern, the boat came very near the ship's propellers. While it is unlikely they were fully out of the water at this time, Boxhall later said, 'I'm not certain if I didn't pass underneath them.'

After rounding the stern and heading a little way up the starboard side of the ship, Boxhall began to feel uneasy about getting closer. Some of the passengers in No. 2 protested about going closer as well. Even though there were just seventeen people in the boat, Boxhall was very concerned about 'being rushed and losing everybody in the boat'.

In the inquiries following the sinking, Boxhall stated that he never saw an open gangway once he got to the starboard side. However, in 1962, Boxhall gave an interview where he claimed that the gangway doors had been open and there was such a crowd in them that he dared not go too close, for fear of being swamped. Whether the latter-day recollection was a confession or merely a confused memory is uncertain, but either way Boxhall had No. 2 turned away from the ship and outward from the stern, and the occupants began to row away in earnest.

Greasers Frederick Scott and Thomas Ranger had seen a group of engineers – including Senior Second Engineer Farquharson, as well as trimmers and firemen – come topside at 1.20 a.m. They had stayed close to them, but now, just before 2.00 a.m., they separated from this group.

Crossing over to the port side, they looked over the edge of the deck and saw Boat No. 4 rowing aft along the port side of the ship. Both men decided to climb out on the davits used to lower Boat No. 16, and attempt to lower themselves into No. 4 as it rowed past. Ranger managed to make it all the way down to the boat safely, while Scott made it halfway, before falling and plunging into the 28°F water. Thankfully, he was pulled into No. 4 within a few minutes.

Back at Collapsible C, First Officer Murdoch was finishing the loading process. In contrast to earlier, it was a very calm scene, with very few passengers nearby. By this time, Bruce Ismay had arrived, and was lending a hand in the loading process. Murdoch called out for any more women and children, but there were no more in the immediate area. With no more passengers forthcoming, Ismay and First Class passenger William Carter both boarded the boat. Murdoch could not have helped but see them, yet there is no evidence that he protested or was upset with this.

This was actually in keeping with Murdoch's process throughout the night, which was women and children first, but permitting men to board the boats when there were no more women. Ismay would later be lambasted by the yellow press for saving himself, which seems unfair considering that there is no evidence that he pushed ahead of anyone or used undue influence to get in the boat. He had a right to survive as much as anyone else that night.

At 2.00 a.m., Murdoch ordered C lowered away. There were forty-three occupants aboard, more than the capacity of the boat, which was rated for forty. Among these were four Chinese passengers who had hidden themselves under the seats until the lifeboat rowed away from the ship.

As Collapsible C lowered the short distance to the water, it kept getting hung up against the ship's side due to the port list, and the occupants had to keep pushing the boat away from the hull to prevent it from catching on the rivets. The slow speed of the lowering had allowed some aboard, including Quartermaster Rowe, to actually witness the sinking accelerate. When Collapsible C started descending, Rowe had noticed that the Well Deck was 'awash', but by the time they reached the water and cast off, the Well Deck had 'submerged'.

Once the ocean reached the Forward Well Deck, this allowed water into the ship through more and larger openings such as the holes for the anchor chains, the hawsepipe and the forward cargo hatches. This quickly accelerated the rate of sinking, which is why witnesses could actually see the process speeding up.

Around 2.00 a.m., the ship's band, which for nearly two hours had been playing inside the First Class Entrance due to the cold, finally emerged on deck. They had been playing ragtime and other upbeat tunes for most of the night, but according to most accounts would soon switch to hymns and more sombre tunes as the end approached.

Popular belief has often placed the band on the port side of the Boat Deck during the night, in the alcove in between the First Class Entrance and the raised roof over the First Class Lounge. However, there are no actual eyewitness accounts that place them there. In fact, more recent research by historian George Behe indicates that the band actually played on the starboard side of the Boat Deck, since all of the witnesses who described seeing them at the end were at that location when the ship plunged under.

Meanwhile, near Collapsible D, Second Officer Lightoller had looked down the stairwell just abaft the Bridge. These stairs led to a landing on A Deck, and then proceeded down to B Deck, providing a direct line of sight. Owing to the port list, water had reached B Deck on that side of the ship, and the Second Officer watched as it climbed the stairs towards A Deck.

The crew working at D called out, 'Any more women and children?' Chief Officer Wilde had noticed that there were no more sailors available to man the boat and turned to Lightoller and told him, 'You go with her, Lightoller.' Lightoller refused, saying, 'Not damned likely!'

At 2.05 a.m., Chief Officer Wilde ordered Collapsible D away, with Quartermaster Arthur Bright in charge. There were just twenty of its forty seats occupied as it lowered away. As it descended towards the water, now just 10ft below, Bright looked forward and saw the sea covering the Forecastle. As the boat passed A Deck, the aft fall got caught up and Bright shouted out to stop letting out the forward fall until they got the aft one sorted out.

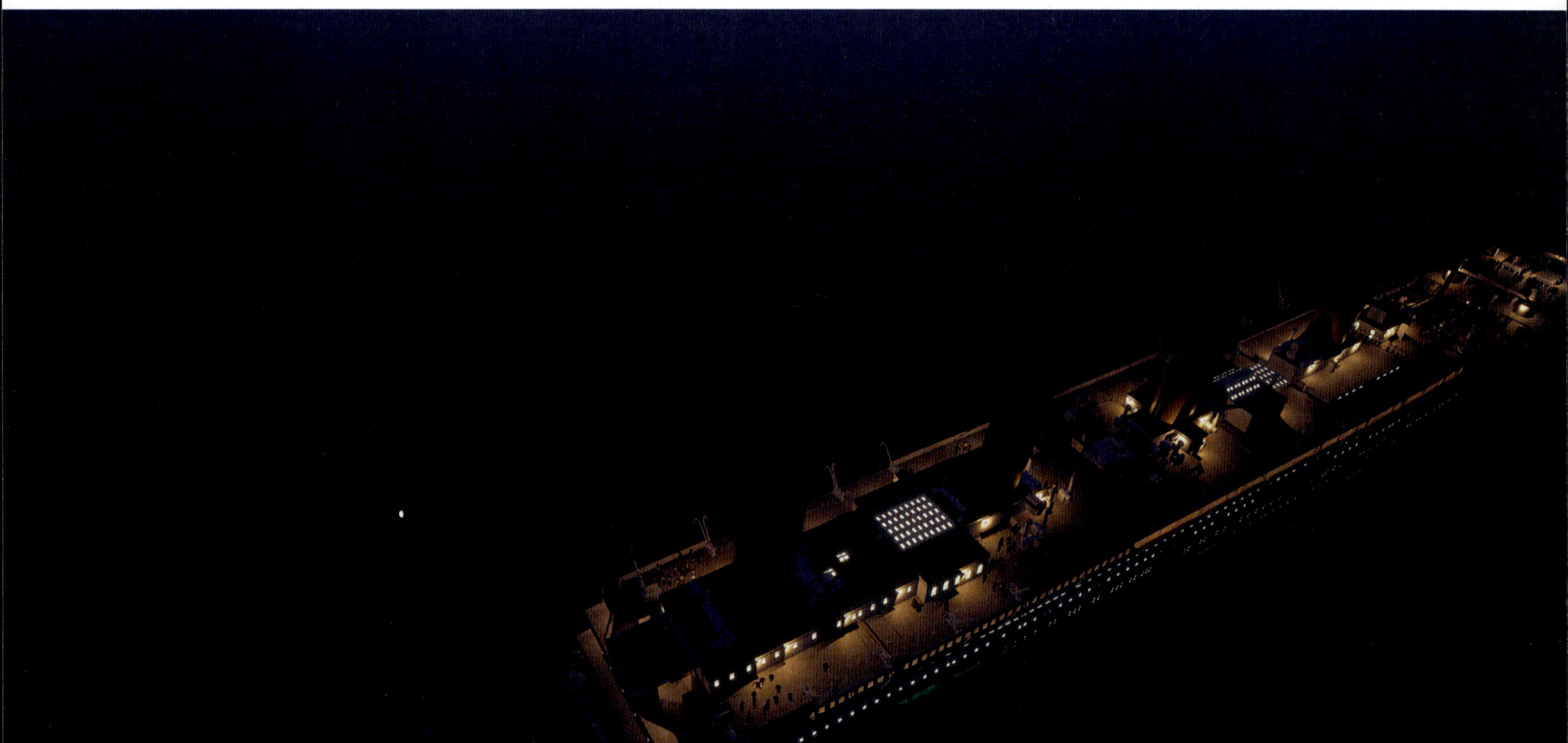

This overhead view at 2.01 a.m. shows Boat No. 4 casting off and Collapsible D being loaded. The Forward Well Deck is submerging, but the Forecastle is still dry. (HFX Studios)

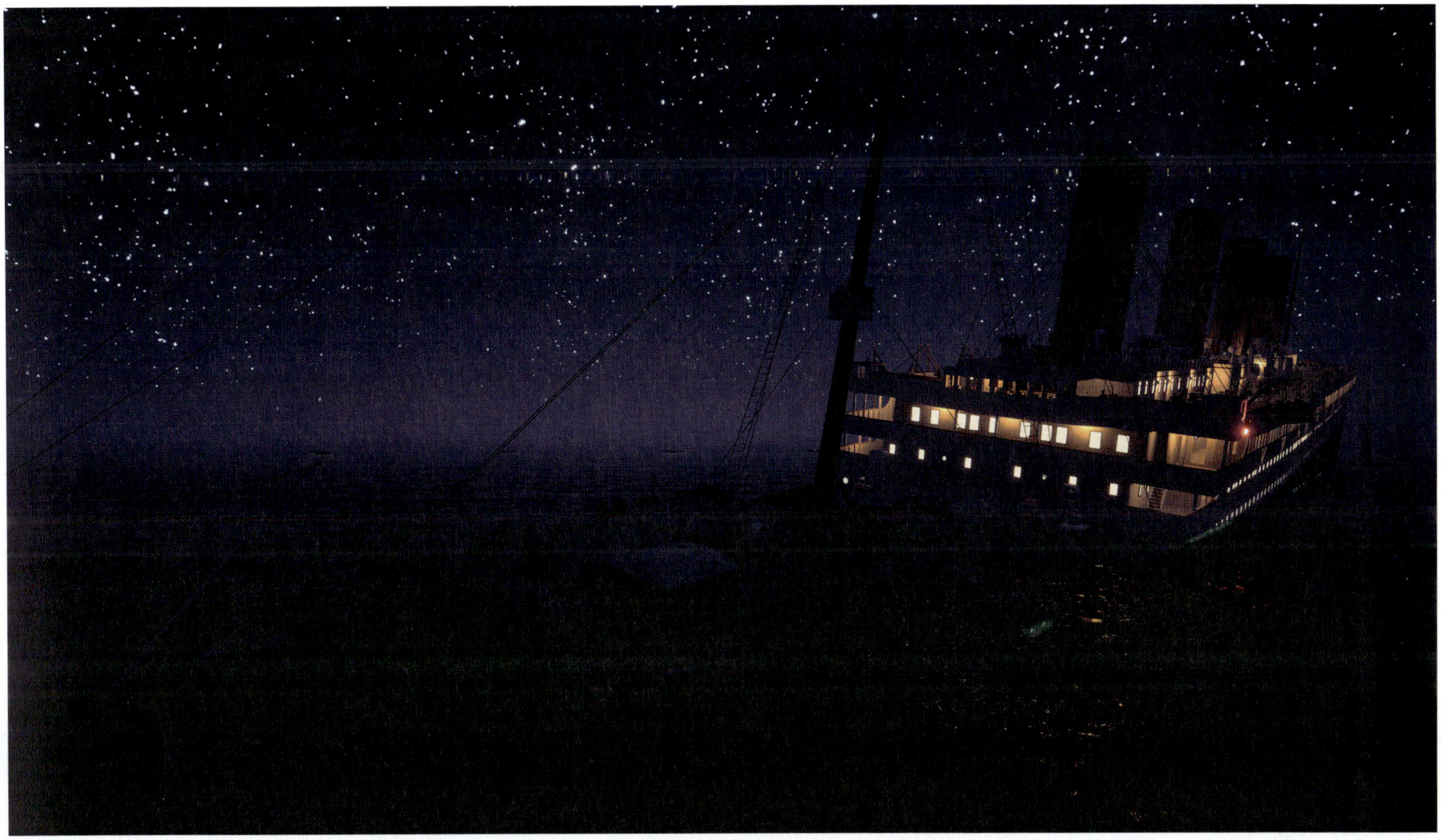

At 2.04 a.m., the port side of the Forecastle is slipping beneath the surface. Collapsible D will begin lowering in about a minute. Boat No. 2 is nearing the stern and will soon cross to starboard. Boat No. 4 is also heading aft, while Boat No. 10 rows away. (HFX Studios)

On the Boat Deck, Captain Smith saw First Class passenger Frederick Hoyt standing nearby. The Captain advised Hoyt to go down to A Deck and 'see if there were not a boat alongside'. He suggested that Hoyt would have to jump and added, 'You'd better do it soon.' It was clear that Smith knew the ship had little time left.

Hoyt descended to A Deck, where, owing to it getting hung up, Collapsible D was still hanging alongside the side of the ship. He climbed over the rail and plunged into the icy water.

Meanwhile, First Class passengers Hugh Woolner and Mauritz Håkan Björnström-Steffansson went down to A Deck. According to Woolner, this was to help any remaining passengers there to a lifeboat. However, Björnström-Steffansson later admitted that they had seen that Collapsible D was still not lowered away and saw a chance to save themselves. The lights were dim and glowing red as they arrived on the port side. Water had begun spilling over the forward bulwark railing on the port side and was soon rushing around their feet.

Fearful that they would end up boxed in against the ceiling, Woolner and Björnström-Steffansson took advantage of the halt in Collapsible D's descent and climbed onto the A Deck railing. The boat was hanging approximately 5ft from the hull due to the port list and both men launched themselves outward. Björnström-Steffansson landed roughly in the boat, but Woolner came up short, striking the gunwale with his lifebelt and chest. As some of the occupants pulled him aboard, his feet dangled in the icy water.

In Collapsible D, Jane Hoyt was unaware that her husband had jumped into the sea. Just after the boat reached the water and began rowing away, the crewmen noticed a man swimming towards them, and he was quickly hauled aboard. As the man lay shivering in the bottom of the boat, he looked up at his wife, who had not recognised him, and called out, 'Jane!' Mrs Hoyt exclaimed, 'My God! It's my husband!' The ladies in the boat then covered him with a fur coat and shawl to keep him warm.

Back on deck, the remaining crewmen turned their attention to the last two collapsible boats aboard, which were located on top of the Officers' Quarters: Collapsible B on the port side and Collapsible A on the starboard side. The proper way of getting the boats down to the Boat Deck involved using a block and tackle that could be hooked through a special eyelet in the stays of the forward funnel. However, it was now clear that there was simply not enough time for this procedure to be completed and they were dangerously close to having the ship sink under their feet. Second Officer Lightoller began working with a group to get Collapsible B down while, simultaneously, First Officer Murdoch and Sixth Officer Moody would try to prepare Collapsible A.

In the Marconi Room, Captain Smith had returned a short time after his previous visit and released Phillips and Bride from their duties, saying they had 'done their duty and could do no more'. Instead of leaving, Phillips doggedly stuck to his work, despite the fading power. Among several other attempted communications, Phillips tried to tell the *Carpathia*, which was racing towards them, that they 'were abandoning the ship, or words to that effect'. Phillips tried to send once or twice more, but was not sure his messages were being heard, or that they were getting a spark, since there were no replies.

Bride had already put on his own lifebelt, but Phillips had not. Bride had to put Phillips' lifebelt on him as his colleague worked. After briefly leaving the room and returning, he found a stoker had snuck in and was literally slipping the lifebelt off Phillips' back. Phillips was so focused on what he was doing that he didn't notice. This sent Bride into a rage, and he attacked the man, possibly hitting him with an object, since he later said, 'I don't know what it was I got hold of.' He added, 'I did my duty. I hope I finished him.' Phillips and Bride then cleared out of the room, leaving the motionless stoker to his fate.

At 2.11 a.m., Collapsible D is pushing away from the ship as the forward end of A Deck on the port side is submerging. (HFX Studios)

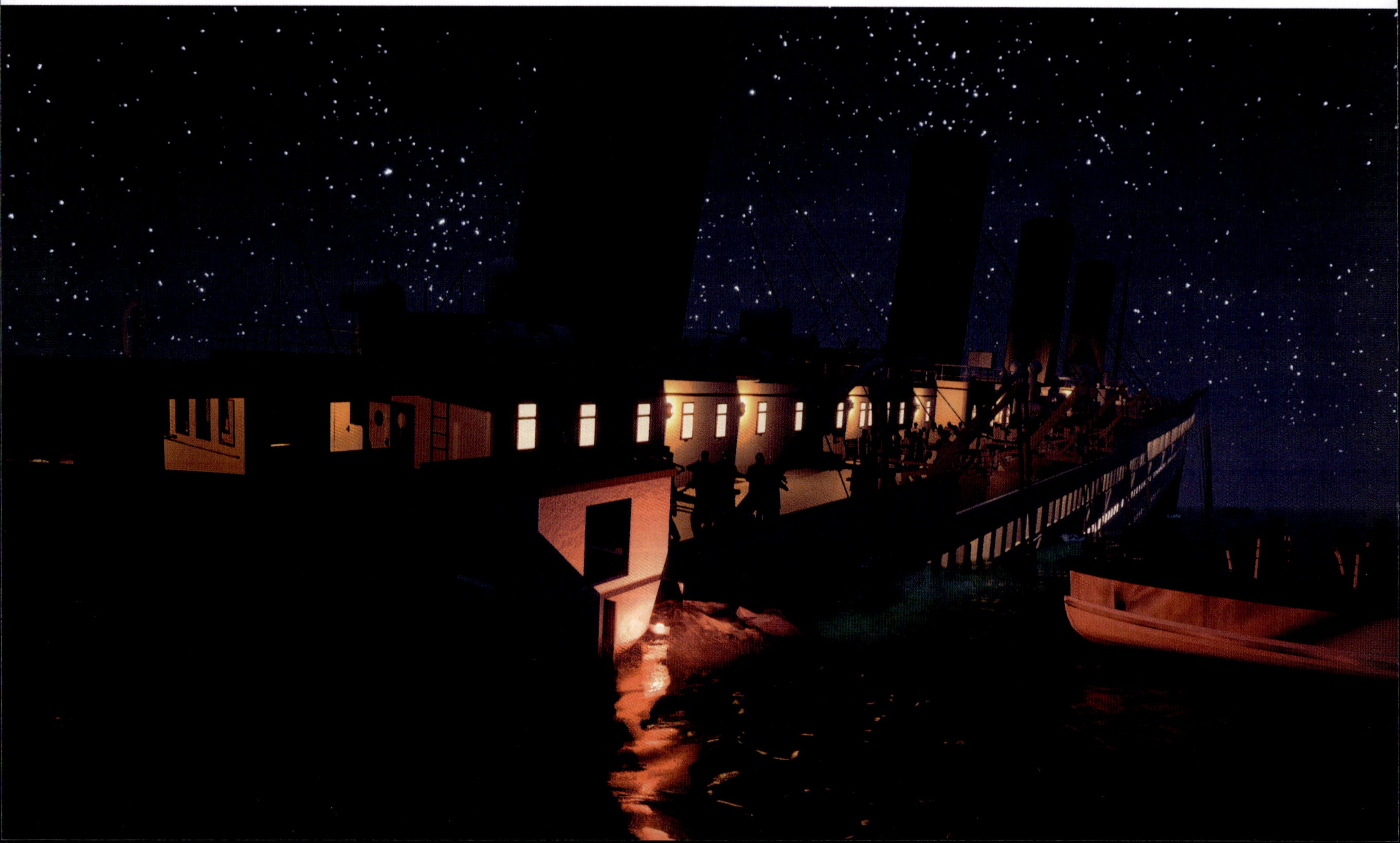

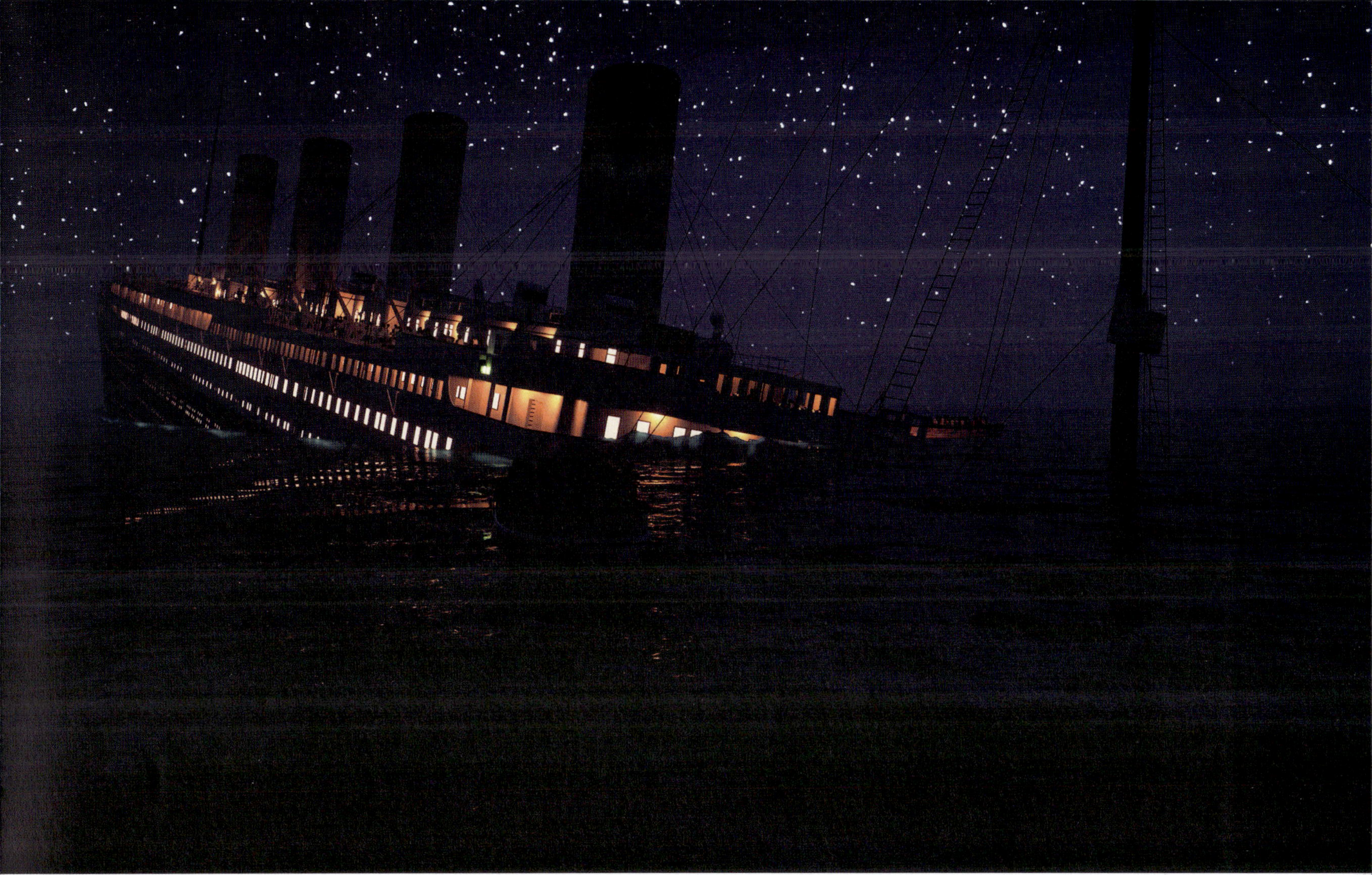

This view, also at 2.11 a.m., shows Collapsibles C and D both moving away from the ship as the forward superstructure is settling into the sea. (HFX Studios)

Around this time, a large crowd of Third Class passengers suddenly began to emerge from below decks via the First Class Grand Staircase and First Class Entrance. Why it had taken them so long to reach the deck is unknown, although it may lend credence to the stories that some steerage passengers were intentionally held back. Many of the crewmen had assumed that the majority of women and children had already left the ship, and were horrified to see so many of them among this crowd, desperately looking for a place in a lifeboat.

On the port side, Second Officer Lightoller and the group of crewmen with him continued to struggle to free Collapsible B. The canvas cover spars from the boat had been set up against the walls of the deck house in the hopes that they could use them as a slide to get the boat down.

Realising that Lightoller's group needed help, Harold Bride climbed up to the roof and aided in the effort. Jack Phillips ran aft, and that was the last time Bride would see him alive.

Meanwhile, on the starboard side, the ship's band began to play their final pieces. Several survivors mentioned the song 'Autumn' as one that was played. Modern speculation has mistakenly led to the belief that when the survivors mentioned 'Autumn', they actually meant the popular waltz '*Song d'Automne*'. However, not a single survivor ever mentioned that song, and some of those who mentioned 'Autumn' specified that it was the hymn by that name.

The majority of survivors in 1912 stated that the final song was actually 'Nearer My God, to Thee'. These stories were being repeated aboard the *Carpathia* even before the ship reached New York. Historian George Behe's recent research strongly supports the conclusion that band leader Wallace Hartley selected the *Propior Deo* version of 'Nearer My God, to Thee' as the final song. Regardless, the evidence is clear that the band played until right before the end, never ceasing in their efforts to calm passengers during the evacuation. Every member of the group would perish.

It must have been a surreal scene as the sounds of string music echoed across the ship, even as officers and crewmen frantically struggled to get the last two lifeboats away. A large group of passengers, including women and children, began to wait near the last two collapsibles, particularly on the starboard side. Many other passengers left aboard, seeing the water so close to the Boat Deck, began to head aft and congregate on the stern of the ship. This growing crowd consisted of crewmen and passengers alike, from all walks of life.

On the stern, seeing that nearly all hope was lost, Second Class passengers Father Thomas Byles and Father Joseph Peruschitz stood together amidst a ring of passengers and began granting absolutions and leading them in prayer. Both men had assisted passengers throughout the sinking.

August Wennerström had a different perspective when he saw the ring of passengers, observing that they were 'asking God and Mary [*sic*] to help them ... They just prayed and yelled, and never lifted a hand to help themselves. They had lost their own will power and were expecting God to do all the work.'

Around 2.10 a.m., on the Boat Deck, Thomas Andrews had just finished tossing deck chairs overboard, to give anyone who ended up swimming in the water something to cling to, before heading towards the Bridge. Near the collapsibles, Captain Smith, with a megaphone in his hand, instructed the crewmen labouring there, saying, 'Well boys, do your best for the women and children, and look out for yourselves.' Then he, like Andrews, headed towards the Bridge.

A few moments later, both men were seen climbing over the Bridge rail on the port side and entering the sea, just as water was reaching their location. There was little chance of being rescued, but both men must have known that choosing to remain behind, when there was nothing left to be done, would be futile.

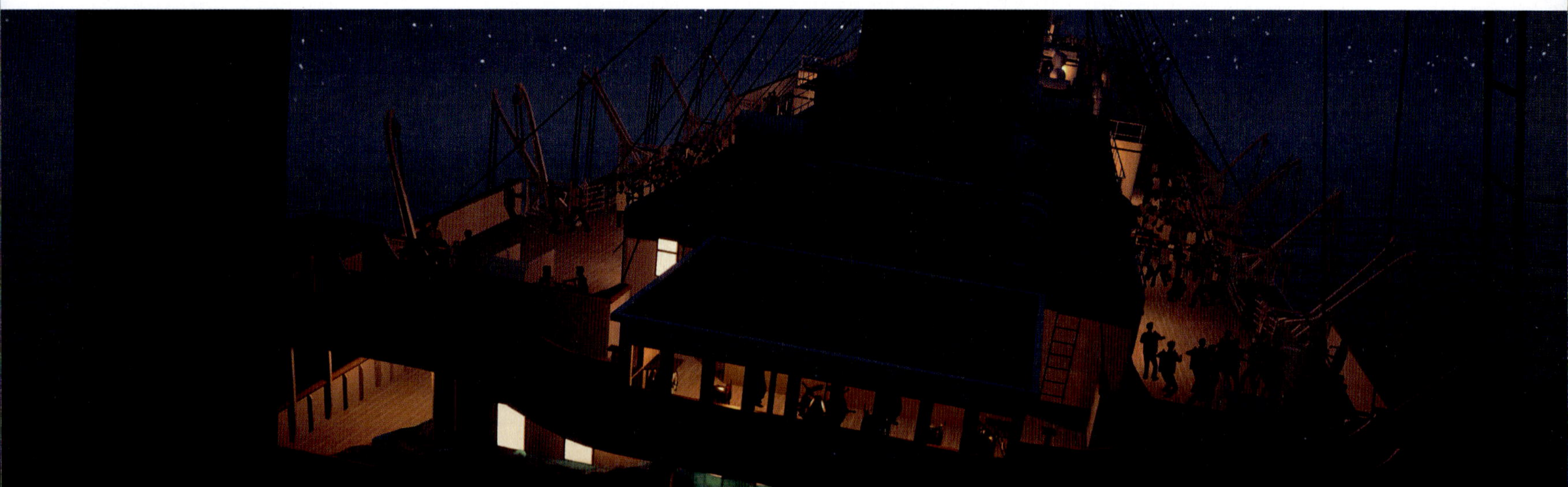

Another perspective at 2.11 a.m., from close to the deserted Crow's Nest, looks down on the water rolling over the front and port side of A Deck.
The port side of the Bridge wing is about to slip under and Captain Smith and Thomas Andrews are about to abandon ship. (HFX Studios)

## THE FINAL SIGHTING OF THOMAS ANDREWS

For over 100 years, the general perception has been that Thomas Andrews, one of the designers of the *Titanic*, was last seen in the First Class Smoking Room, staring at the painting of Plymouth Harbour in a daze, waiting for the ship to sink around him.

Investigation into the origins of this perception led to Shan Bullock's 1912 book, *Thomas Andrews: Shipbuilder*. Bullock stated that, about fifteen minutes before the end, an assistant steward 'saw him [Andrews] alone in the smoking room, his arms folded over his breast and the [life]belt lying on a table near him'. Upon asking whether Andrews was going to make a try for it, there was no response. Andrews 'just stood like one stunned'. Although Bullock did not identify this 'assistant steward', it is generally believed to be Verandah Café Steward John Stewart. However, other evidence says that Steward John Stewart left the ship in Boat No. 15. He is specifically identified in this lifeboat in the British inquiry testimony.

We know that Boat No. 15 left the *Titanic* at about 1.40 a.m., long before Bullock's fifteen minutes before the end. Assuming the Smoking Room account was Stewart's, then the sighting of Andrews had to be long before the sinking, before 1.40 a.m. Andrews had much time to move on from the Smoking Room and to continue to help other people off the ship, as he had been earlier.

In fact, Bullock himself did not claim that the sighting of Andrews in the Smoking Room was the last time he was seen. He mentioned subsequent sightings of him, including an unnamed officer saying that Andrews was seen throwing deck chairs overboard, and he was seen by an unidentified witness heading towards the Bridge to speak with Captain Smith, with a lifebelt in his hand.

Importantly, we do have a very credible eyewitness who recalled a much later sighting of Andrews. A 21-year-old Mess Steward, Cecil Fitzpatrick, gave an account in the *Western Daily Mercury* of 29 April 1912. In it, Fitzpatrick said he passed through the Bridge from the port side and saw Captain Smith talking to Andrews. He heard Smith say, 'We cannot stay any longer; she is going!' Fitzpatrick collapsed against the starboard side of the Bridge entrance, then recovered his senses and realised he needed to get to a boat. He tried to help launch Collapsible A, but found it was wedged between the davits and the spar. Then the ship dipped, and he ended up in the water with many others, eventually finding his way to Collapsible B. Lillian Bentham corroborated Fitzpatrick's account when she placed the young Mess Steward first in Collapsible B, and later in Boat No. 12.

Fitzpatrick's account places both Andrews and Captain Smith on the Bridge toward the very end of the sinking. For the next part of what Fitzpatrick saw on the Bridge before being washed off the ship, we must look at a private letter written by Andrews' good friend, David Galloway. It was addressed to Andrews' uncle, Lord Pirrie, dated 27 April 1912. Galloway had interviewed some of the crew on the *Lapland* when they were returning to England, attempting to ascertain how Andrews met his end.

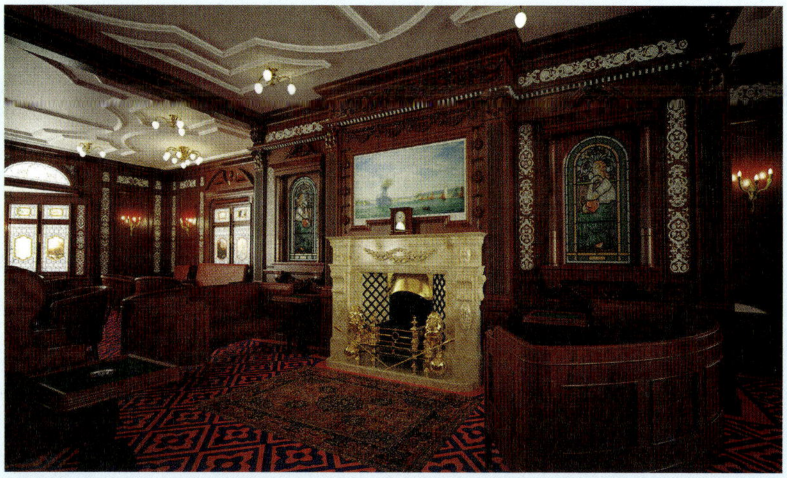

Thomas Andrews is often depicted as standing here by the Smoking Room fireplace just minutes before the end. In reality, the sighting had taken place much earlier and he was seen in several other places afterwards. (Chris Walker)

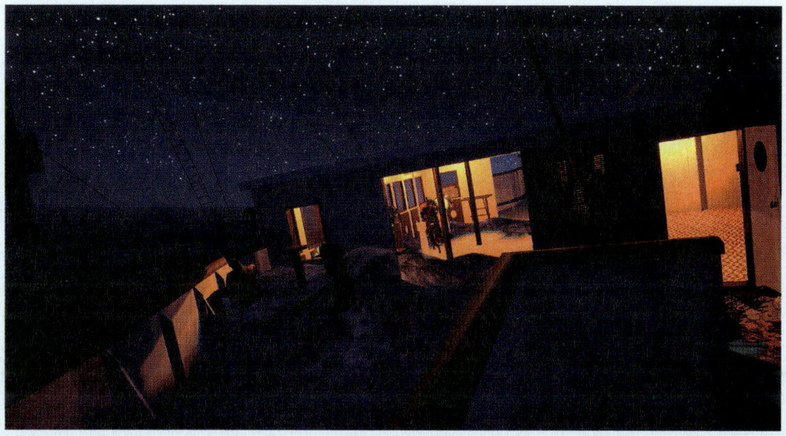

Thomas Andrews and Captain Smith climb over the railing as the Bridge goes under. (HFX Studios)

Galloway reported to Pirrie that a 'young mess-boy' saw both Andrews and Smith on the Bridge, and overheard Smith say, 'It's no use waiting any longer' – notice the similarity to the wording in Fitzpatrick's press account – then 'both men entered the sea together'. Though Galloway does not specifically name the survivor, his description of this person being a 'young mess-boy' and being on the ship so late in the sinking means that the only likely candidate is Fitzpatrick.

It seems obvious to us that Thomas Andrews made it to the Bridge before the ship sank and dived overboard with Captain Smith at the very end.

At that very moment, on the port side of the Officers' Quarters, Second Officer Lightoller, Harold Bride and the rest of the group working at Collapsible B were finally able to push it over the edge of the deck. Unfortunately, the boat flipped over as it fell, landing upside down on the Boat Deck. Owing to the port list, water had already begun pouring onto the deck there, and the collapsible landed in it. At about the same time that the collapsible fell, Bride had caught a glimpse of Captain Smith jumping into the sea. Several of the men who had tried freeing Collapsible B scrambled down, including Harold Bride. With water washing across the deck, there was no chance to right the boat, much less hook it up to the falls.

On the starboard side, First Officer Murdoch's group was desperately trying to get Collapsible A down. The Boat Deck was still dry on the starboard side, because of the strong port list. Murdoch had the forward set of davits cranked back in to receive the collapsible.

Lamp Trimmer Hemming had just untangled the falls that had been used to lower both Boat No. 1 and Collapsible C, and then went to hand the block up to Sixth Officer Moody and the men on the roof of the Officers' Quarters. Moody said, 'We don't want the block. We will leave the boat on deck.'

The crew then managed to push Collapsible A over the edge of the deck, where, unlike Collapsible B, it slid down two planks propped against the wall and landed upright. The fall appeared to damage the boat, though, and the crew were unable to raise its canvas sides as a result.

Evidently, Moody's opinion that the collapsible should be left on the deck was overruled, as a group of '6 or 7' men – including Murdoch and Moody, Steward Edward Brown, Saloon Steward Thomas Whiteley, Mess Steward Fitzpatrick, Second Class passenger William Mellors and Third Class passenger Eugene Daly – slacked and hooked up the falls to the boat, then attempted to push it uphill against the list to the edge of the deck.

Water is washing across the Bridge and gurgling up the starboard stairs from A Deck below in this view at 2.14 a.m. (HFX Studios)

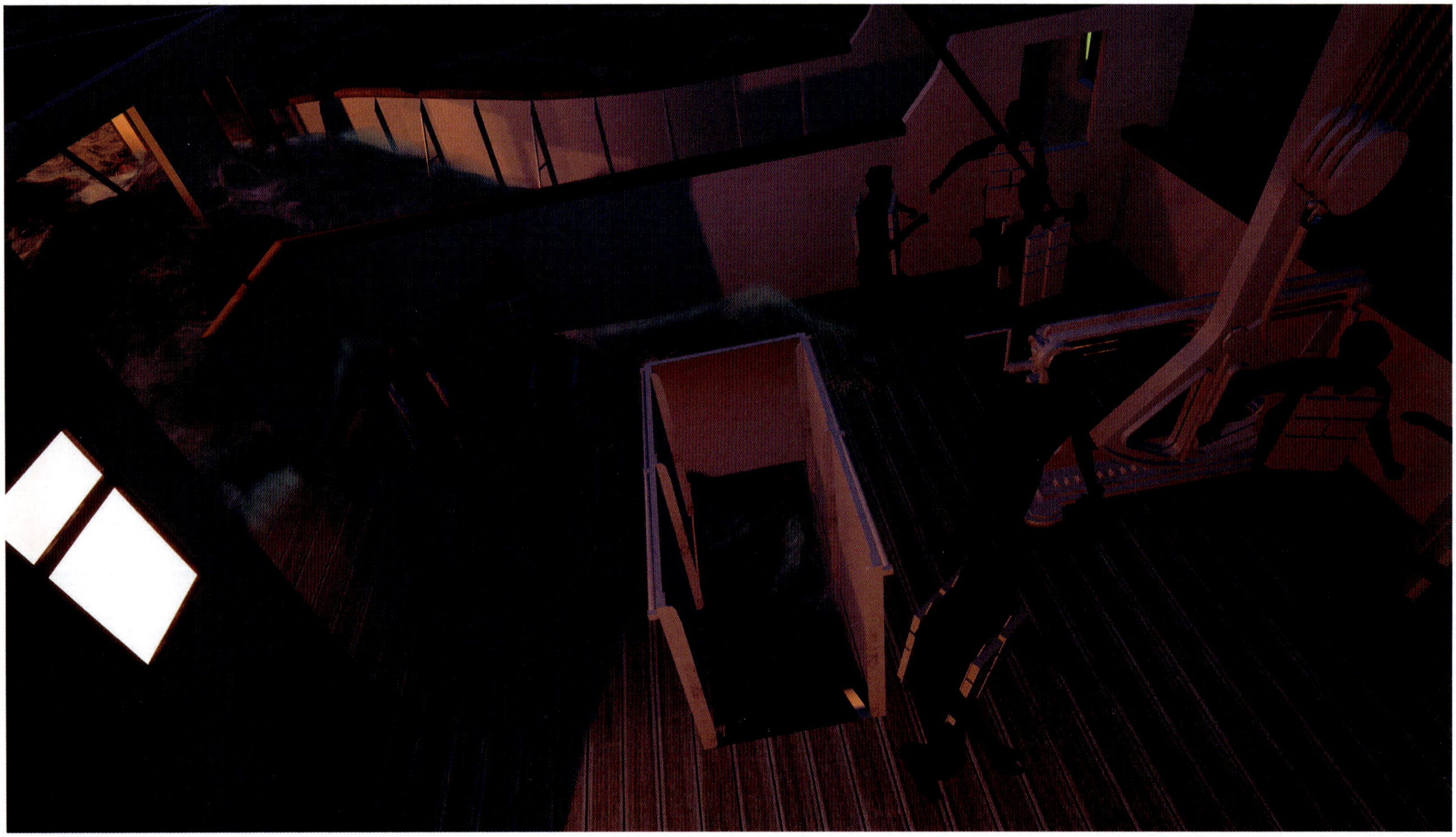

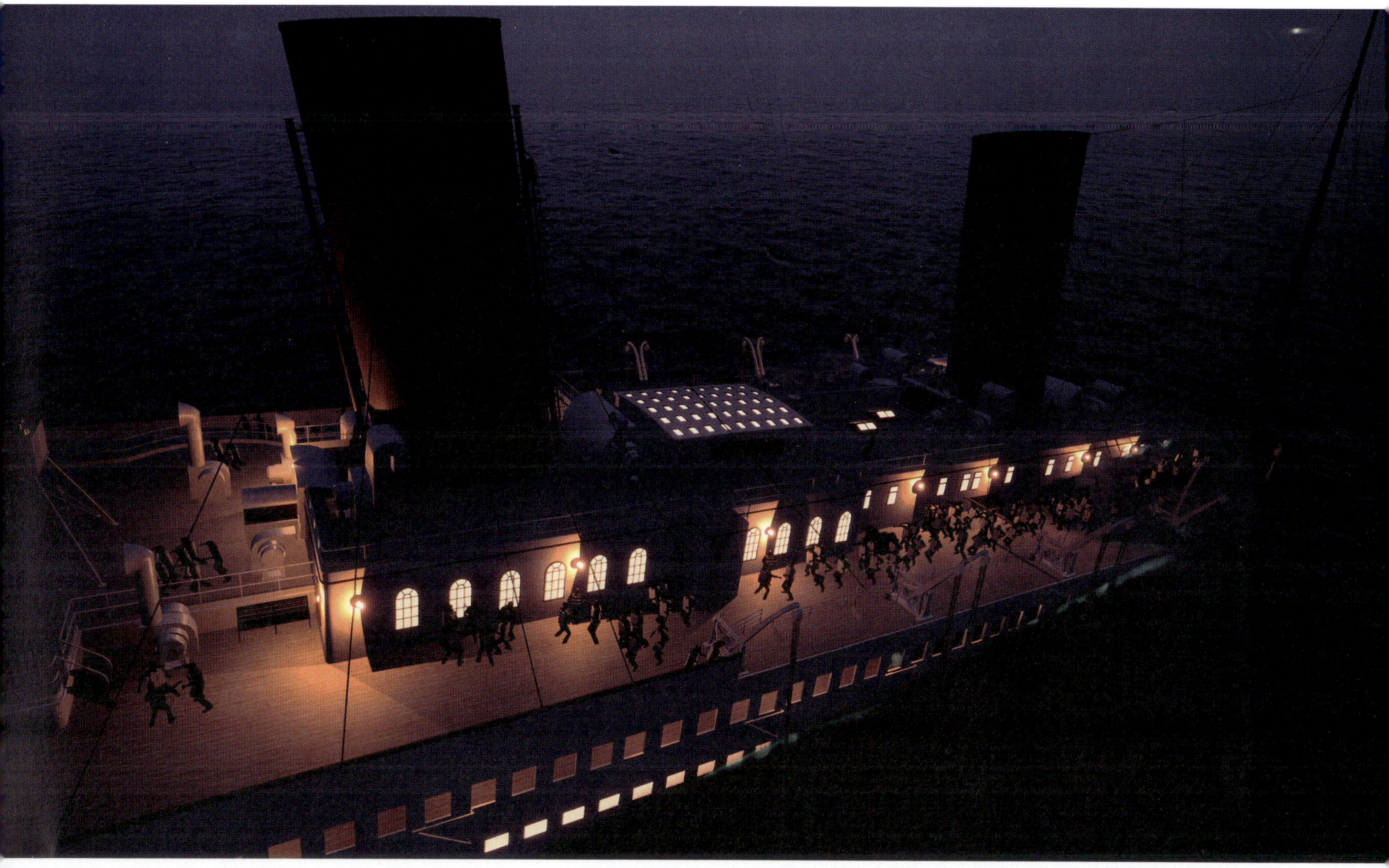

Among the large crowd gathered nearby were a number of women, who were waiting to board the boat. As the group of crewmen laboured in desperation, water began gurgling up the forward hatchway onto the Boat Deck. Fireman John Thompson leapt into the boat and, despite having his hands burned in a steam-related accident in the boiler rooms earlier that night, frantically but unsuccessfully tried to put the plug in.

At 2.15 a.m., as water began running aft along the deck, *Titanic* took a 'slight but definite plunge', which sent water surging aft along the deck like a wave. The crowd nearby began running aft in a frenzy, most heading towards the stern. Others turned and jumped overboard, while those right next to Collapsible A scrambled into it. Steward Brown leapt into the boat and managed to cut the aft falls.

Multiple eyewitnesses who were forward on the starboard side stated that after this initial plunge, the Boat Deck rose again slightly. The exact reason for this motion is unknown, but enough eyewitnesses consistently described this to suggest that it really happened.

This brief righting action allowed the men still near Collapsible A to free it. William Mellors and Eugene Daly managed to cut the forward falls, and the boat began to float free. As *Titanic* continued to plunge down, the wave rolled aft, swamped the Bridge and washed many passengers and crew overboard. Those who were still on their feet continued to rush aft, where the crowds continued to swell in number. *Titanic*'s port list eased as it sank further down, the ship settling onto a more even keel.

*Titanic* making her 'slight but definite plunge' and coming up on an even keel at 2.15 a.m. Collapsible A floats free. Amidships, Jack Thayer and Milton Long go overboard, while Colonel Gracie and his friend, J. Clinch Smith, rush aft and will shortly be overtaken by the wave rolling aft. (HFX Studios)

## THE CHRONOMETER

A ship's chronometer was used to keep time. However, unlike a Magneta clock, which was used to keep track of time on the ship itself as she moved east or west across the ocean – similar, in a way, to time zones on land – a chronometer was used to keep track of time at a fixed point on land.

Marine chronometers were developed during the eighteenth century to solve the problem of determining east/west longitude at sea. For an explanation of how this was done as ships crossed the North Atlantic, please see our previous section, 'Keeping Track of *Titanic*'s Maiden Voyage' on page 68.

*Titanic* had two chronometers, a primary and a back-up. The 1996 *Titanic* expedition recovered one of these. This chronometer was restored and eventually displayed, complete with its glass cover, at the Science Museum in London. Photographs of it have appeared in print and online. One photo, taken in May 2003, shows the chronometer's hands showing approximately 12.39. But is this time anywhere near accurate? (This photo can be viewed by searching online for 'Science Museum London *Titanic* Chronometer'.)

Since we know *Titanic* sank at about 2.20 a.m., the Bridge went under water about five minutes earlier, or around 2.15 a.m. As the rules required that the chronometer maintain Greenwich Mean Time (GMT), what would that be? The proven time difference between the *Titanic* and New York was two hours and two minutes. This would mean the difference between *Titanic* and GMT would be two hours and fifty-eight minutes, and the time of the sinking in GMT would be 5.18 a.m. If we adjust this by the five minutes earlier when the Bridge went under, the chronometer should have shown somewhere close to 5.13, definitely not 12.39.

Examining other photos of the chronometer reveals slightly different settings of the hands. For example, one photo showed a time of 12.38. A third photo, taken in 2010, showed both hands dismounted and lying across the clock face.

Examination of other chronometers of that era shows that they all used a retaining nut to hold the hands to the pinion in the centre of the clock face, connecting to the clock's innards. A close look at the May 2003 photo shows no retaining nut at all. Also, the square hole in the centre of each hand shows nothing holding them in position – in other words, no pinion either. With no nut and no pinion, the hands of the chronometer are just resting on the face, however the curator set them, and have nothing to do with whatever time the chronometer was actually showing when the Bridge went under.

A picture of the chronometer face immediately after recovery, or definitely before restoration, was needed. Luckily, while the 1998 *Titanic* Exhibition in Zürich, Switzerland, did not display the chronometer itself, a booklet printed for the exhibition contained a photograph of the chronometer before its conservation. Because of copyright rules, we cannot display the actual photo, but researcher Samuel Halpern was able to recreate the image, shown at left.

As can be seen, before restoration much of the glass cover of the chronometer was covered in some sort of rust-covered staining. The dial itself was blackened over much of the face, though most of the maker's name, 'J.W. Ray & Co.' and most of 'Liverpool', can be discerned. Note that the minute hand is completely disconnected and was lying across the face of the clock. The hour hand is pointing somewhere between 6 and 7, indicating a time of approximately 6.45. This has no meaning at all, comparing it to the GMT time we referenced above of 5.13.

In all likelihood, when the chronometer was torn loose from the deck of the ship during the sinking and break-up of *Titanic*, the innards and dial lost any attachment to the real GMT time of the sinking. Since the hands were not attached, we will never be able to prove anything about it, other than that the 12.39 time displayed on the chronometer for the Science Museum showing of May 2003 – and used to argue that the time difference between *Titanic* and New York was one hour and thirty-three minutes rather than the actual time difference of two hours and two minutes – had no basis in reality at all.

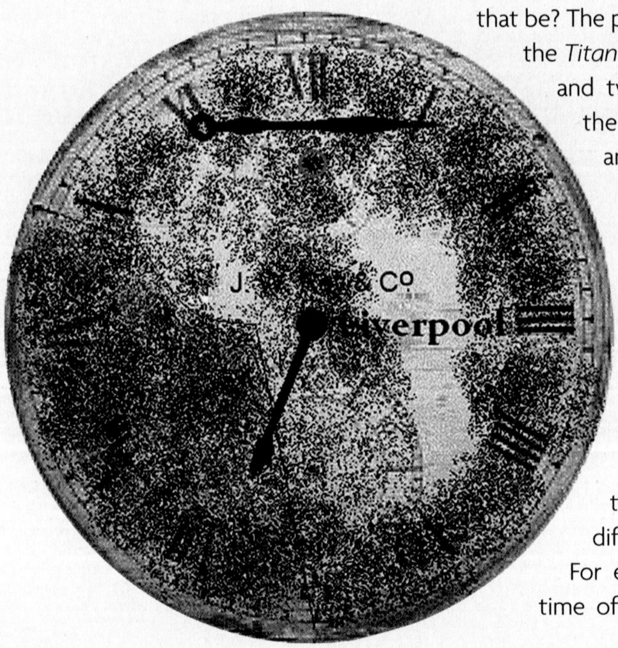


This artistic reproduction based on a real photo shows the face of the recovered chronometer immediately before restoration. The dismounted hands and missing mechanism show that the chronometer cannot be used to prove any times on the night of the disaster. (Samuel Halpern)

On the roof of the Officers' Quarters, Second Officer Lightoller thought of heading aft as well, only to realise the futility of this. Instead, he walked forward onto the roof of the Bridge and dove into the water. He described the cold shock of the 28°F water as being like 'a thousand knives being driven into one's body'. Noticing that the Crow's Nest was coming level with the sea, his first instinct was to swim for that, before he realised that there was no point in grabbing onto anything still attached to the ship. At that moment, the force of water rushing down the intake for a stokehold ventilating fan pinned him against the fidley grate, forward of the first funnel. He felt he was going to be drowned until a rush of hot air from below freed him. He came to the surface sputtering, found himself right next to the upturned Collapsible B, and quickly grabbed a rope and held on.

As the ship's downward plunge continued, water soon penetrated the First Class Entrance and forward Grand Staircase. The staircase may have broken up in the rush of water, with the remains of it floating out of the destroyed glass and wrought-iron dome. Somehow, as the Bridge and forward Boat Deck continued to submerge, Collapsible B was pushed over to the starboard side of the ship, ending up near the starboard side of the first funnel as water began to reach its base.

Suddenly, the funnel collapsed, falling forward and slightly to the starboard side, nearly crushing Collapsible B and landing on a mass of people who were struggling in the water. The wave from this impact pushed Collapsibles A and B further away from the ship, but also washed many of the individuals standing in Collapsible A out of the boat.

Down and down *Titanic* plunged. Screams and shouts echoed across the water as individuals attempted to flee for their lives. When water reached the base of the second funnel, it also collapsed, shooting out a fan-shaped spray of sparks and coal in the process. In the water, Eugene Daly said, 'Those people that covered the water were sucked down in those funnels ... like flies.'

The lights of the ship continued to burn steadily, although even more dullish-red than before, as the power supplied by the emergency dynamos began to fade. The stern of the ship rose higher and higher, until it reached an angle that some believed was nearly 30 degrees.

Next, horrific roars and groaning sounds began to echo out over the water. Some compared these to actual explosions and believed that the boilers had burst. In fact, it was the sound of rending and tearing as *Titanic*'s hull was being subjected to far more stress than it was ever designed to withstand. At about 2.17 a.m., nearly all of the lights on the ship suddenly shut off as the ship violently broke in half.

The stern, still attached to the bow by some fragments, settled back slowly, nearly to an even keel. As it did so, the third funnel collapsed forward and the fourth funnel fell backwards onto the deck. Miraculously, some of the ship's emergency lights still burned dully, even after the break-up.

With the stern section pausing after settling back, it momentarily gave some false hope that it might stay afloat. Atop Collapsible B, Second Officer Lightoller was heard to yell out that the stern would float, and those in the water should swim toward it.

However, the broken end of the stern soon began to flood and the stern was pulled upright. As the Poop Deck and stern rose upwards, the stern section corkscrewed around to port. To some of the individuals who had scrambled atop Collapsible B's overturned hull, it appeared as if the propellers and stern would topple over on them, but they did not. To those in the lifeboats, the nearly 1,500 people stranded on the stern looked like swarming bees, 'clinging in clusters or bunches ... only to fall in masses, pairs or singly', as the stern rose higher.

When the stern reached a near vertical position, it paused for several moments, until enough water flooded in that it began diving down with ever-increasing momentum. As water engulfed the last portion of *Titanic*'s fantail, there was a slight gulp. The dreaded suction proved to be a non-factor, as proven by Chief Baker Joughin, who claimed his head did not even go under water as he rode the stern down to the surface.

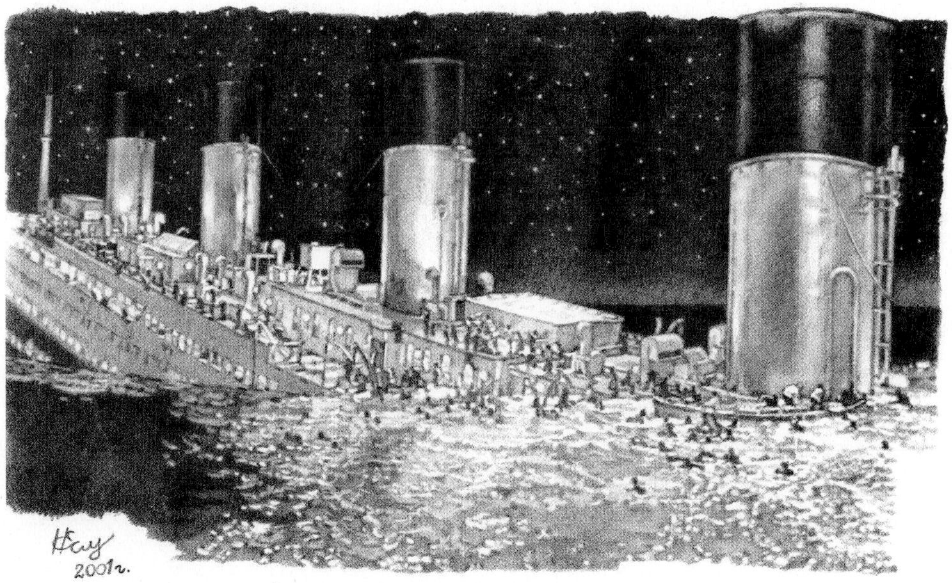

As the base of No. 1 funnel begins to submerge, hundreds of people struggle for their lives in the maelstrom. (Eugene Nesmeyanov)

## THE 'SLIGHT BUT DEFINITE PLUNGE', THE COLLAPSE OF THE FUNNELS AND THE BREAK-UP

### THE 'SLIGHT BUT DEFINITE PLUNGE'

Late in the sinking, at about 2.15 a.m., *Titanic* was poised with her Bridge and forward Boat Deck just above the sea. Jack Thayer, standing on the starboard Boat Deck behind the empty davits for Boat No. 7, just forward of amidships, recalled that the 'list to port had been growing greater all the time'. Due to the list, the water seems to have been on the port Bridge wing and forward Boat Deck, but not on the starboard side.

Just then came a 'tremble' that ran 'through the whole of the ship', followed by 'loud reports inside' that some likened to a volley of musketry. Some eyewitnesses guessed that either a watertight door or a bulkhead had failed. Jack Thayer recalled that suddenly the ship 'straightened up on an even keel and started to go down fairly fast'. Others described her movement as a 'plunge' or 'lurch', and a wave washed aft along the Boat Deck, swamping everyone in the area and washing them into the sea. Clearly, water had shifted suddenly inside the ship, changing her centre of gravity.

Then came an odd phenomenon that has been all but forgotten but was clearly reported by those on the starboard Boat Deck forward near Collapsible A. The forward Boat Deck rose out of the water very briefly. This exact phenomenon was described by Richard Norris Williams, William Mellors, Eugene Daly, Cecil Fitzpatrick and others.

Initially these reports make little sense, but *Titanic* was then balanced precariously like a see-saw. The stern remained buoyant while the flooded bow had just taken a quick nosedive as her balance shifted. It seems that the ship had temporarily found her equilibrium, recovering very slightly before further flooding took back over and the downward plunge resumed. A good analogy might be when one brakes heavily in a vehicle: the nose of the car dips, but if the brake pedal is suddenly and completely released, the nose bobs back up. Whatever the cause for this slight recovery, it did not last long, and the bow began to plunge once again.

### THE COLLAPSE OF THE FORWARD FUNNELS

Recent depictions of the *Titanic*'s final minutes have shown the ship with a strong port list from the final five to ten minutes right through to the stern's final dive. However, these depictions contradict eyewitnesses like Thayer, who said the ship returned to something like an even keel at about the time of the plunge. Indeed, the fact that the ship was more or less on an even keel played a pivotal role in how the next events in the disaster played out.

The collapse of *Titanic*'s four towering smokestacks actually killed some people in the water. But questions have long surrounded the way that each funnel fell. We can establish what led to the collapse of the forward two funnels. Neither of the two after funnels fell because of the angle of the ship, so there must have been a common influence on both forward funnels to cause their collapse. Harland & Wolff's Edward Wilding testified that when the upper decks:

> … became submerged and the water got inside the house, the water would rise outside much faster than inside, and the excessive pressure on the comparatively light [funnel] casings which are not made to take a pressure of that kind would cause the casing to collapse; would take the seating from under the funnel and bring the funnel down.

Looking at the shreds of tangled funnel remnants on the wreck, there seems to be some support for this notion. In truth, the parting of the funnel stays was caused by the funnels' collapse, not the other way around.

Which way did the forward two funnels fall? Recent depictions have them falling toward the port side. A pair of davits on the port side of the Bridge on the wreck have often been referred to as forensic evidence of this, but there are many davits scattered around the wreck site and we simply cannot say what happened to those davits after the ship left the surface. The davits cannot be used as reliable evidence that the funnels fell to port. We also know that a portion of a funnel steam pipe became lodged on the aft railing along the starboard side of the Forecastle. This should not surprise us, because multiple eyewitnesses actually reported that the forward funnel fell toward the starboard side rather than to port, and that it also fell toward the bow of the ship. How can we have such a clear picture?

We know that it nearly took out Collapsible B, which had managed to work its way around to the starboard side of the ship in the turbulent waters. To do this, the boat had to cross from one side to the other from the port Boat Deck where it had fallen upside down. Victor Sunderland, who was on the overturned boat at the time, confirmed this. He said, 'We passed under the forward funnel and just as we were clear, it fell.' Fireman John Thompson was on half-swamped Collapsible A, washed off of the starboard side of the deck, and he recalled that, as the funnel came down, it caused 'so great a wash that our boat was sent spinning and I was knocked violently against one of the davits'.

We also know that when he jumped overboard, Lightoller first swam forward toward the submerging Crow's Nest, then turned to starboard and swam away from the ship. Further, he specified that the forward funnel fell to starboard, nearly hitting him as it came down, landing right beside Collapsible B, and throwing him and the boat clear of the ship.

Similarly, Seaman Albert Horswill later reported that the forward funnel had 'fallen into the water on the starboard side with a terrific splash, unquestionably doing a great deal of injury to people struggling in the water'. Richard Williams, who had gone into the water with his father from the starboard deck, watched as this same funnel crushed his father, just a few feet away. Without any doubt, the forward funnel fell to starboard and forward.

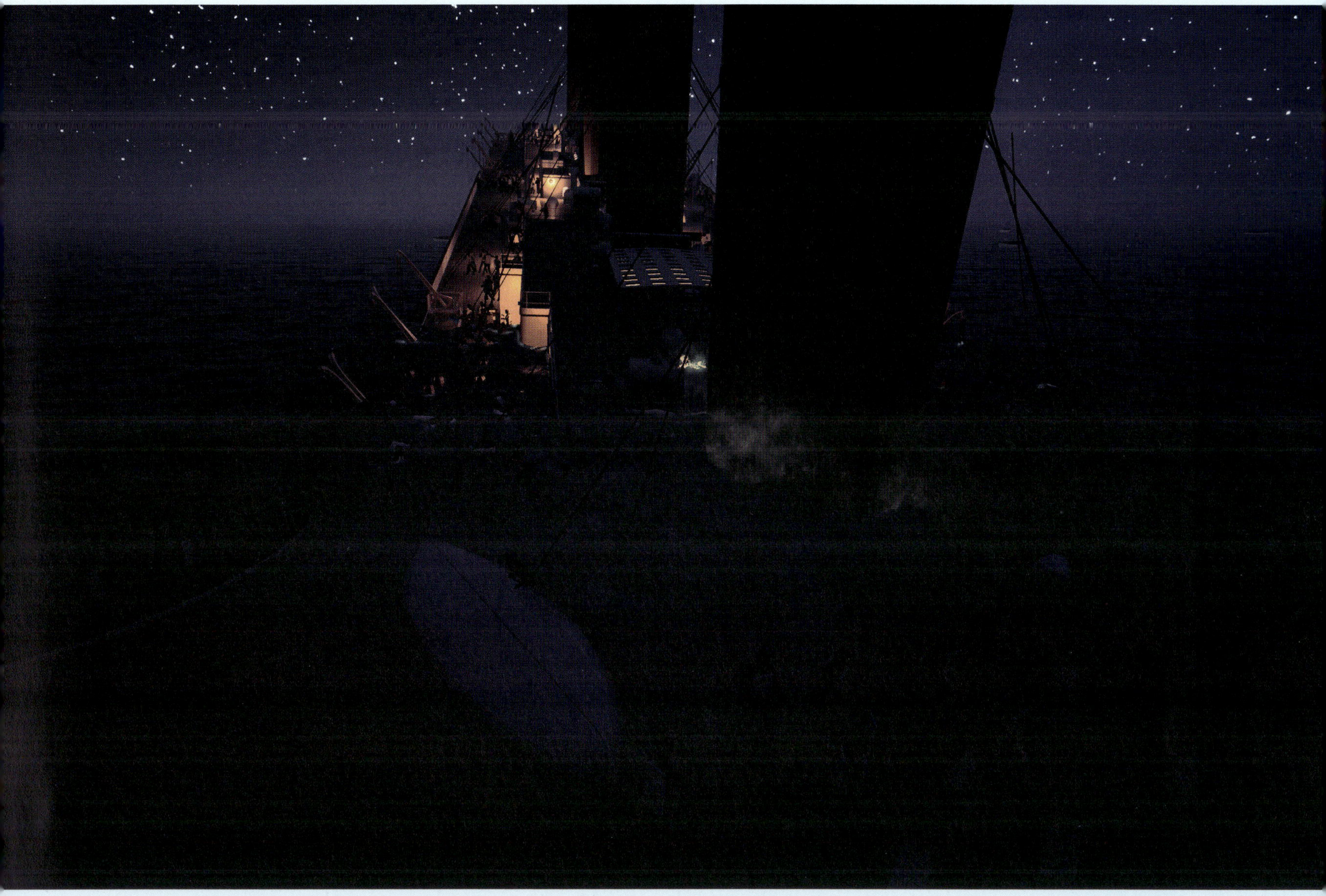

The second funnel is even easier to understand: Jack Thayer had gone over the starboard rail of the Boat Deck near midships. As he was trying to get away, he looked back, 'and the second funnel fell and missed me by about ten yards. This funnel … made a tremendous additional wash and suction. I was drawn down again.' He also recalled that it came down 'emitting a cloud of sparks'. Steward Percival Keen also saw the second funnel break off, noting that it killed a number of people in its fall. If the second funnel had fallen to port, it would not have nearly crushed Jack Thayer.

We can also say that the two forward funnels did not fall simultaneously, but rather sequentially as water pressure around the base of each funnel built to the point where it collapsed the thin plating of each and more or less brought them down under their own weight. Since the forward funnel was both further forward and also deeper in the water than the second, because of the ship's forward trim, there is no question that the forward funnel fell first, followed soon by the second. However, this does not mean that a great deal of time passed between the two events, because the forward decks were then submerging rather quickly.

As the forward funnel begins to fall forward and to starboard, Collapsible B has already been washed around to the starboard side, just clearing the funnel before it comes down. (HFX Studios)

*Right:* This terrifying view shows the perspective from Boat No. 2 as *Titanic*'s stern rears high into the night sky at about 2.16 a.m. The No. 2 funnel has just collapsed into the sea. (HFX Studios)

*Below:* Also at about 2.16 a.m., this view shows the Poop Deck crowded with passengers and crew from all walks of life as they unite to face a common catastrophe. Forward, the remains of the No. 2 funnel can be seen slipping beneath the waves. (HFX Studios)

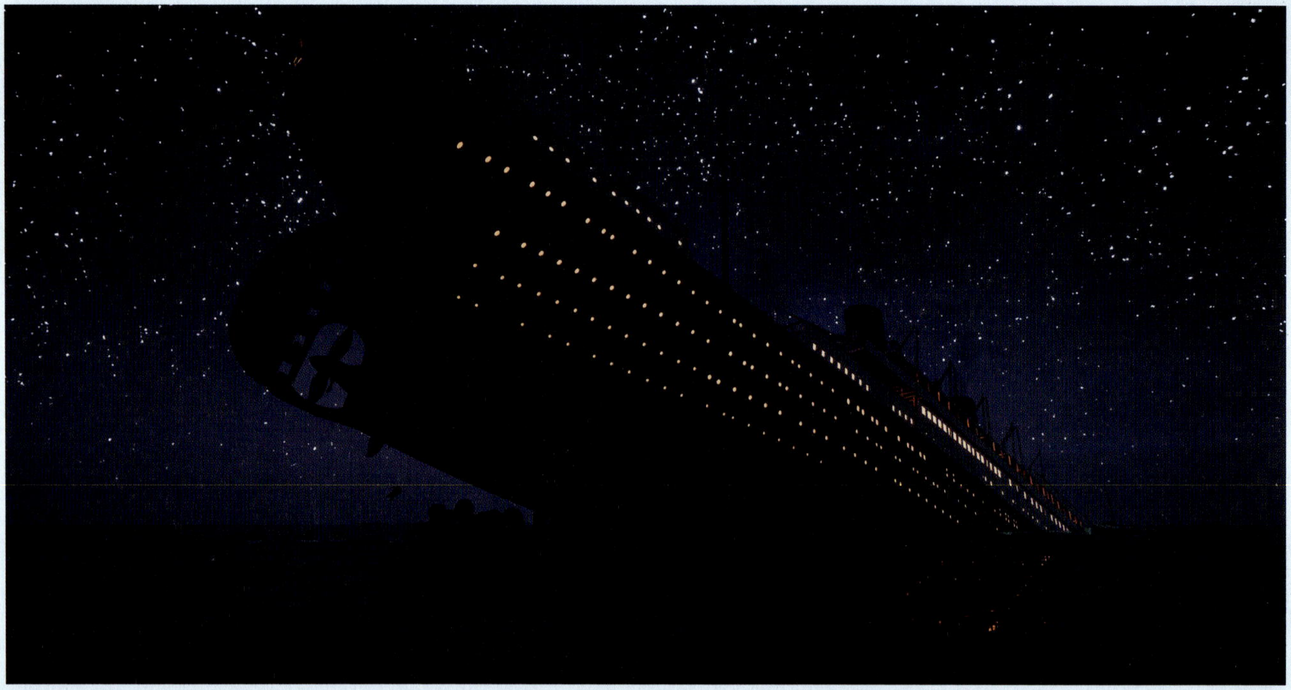

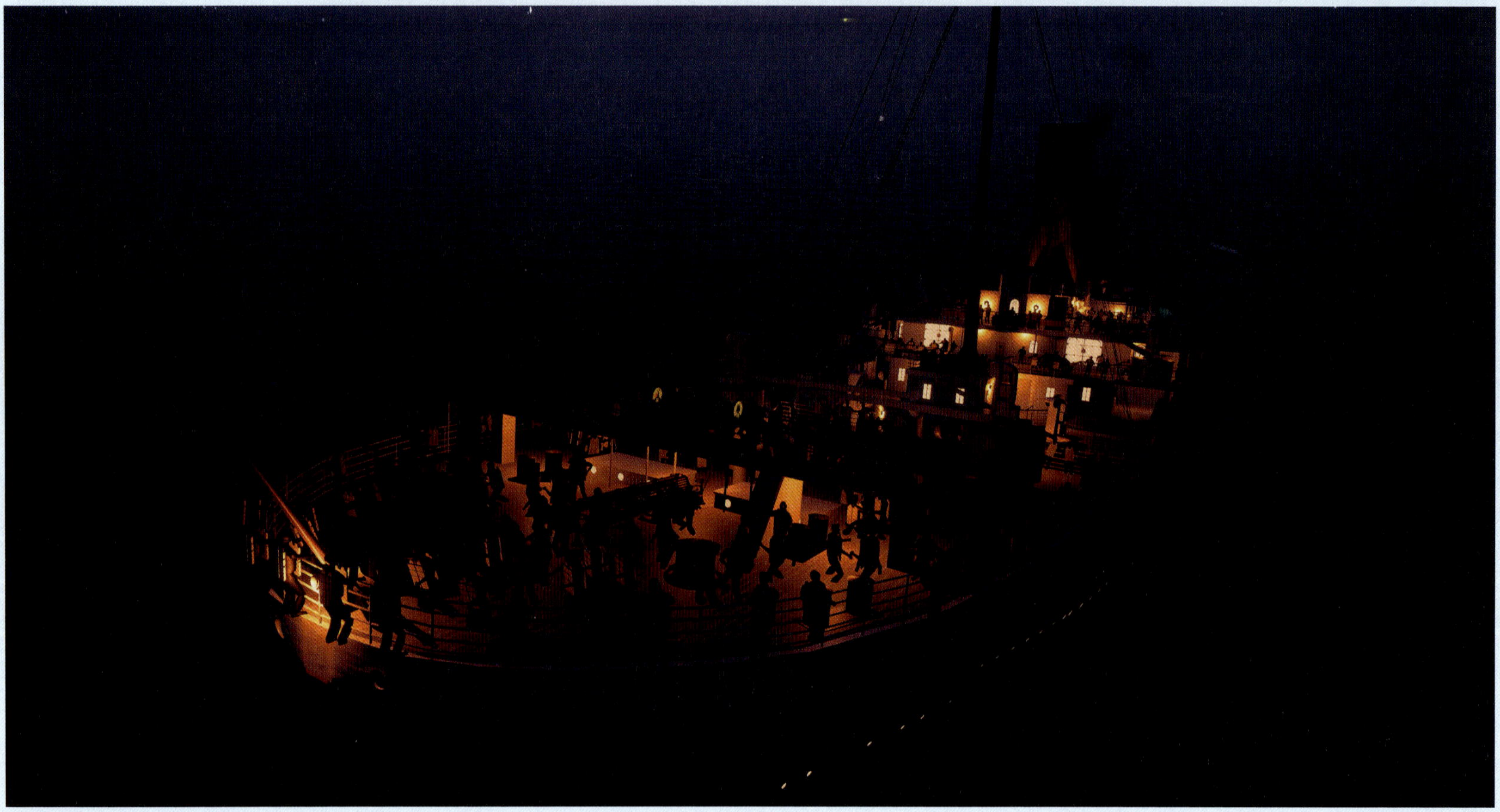

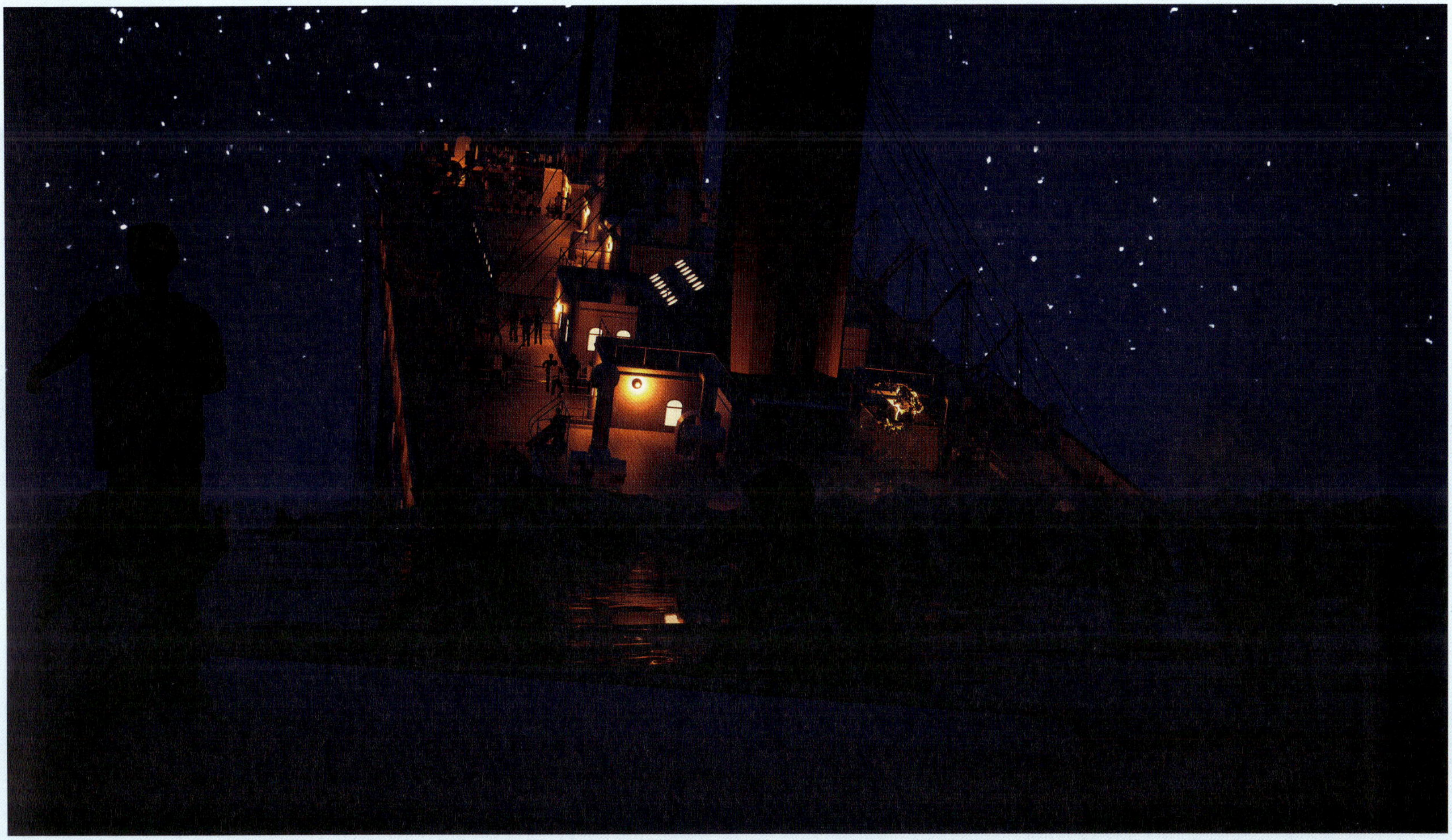

This is the perspective that Second Officer Lightoller and others on Collapsible B had just moments before the ship broke. Lightoller correctly remembered that the No. 3 funnel was then out of the water. (HFX Studios)

**THE BREAK-UP**

Now that we have established how the forward two funnels fell, we can discuss the break-up. It is amazing how many survivors said that the ship broke on the surface while it sank. During the last two decades, we have been compiling survivor accounts, tabulating their reports and comparing the eyewitness evidence with other data.

Since the wreck's discovery, there have been numerous ideas and theories on how the ship broke up and sank. Forensic evidence at the wreck site alone is easy to misinterpret because it is difficult to tell what happened to the pieces of wreckage as they sank to and collided with the sea floor versus what damage was done to them during the ship's break. Cherry picking eyewitness accounts to support preconceptions is also a terrible idea, as people tend to find only what they are looking for to support their pet theories.

Computer simulations, forensic analyses and stress calculations, on their own, are subject to the old axiom of 'garbage in, garbage out', and frequently do not take into account the complete picture. The only sort of analysis that has any hope of accuracy is to take all of these disparate lines of evidence and begin to allow the pieces to fall into place, and then to accept the preponderance of evidence.

As the ship continued its downward plunge, the buoyant and unflooded stern was correspondingly lifted higher and higher from the water, exposing all three propellers, the entire rudder and some of the ship's bottom hull at the stern. How high an angle did the ship attain? Quite simply, we can't say for sure.

Many survivors spoke about high angles, with some specifically referring to angles of 30 degrees or greater. However, it can be very difficult to judge such angles unless one is practically broadside to the ship as it sank, and most of the lifeboats were at oblique angles either fore or aft and to one side or the other.

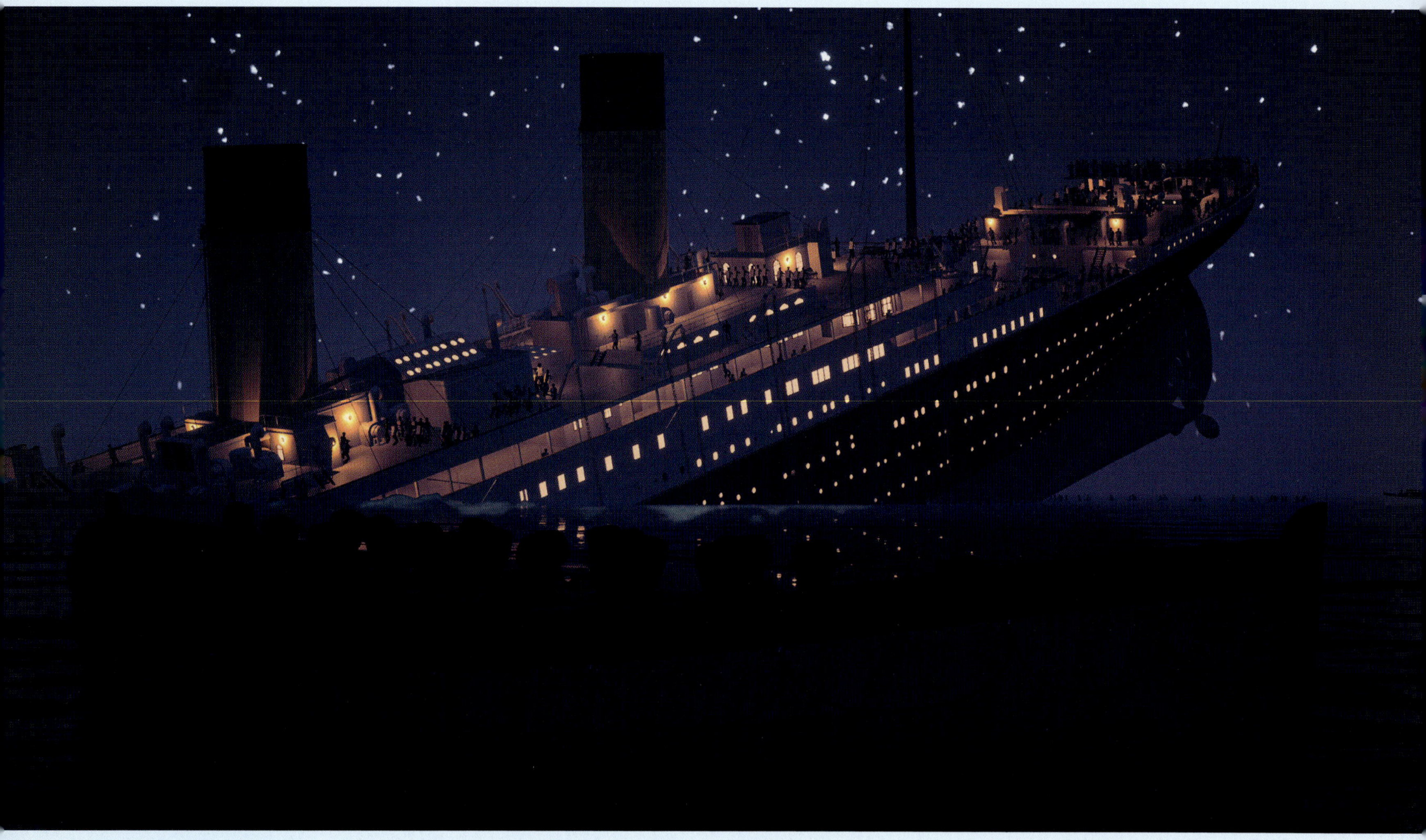

Boat No. 6, carrying First Class passengers Margaret Brown and Major Arthur Peuchen and crewmen such as Lookout Frederick Fleet and Quartermaster Hichens, lies in the foreground as they face the horrifying spectacle of *Titanic* sinking. (HFX Studios)

Stress calculations indicate that the maximum stresses imposed on the ship's hull form would have been attained at a forward trim of about 15 degrees. After that point, increasing angles would have brought a gradual reduction in stress on a descending curve. However, we also know that for a number of minutes prior to the break-up, the ship's hull was giving audible signs of suffering high stresses, even before the ship attained that angle of forward trim. Quite simply, no ship was ever designed to withstand such tortures. As Boat No. 4 moved aft along the port side of the ship, some of the boat's occupants reported hearing 'a cracking noise resembling china breaking'. At a similar time, Chief Baker Charles Joughin was in the A Deck Pantry, just aft of the Lounge, and he reported a 'kind of crash as if … part of the ship had buckled', which he thought sounded 'as if iron was parting'.

In all honesty, there is plenty of unbroken china left stowed in the cupboards within the wreck. Although some of this crashing could have been china or other breakables within the ship, one would not expect eyewitnesses in such different locations to have remembered this sound so clearly if it were merely dishes breaking. More likely, Joughin was correct when he said it seemed as if something within the ship was giving way. Even more interestingly, Joughin was then standing in the area that was about to be torn to shreds as the ship broke up.

Clearly, *Titanic*'s structure was suffering from the strains being imposed on it minutes before she actually broke. The visible break-up was only the grand finale of a series of ongoing stresses and failures of certain components of *Titanic*'s structure. The strain was being taken up by other structural members, but then, as the stern's angle reached above that peak stress angle of 15 degrees, the strain began to decline. However, things were already past the point of no return.

What happened next was astounding for eyewitnesses. The main structural failure took place 'just in front of the third funnel', as Jack Thayer said. Thayer was remarkably observant, for even now on the wreck we can see the precise location of this break at the aft end of the bow section. It took place just in front of the No. 3 funnel uptake, opening the top decks wide and tearing through the aft section of the First Class Lounge, and this break went nearly straight down to the double bottom.

Even though he later claimed that the ship sank intact, Lightoller did recall, with his uncannily selective memory, that 'the whole of the third funnel' remained visible at about that time. We also know that a number of eyewitnesses said that the break occurred in the vicinity of the Lounge, or just abaft of amidships. As the upper decks here opened up widely, a fan-shaped shower of sparks flew into the night sky, brilliantly illuminating the scene.

However clean this break was, there was a lot that was going on behind it as part of the break-up. The primary hull aft of the main break was clearly coming apart as well. For example, the 'Big Piece', a large section of shell plating on the starboard side across from the Reciprocating Engine Hatch and some distance behind the main hull break, could only have been shredded from the surrounding structure as a function of the ship's break-up. Nothing else could explain the way the incredibly strong, double-thick steel plates were torn like a sheet of paper. This helps demonstrate that, in simple terms, different portions of the structure were being pulled this way and that, in essence, deciding which half of the ship they were going to stay with, or whether they were simply going to peel away from both structures entirely.

Furthermore, while many eyewitness accounts place the main break of the ship around amidships, some reported the ship broke by the aft expansion joint, or even as far aft as behind the fourth funnel. At first glance, these disparate statements seem to make little or no sense. However, they actually lead to a reliable conclusion supported by forensic data on the wreck: the ship's break-up on the surface was far messier than the main break alone would indicate.

Structurally speaking, *Titanic*'s primary hull was something like a box girder formed by the keel, the ribs and side shell plating, and at the top, B Deck formed the 'strength deck'. Where this deck met the sides, special reinforcing 'doubler' plates added extraordinary strength to the shell plating and helped to hold the entire box together.

Everything that sat atop the floor of B Deck was referred to as 'superstructure'. It was made of comparatively lightweight steel, more or less riding atop the hull, and carried none of the stresses that the hull encountered as the ship worked through heavy seas. To allow the superstructure to 'flex' as the hull beneath it flexed at sea, it was built in three segments separated by expansion joints, rather than in one long structure. These joints ran from port to starboard across the ship and on each deck. As *Titanic*'s primary hull came apart beneath the superstructure, the comparatively light upper works shredded in a breathtaking frenzy of destruction.

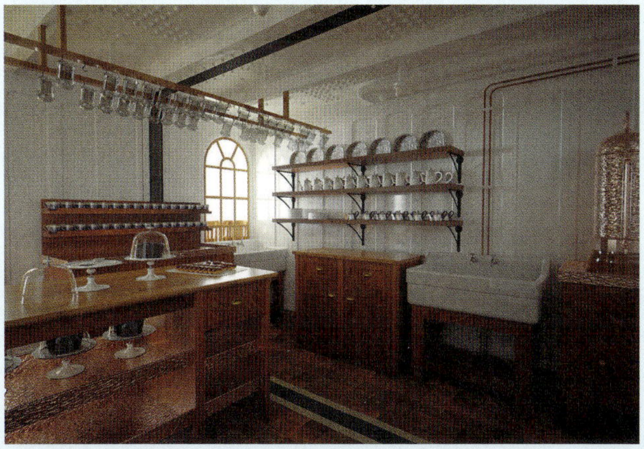

*Left:* Just a couple of minutes before the break-up, Chief Baker Charles Joughin stepped into this pantry on A Deck, just behind the Lounge, for a drink. He heard something metallic give way deep within the ship and left quickly. Moments later, this room was torn apart in the break-up. (Chris Walker)

*Below:* A fan-shaped shower of sparks erupts as *Titanic* splits in two. The stern is settling back, the forward 'tower' is breaking off, the No. 3 funnel collapses forward, while the No. 4 funnel falls astern. Some people at the rails near the break-up plummet into the ship. (HFX Studios)

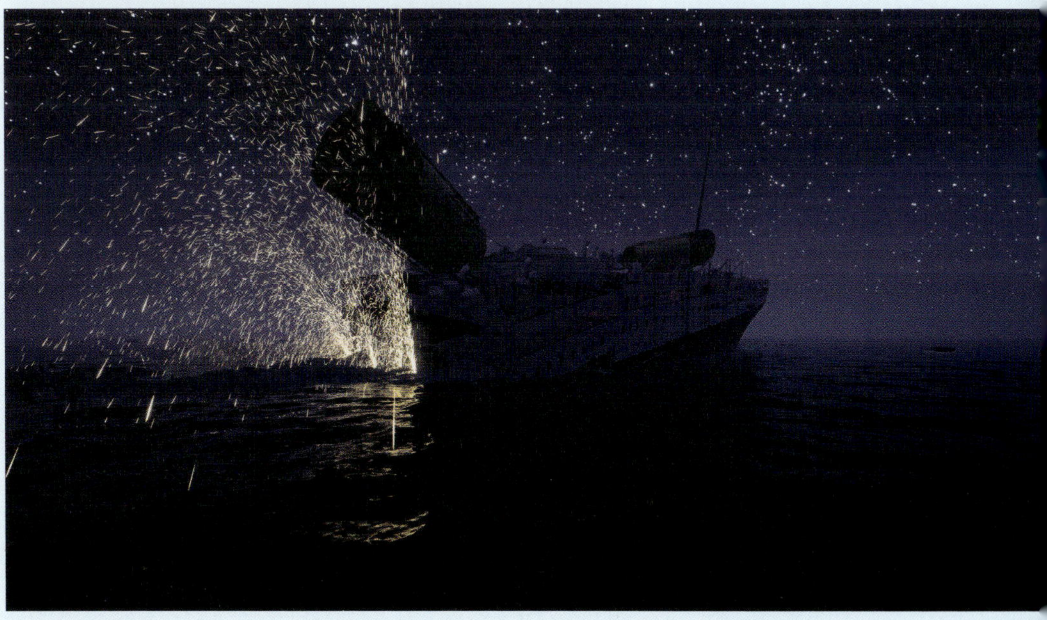

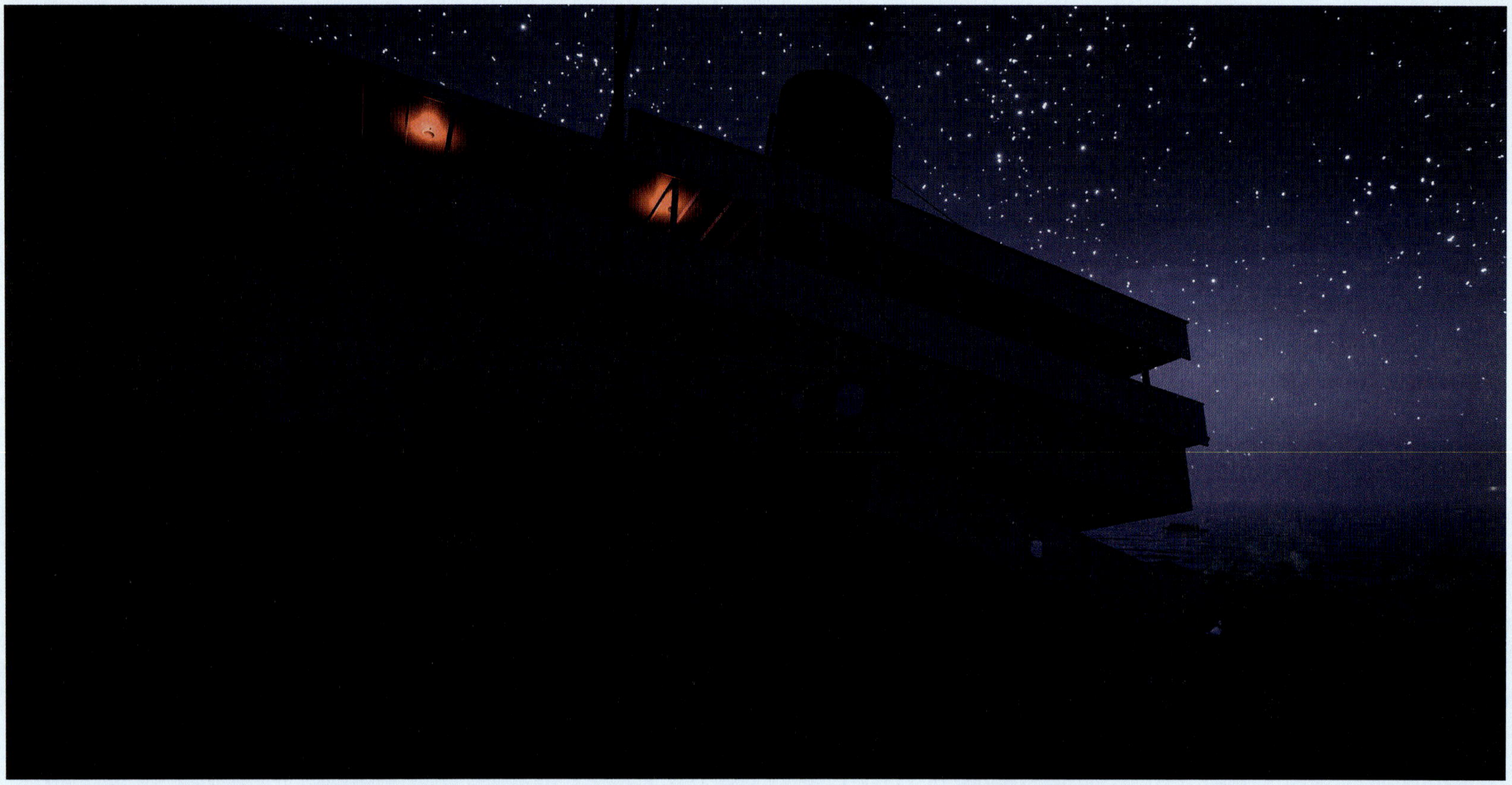

*Above:* On the starboard side, the carnage taking place along the ship's side is clear. An enormous 17-ton section of hull plating from C and D Decks, later termed the 'Big Piece', is shredded from the side. Some emergency lights continued to burn following the break-up. (HFX Studios)

*Right:* An internal and external view of the 'Big Piece' as it looked immediately after being torn from the ship. (Michael Brady)

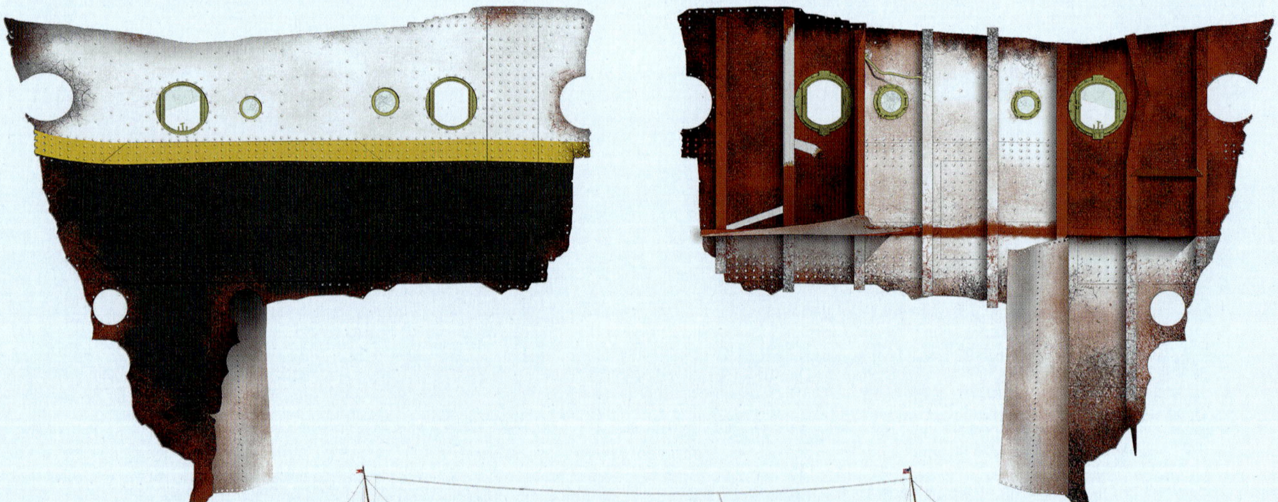

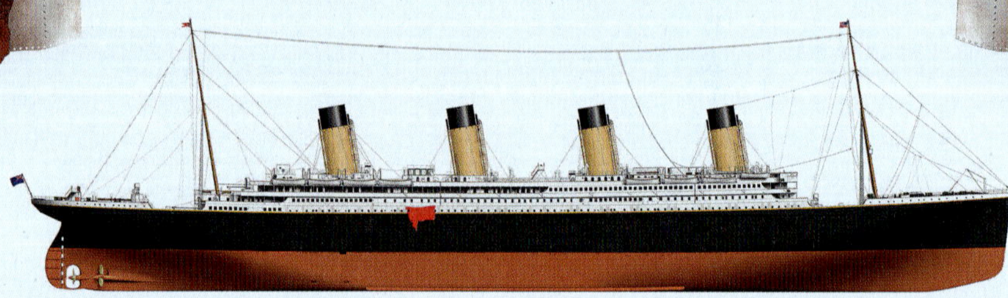

The superstructure between the bow and stern portions of the wreck came apart in two main segments, or 'towers'. Their dividing point was close to, but not precisely located at, the aft expansion joint. Each section fell away from the ship as enormous chunks; the after one is rather intact today, about 300ft from the stern. Debris from the forward section, just behind the main break in the vicinity of the third funnel, is further away from the stern. Because eyewitnesses reported seeing the ship break in the area that divided these two towers, and some of them also reported some odd visual phenomenon that would not make sense if the entire ship's structure was making that kind of motion, it seems that the forward tower came apart at the surface and visibly detached from the stern. Nothing else would explain the forensic data and the eyewitness reports.

It is also very likely that the aft of these two towers was similarly 'displaced' and badly damaged, enough so that people reported that the break occurred closer to or even behind the fourth funnel, yet it seems that it was better connected to the main aft portion of the ship and remained with it for some distance after leaving the surface. This tower is located closer to the stern than the other debris.

Meanwhile, the forces being exerted on the ship's stern hull were literally shredding it. Two enormous sections of the double bottom and keel tore apart from the main sections of the wreck, likely during the main break. This event dislodged the boilers in Boiler Room No. 1, which fell more or less straight to the bottom. Their location on the sea floor today, 2.5 miles below where the ship broke apart, helps us to pinpoint where the ship was when this event happened even more closely than the main wreck sections on their own could. As the aftmost of these two pieces tore away from the stern section, it broke right through the reciprocating engines, ripping the forward cylinder off of each.

It is clear that during this horrendous carnage, the third and fourth funnels fell. The third funnel almost certainly fell forward during the separation of the forward 'tower', and the fourth funnel was reported to have fallen astern but remained close to its original location until the area sank from sight. It remains very near to the stern on the sea floor today.

The stern settled slowly back into the water and the sequential loss of support seems to have acted as a 'brake' on its collapse into the sea. As it came down, one – and only one – survivor reported that it canted heavily to port, flinging many of those still on the ship against the port rail.

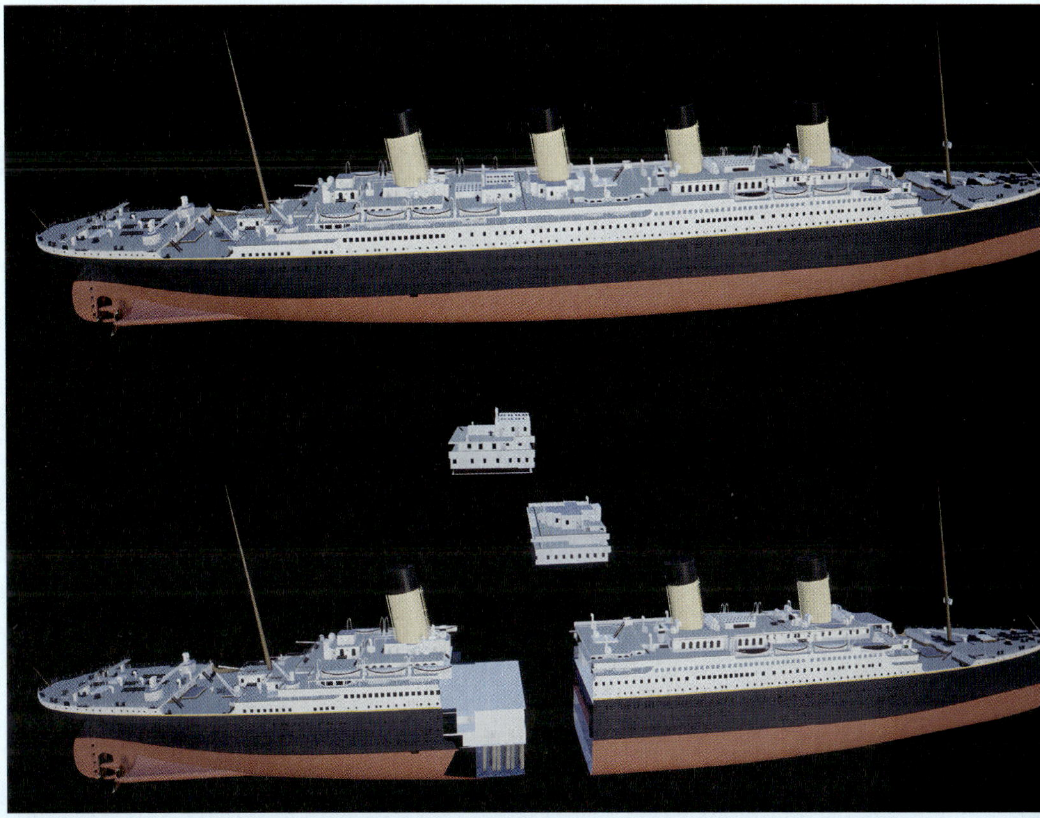

This illustration shows *Titanic* intact and the two major sections after the break-up. The two tower segments that came apart at the surface are shown as separated. (HFX Studios)

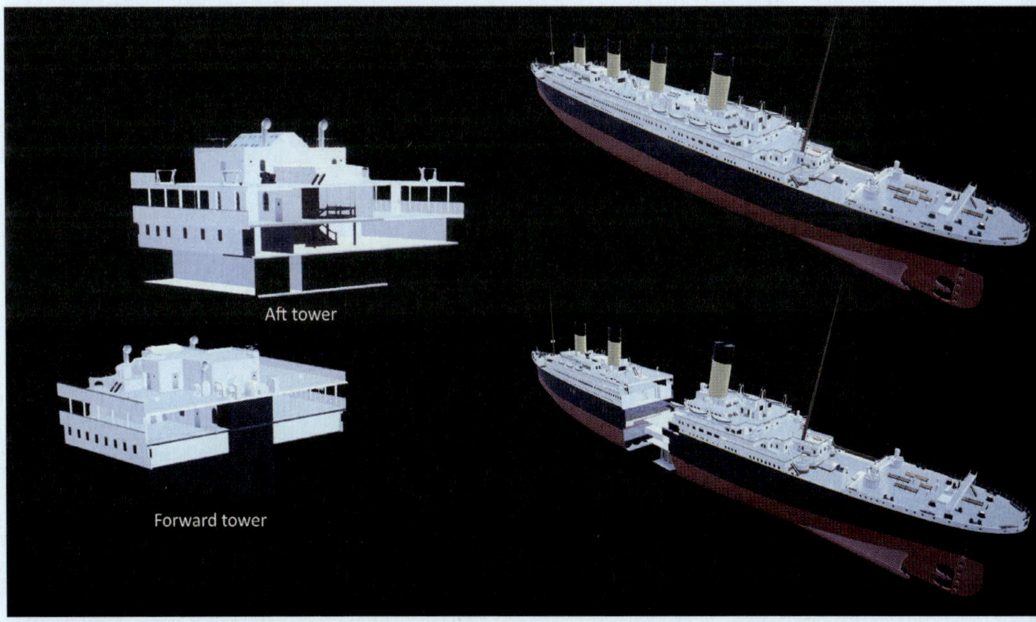


This view on the port side looking forward shows the two tower segments, in rough terms, and where they came from. (HFX Studios)

This was Chief Baker Joughin. However, other survivors, some of whom were also on the stern at the same exact time, did not report a strong port list just then. Joughin also reported that the stern never recovered from this port list, whereas a strong port list would be ruled out by what survivors reported happened next.

**THE FINAL PLUNGE**

At this point, with the stern settling back to something like an even keel, many believed that it was going to float. However, that was not to be. After it had remained level for a brief period, it began to tip forward toward its mangled front end.

Not all of the factors are known as to why the stern tipped forward and sank as quickly as it did. Certainly, the weight of the remains of the reciprocating engines and the turbine likely helped speed that motion up and what was left of the Reciprocating Engine Room would have flooded nearly instantly once its forward end broke open. Considering the trauma that the hull was experiencing, it is likely that water would soon have made its way aft, one way or another.

However, the speed at which the stern tipped forward and down indicates something else was also happening. Considering the fact that it also began corkscrewing around in an anticlockwise motion, if viewed from above, it seems that the stern may still have been connected to the flooded bow section, which had disappeared immediately after the break. What held them together is rather difficult to say with certainty, because looking at the wreckage today, it is often very difficult to interpret what damage happened to certain pieces during the descent to the sea floor and impact with it, and what damage had occurred as a result of the break-up of the ship. We also know that the bow did not rotate in the anticlockwise manner that the stern did, since it remains pointed largely in the direction it had been pointed in when it left the surface.

As the stern raised the propellers and rudder high into the sky and began to rotate anticlockwise, if there had been a strong port list that was maintained through the final plunge, it would have become very obvious to multiple eyewitnesses watching from a variety of vantage points. Furthermore, if the stern's angle of final descent was not very steep, then not everyone would have reported that it attained a nearly vertical position.

By all accounts, the stern sank rather slowly in a vertical dive, but some remembered that its downward momentum began to build as more and more of it became submerged. Footage of sinking ships suggests that these final moments are usually accompanied by jets of air escaping from every conceivable open port, hatch or vent, because air is literally being forced out of the hull under pressure as the seawater rushes in.

In the end, it seems likely that the stern rotated more than 90 degrees to port (anticlockwise, as viewed from above) as it rose up and then dove into the depths. After the ship left the surface, there were reports of one or more sounds that seemed like underwater explosions. Frighteningly, it seems that these could have been the sounds of portions of the stern imploding as water pressure increased around pockets of trapped air within the wreck.

What exactly happened to *Titanic* when she left the surface is a forensic mystery of outstanding proportions. There are many hotly debated theories out there. However, we can paint a very clear scene of what happened on the surface because, in addition to the forensic data, we have strong eyewitness testimony from multiple vantage points.

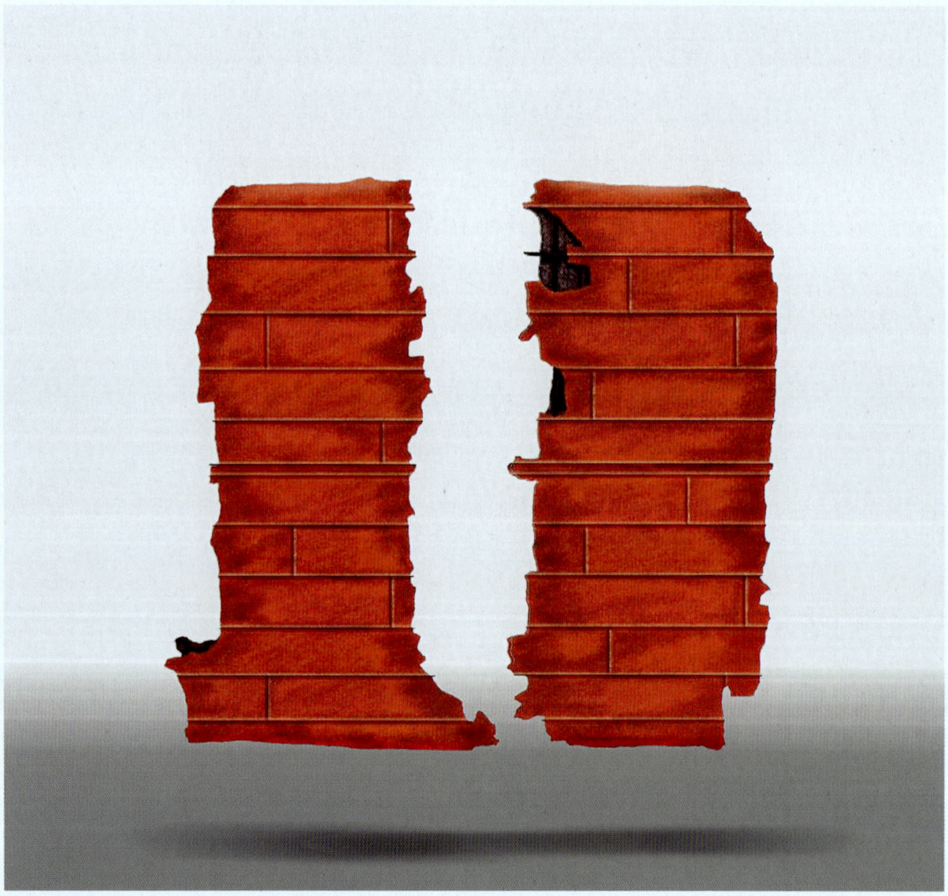

A rendering of the two 'missing' segments of the double bottom. When the aft segment tore off, it took the forward cylinder of each of the four-storey, 1,300-ton reciprocating engines with it. Boiler Room No. 1 disintegrated, and its five single-ended boilers fell almost vertically to the sea floor. (William Barney)

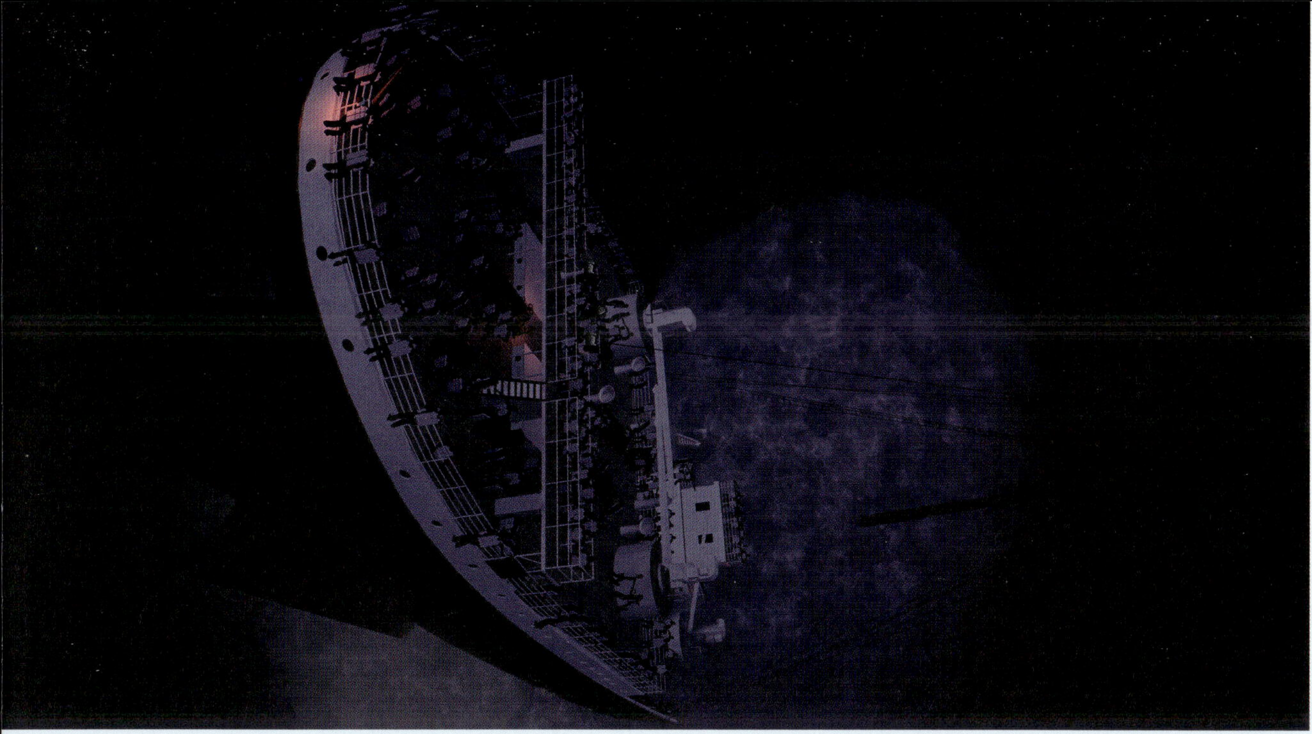

*Left:* Despite temporarily giving hope that the stern would float, it soon heaved back up into the night sky. This view looks down as the Second Class Decks slip beneath the sea. (HFX Studios)

*Below:* This selection of six images, from the perspective near Collapsible C, shows the stern lifting up and rotating anticlockwise before finally disappearing at 2.20 a.m. (HFX Studios)

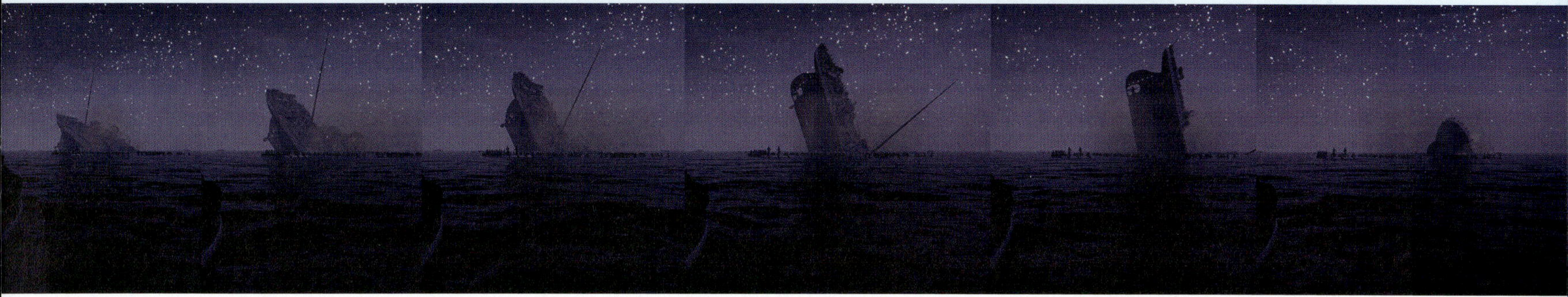

It was 2.20 a.m. As the ship disappeared, a mushroom cloud-like blanket of steam or smoke billowed outward, engulfing those close to where the stern went under in a supernatural haze. Those in the lifeboats sat in silence, stunned that the ship was gone. In Collapsible C, Bruce Ismay had turned away from *Titanic* in its final moments, as he 'did not wish to see her go down'. The quiet lasted for only a few moments before the roughly 1,500 individuals who were in the water began to scream and yell for help in a terrifying cacophony of noise.

The heartrending cries and pleas from the water continued for some time. With the water only being 28°F, all but the hardiest individuals, whether they were buoyed up by lifebelts or not, would last only fifteen to forty-five minutes before exhaustion or even unconsciousness set in. For those in the lifeboats listening to the chorus of cries, it must have seemed an eternity.

The sound of those freezing to death and drowning in the waters would haunt everyone who heard it for the remainder of their lives. Within an hour, nearly everyone in the water had perished, and an eerie silence spread across the ocean, which was as still as glass. The only noises at this point were the cries of those in the boats, who suddenly realised both the enormity of what had happened and what the growing silence meant.

In the immediate aftermath of the sinking, there were some who protested or argued in favour of rowing back to the site of the sinking, to attempt to rescue some from the water. In Boat No. 5, Third Officer Pitman suggested this and was barraged with panicked pleas not to do so. Similar scenes played out in other boats, such as No. 1, which had just twelve occupants, as well as No. 2, with just seventeen aboard, and Nos. 8 and 9. In Boat No. 6, Quartermaster Hichens continued to be belligerent with the other occupants of the boat. When they suggested rowing back, he was unyielding in his refusals, insensitively saying there were only a 'lot of stiffs there'.

Boat No. 12 would attempt to row back for survivors, but when they got close to the site of the sinking, they only saw 'a couple of hundred deck chairs'. The occupants in Boat No. 4 were spared the decision about whether to act or not. The boat had been so close to the ship when it went down that some individuals, such as Greasers Ranger and Scott, had dropped down the falls to get into the boat, while some others, seeing the boat just off *Titanic*'s side, had jumped for it and swam towards the craft. The occupants in No. 4 rowed back towards them and were able to pull around eight individuals from the water. Even though they had only been in the water for a brief period of time, several of these individuals passed out from hypothermia, and some died.

Fifth Officer Lowe proved himself to be the hero of the night when he assembled a flotilla of lifeboats consisting of Boat Nos. 14, 12, 10, 4 and Collapsible D. He would redistribute occupants between the boats, and then, after approximately an hour, rowed back and was able to pull three to four people from the water, one of whom later died. By Lowe's own estimate, he had waited too long.

Lowe and the other men manning Boat No. 14 experienced a horrific scene as they attempted to rescue more survivors, passing through a large field of floating dead. Lowe proved himself a very resourceful sailor during the night, being the only crewman who managed to get his lifeboat's sail erected, sailing around the scene as morning approached and towing Collapsible D.

Besides those who were actually in the water, those who clambered into Collapsible A or who climbed aboard upturned Collapsible B had by far the most difficult time during the night. When Collapsible A was washed free from the ship, it initially had around twenty to thirty individuals aboard. However, because the canvas sides of the boat could not be raised, icy water sat in the bottom of the boat, up to the knees of those aboard. Every time more individuals tried to climb aboard, the instability of the boat caused it to tip and dump most of the occupants overboard, and each time this happened, fewer and fewer had the strength to get back aboard. Some individuals in the boat succumbed to hypothermia and fell or were tossed overboard during the night to lighten the boat. The survivors in this boat were eventually rescued by No. 14, by which time there were only around twelve or thirteen still alive.

Similarly, over thirty individuals managed to climb atop Collapsible B's keel during the night. At first, some of the individuals on the craft prevented others from climbing aboard, fearful of sinking. At least one individual was hit over the head with an oar when he tried to climb up. As with Collapsible A, multiple individuals succumbed to hypothermia during the night, eventually dropping overboard.

Second Officer Lightoller took charge of the boat, however, and when waves began to pick up in the morning and caused the collapsible to rock and begin to settle, he organised the survivors to lean this way and that, against the waves, attempting to keep them level and afloat. Harold Bride and Eugene Daly were among the twenty-eight to thirty individuals who survived until Boat Nos. 4 and 12 rescued them.

There is some compelling evidence, although not definitive, that Captain Smith approached Collapsible B during the night, and seeing that there wasn't room to climb aboard, stayed nearby, encouraging those aboard, saying, 'Hello boys, keep calm, boys.' Before swimming away and becoming still, he reportedly said, 'All right, boys; good luck and God bless you.'

## LIFEBOAT DISTANCES AND LOCATIONS

As we were working on our videos for the 2021 anniversary of the sinking, we created a chart (opposite) showing the locations of all the lifeboats in comparison to *Titanic*, and where they would have been at the time the ship sank.

As we worked at the details, we found many discrepancies between the survivors' accounts. One person would say a particular lifeboat was ¼ mile away from *Titanic*, another person in the same lifeboat would say ½ mile. We found cases where a lifeboat was having difficulty being rowed decently, but was still said to have moved quite a way from the sinking ship. We decided we needed to pay more attention to what a reasonable distance was, and mainly concentrate on how the lifeboat moved before the sinking, and in which direction, particularly since estimates of distances at night, without any specific frames of reference, are very subjective.

All lifeboat lowering times are approximate, and reflect when the lifeboat started to lower from the upper decks.

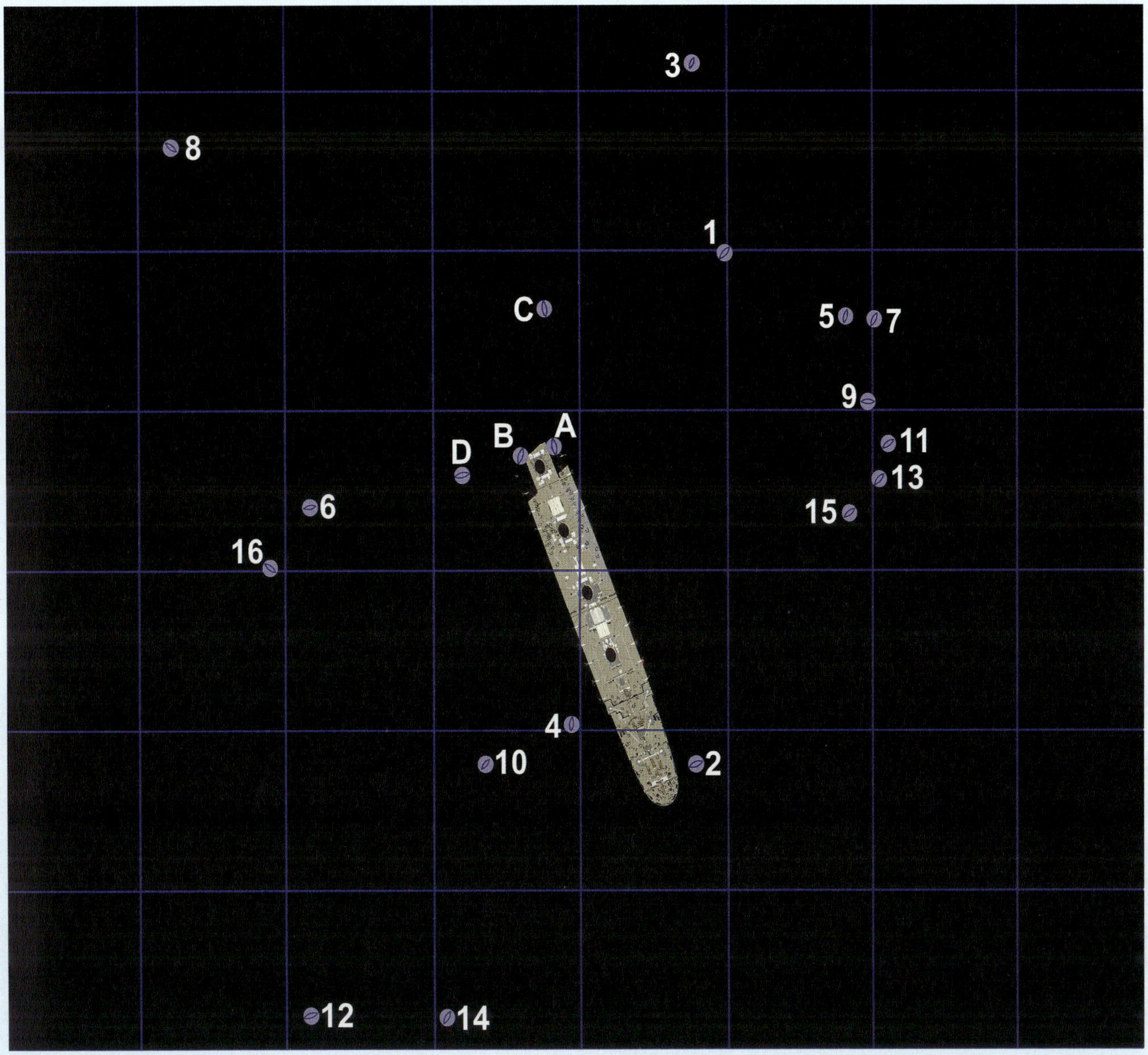


This chart shows the approximate positions and distances of *Titanic*'s lifeboats at 2.15 a.m. Each square on the grid represents 100 square yards and due north is straight up. (HFX Studios)

**NO. 1**

When lowered at around 1.05 a.m. under the command of Lookout Symons, they were ordered to 'stand off a little ways and come back when called'. However, they rowed toward the light on the horizon, off the *Titanic*'s port bow, and away from the ship. According to crewmen Hendrickson and Taylor, they were about 150–200 yards away when the ship went under. Our calculations agree with the estimate of 150 yards.

**NO. 2**

Lifeboat No. 2, under Fourth Officer Boxhall's command, was lowered with Seaman Osman assisting, at around 1.45 a.m. Boxhall said the lifeboat pulled 100ft away, and then he heard someone (Smith) say to come around the starboard side. They had to be pulling fairly close to *Titanic*'s hull to make the distance they needed to make, and with only four oars aboard. Boxhall was using a stroke oar, while passenger Mahala Douglas was at the tiller.

Boxhall claimed they came very close to the propellers, before turning around them to go along the starboard side to look for a gangway door. Boxhall said the door was closed, and Osman stated they could not get close in at that point, due to the ship's list, and they subsequently pulled astern. He said the *Titanic* went down when they were about 60–100 yards away.

But No. 2 had only four oars, and they were having trouble making headway through the water. Did they really have time to make 60–100 yards before the ship sank, considering how close they had just been to the ship looking for that gangway door? We do not think so. It's possible they were far closer when *Titanic* sank. They must have passed close around the propellers, as they were partially out of the water, and before the actual break-up started.

Our animation shows this happened around 2.15 a.m. If they had just a minute or two more to come around to starboard and forward to get a look at the gangway door before they pulled away, say 2.17 or 2.18, they had to be right next to the ship when the lights went out and the ship split in pieces. We estimate No. 2 had to be around 10–20 yards from *Titanic* when she took her forward plunge.

**NO. 3**

Lowered at 12.55 a.m., according to Seaman Moore, No. 3 'got clear of the ship as soon as we were lowered in the water', and was just over ¼ mile (440 yards) away when the ship sank. We think she was a bit closer than that, more likely around 250 yards.

**NO. 4**

Boat No. 4 was one of the last boats to leave the port side, at around 1.50 a.m. Loaded and launched from A Deck, it was close enough to the sinking ship to pick up Greasers Scott and Ranger as they slid down the falls that had been previously used to lower No. 16 – Ranger dropped directly into the boat, Scott into the sea just a short distance away. Quartermaster Perkis, in charge, also picked up seven or eight men from the water, several of whom died in the lifeboat. Then they pulled away, found No. 12, and became part of Lowe's flotilla of boats (Nos. 4, 10, 12, 14 and Collapsible D). When it became light, No. 4 headed across the debris and body field, and helped No. 14 rescue the survivors off the overturned Collapsible B.

Since they were right up next to *Titanic* when Ranger and Scott came down from the falls, and then picked up swimmers in the water, they could not have moved too far from the ship by the time she sank. We estimate they were about 20 yards away at first, and then moved farther away to join up with the other lifeboats.

**NO. 5**

The second boat lowered on the starboard side at about 12.45 a.m., No. 5, commanded by Third Officer Pitman, pulled away from the ship and met up and tied to No. 7. Boat No. 7 was somewhat undermanned and Pitman transferred about six people to it to 'even them up a bit'. They then pulled farther off, and by the time of the sinking they were at somewhere between 300 and 500 yards away, according to Pitman, Quartermaster Olliver and Steward Etches' estimates. We think that's a bit too far – our calculations show them around 200 yards away at the time of the sinking.

**NO. 6**

Quartermaster Hichens was put in command of No. 6, which lowered about 1.10 a.m., by Second Officer Lightoller. He was told, 'Pull toward that light', referring to the light seen by many off the port bow. Major Peuchen, who was allowed to slide down the falls into the lifeboat, said, 'I think we must have been five-eighths of a mile' (or 1,100 yards) away when *Titanic* sank. The lifeboat did meet up with No. 16, who transferred a fireman into No. 6 to help row, as Hichens only had two other men aboard, Lookout Fleet and Peuchen. At the time of the sinking, we think No. 6 was actually only 150 yards or so from *Titanic*.

**NO. 7**

The first boat to be lowered, at about 12.40 a.m., No. 7, met up with No. 5, as stated earlier, and received about six people from her, to allow the number of people to even out between the boats. It originally had rowed about 20 yards out, then pulled farther away, about ¼ mile (440 yards) to meet No. 7. They eventually untied and 'laid off, then, until [No. 7] saw the lights of the *Carpathia*'. This 440 yards seems a bit distant, since they were close to No. 5. We think No. 7 was only about 200 yards off the sinking ship.

## NO. 8

Boat No. 8, under the command of Seaman Thomas Jones, was the first boat lowered on the port side at around 1.00 a.m. Captain Smith told them, 'Pull for that vessel; land your people and return to the ship', meaning the light sighted off the port bow.

There are different accounts as to how long they rowed for it. Jones said about two hours, but passenger Mrs J. Stuart White said half an hour. Regardless, at some point they turned around and turned back toward the *Titanic*. Steward Crawford estimated they were 1½ miles away (2,640 yards) when the ship sank. This seems way too far to us – we think they were much closer, probably less than 300 yards off *Titanic*'s bow.

## NO. 9

Boat No. 9 was loaded and launched from the Boat Deck at around 1.30 a.m., with Boatswain Haines in charge. Haines claimed they pulled off 100 yards at first, but when he saw the ship going down by the head, he pulled farther away. By the time *Titanic* went under, No. 9 was likely about 200 yards away, and not too far away from the other aft starboard boats.

## NO. 10

For some unknown reason, No. 10 was readied quite a bit later than the other aft port boats, and was lowered last of that quadrant, at around 1.50 a.m. First Officer Murdoch ordered Able Bodied Seamen Buley and Evans in, to help the passengers aboard. After some issues during the lowering, they pulled away, but how far is questionable. Steward Burke said ¼ mile (440 yards), Seaman Evans, 200 yards, and Seaman Buley, 250 yards.

During the next hour or so, other boats had joined them, becoming what has become known as Lowe's Flotilla – Nos. 4, 12, 14 and Collapsible D, in addition to No. 10. Most of these boats remained floating in this area until the *Carpathia* was sighted after daylight. At the time of the sinking, we think this lifeboat was slightly less than 100 yards away, due to its later time leaving the ship.

## NO. 11

Boat No. 11 was loaded from A Deck and lowered to the sea around 1.35 a.m., with Quartermaster Humphries in charge. According to Able Bodied Seaman Brice and Steward Mackay, they rowed ¼ mile (440 yards) away from *Titanic* and then rested on their oars until sighting the *Carpathia* in the morning. We think 200 yards is a more appropriate figure.

## NO. 12

Lowered at approximately 1.30 a.m., under the charge of Able Bodied Seaman Poingdestre. According to him, No. 12 rowed about 150 yards away from the ship. After *Titanic* went down, and hearing the cries of people in the water, Poingdestre pulled back in to about 100 yards away, but found no one to pull from the water. He did join up with the other boats in Lowe's Flotilla (see No. 10 above) and received a sailor (Able Bodied Seaman Lucas) and two firemen from Collapsible D, as No. 12 was short on men.

Boat No. 12 also was hailed by Second Officer Lightoller on Collapsible B, and Nos. 12 and 4 between them pulled about twenty-eight or thirty survivors aboard. In this case, Poingdestre said it was farther out, then came back into 100 yards. No. 12 likely got as far away as 200 yards at the time of the sinking.

## NO. 13

The order to lower away for No. 13 was given at about 1.40 a.m., and Leading Fireman Barrett took charge. It had problems as it lowered, nearly being swamped by the condenser discharge forward of it, then pushed aft, where it was nearly crushed by No. 15, which was lowered immediately after No. 13.

When they got away from the ship, Steward Ray stated they were 'about a mile off [1,760 yards] when the lights went out'. We know Nos. 13 and 15 were lowered very close to the same time and remained close as they moved away. Regardless of Ray's estimate, they probably were around 200 yards away, and not far from No. 9, which had been launched shortly before.

## NO. 14

Fifth Officer Lowe helped load No. 14 and boarded it as it started to lower. Lowe had the boat rowed about 150 yards from the ship and soon herded Boat No. 10 and Collapsible D together, as well as Nos. 4 and 12, which all ended up nearby. However, Steward Crowe thought it was nearer to a mile (1,760 yards).

In preparation to head back into the debris and rescue those still alive, Lowe transferred most of the occupants from No. 14 into the other lifeboats. After about an hour and a half, Lowe took Buley and Evans, plus several others, from No. 10 to No. 14 to help him go back and search for survivors. They did find three or four people alive (one of whom died in the lifeboat); then, not finding anyone else, they headed away from the area.

They again sighted Collapsible D looking 'sorry', and Lowe approached, attached a line and proceeded to tow D toward the *Carpathia*. The lifeboat also happened upon overturned Collapsible A and took aboard about twelve or thirteen survivors.

At the time *Titanic* broke up, No. 14 was probably about 200 yards off.

**NO. 15**

It appears that No. 15 stayed close to No. 13 during the night. Fireman Dymond was in charge. Fireman Taylor said they rowed about ¼ mile (440 yards) upon reaching the water and were heading toward a light 'with the other boats'. This figure seems a bit too high, and we think it likely that it was near to No. 13, and only around 200 yards from the sinking ship.

**NO. 16**

Under the charge of Master-at-Arms Bailey, No. 16 moved off from *Titanic*. According to Steward Andrews, they moved off about ½ mile (880 yards) from the ship. It did tie up with No. 6 for a period and transferred a fireman to that boat, which needed more rowers. We think around 200 yards is a more likely figure, given that they rendezvoused with No. 6.

**COLLAPSIBLE A**

After being pushed off the roof of the Officers' Quarters, Collapsible A landed right-side up and floated off the starboard Boat Deck as the Bridge was dipping under water at around 2.15 a.m. As such, by the time the stern went under a few minutes later, she was very close to the sinking ship, and a number of people were able to stay on her, or some climbed onto her from the water. Over the course of the next several hours, some of these people died and fell overboard, until Lowe, in No. 14, rescued them as the sky was starting to lighten. Collapsible A was within yards of the *Titanic* as she dove under and broke apart.

**COLLAPSIBLE B**

Also stowed on the port-side roof of the Officers' Quarters, Collapsible B was pushed off the roof and landed upside down on the Boat Deck, just as water was starting to cover the deck. It also floated off the ship around 2.15 a.m., and for some reason, floated to starboard around the front of the Bridge. The first funnel just missed the lifeboat by feet when it collapsed, and pushed it farther away to starboard.

A number of people climbed aboard the overturned collapsible, but as they had no paddles or oars, it just drifted while some of the people standing on top died and fell overboard. In the morning, Nos. 12 and 4 picked up the survivors from B and delivered them to the *Carpathia*. Like Collapsible A, this lifeboat was also very close to the sinking, within yards.

**COLLAPSIBLE C**

The *Titanic* was considerably down by the bow when Collapsible C, with Quartermaster Rowe in charge, was launched at around 2.00 a.m. Rowe steered for a light he thought was 5 miles distant and pulled through the night, but could not get closer. They then spotted a green light to their rear (Boxhall's light in No. 2) and turned around to head toward it. Daylight was breaking and they continued on to the *Carpathia*.

Rowe thought Collapsible C was ¾ mile (1,320 yards) away from *Titanic* when she sank, but since they got away from the sinking ship only ten or fifteen minutes before the sinking, we think this is way too great a distance. Probably they did not go beyond 100 yards from the ship.

**COLLAPSIBLE D**

The last boat to be successfully launched, Collapsible D was lowered 10ft to the sea at around 2.05 a.m., with Quartermaster Bright in charge. As they were lowering, passenger Björnström-Steffansson was able to jump in. His friend, Woolner, tried to jump in, but his lower body ended up hanging into the water. Passenger Fred Hoyt was able to swim the short distance to the lifeboat.

Because of a lack of crewmen to pull the oars, D had difficulty making headway, but they were still able to make 100 yards to meet up with the rest of Lowe's flotilla. After Lowe had headed into the debris to find survivors and exited, he came across Collapsible D, which looked 'sorry', and took her in tow to take it to the *Carpathia*. When *Titanic* sank, they were likely just under 50 yards away.

In Boat No. 2, Fourth Officer Boxhall periodically lit off flares during the night. He had placed a box of these in the boat before boarding it, almost as an afterthought. He felt that lighting these might help keep the lifeboats together once they were afloat. When he lit the first flare, shortly after the sinking, the cries from the water got even louder, probably because those struggling for their lives in the sea thought the light was that of a rescue ship.

Without a doubt, the flares helped save many of the survivors' lives. Captain Arthur Rostron and the crew of the *Carpathia* began seeing the green flares as they drew closer to the scene. Ironically, the distress co-ordinates that were calculated by Boxhall and transmitted by the wireless operators earlier that night were wrong, indicating a position some 13 miles to the west of where *Titanic* actually sank. While coming up from the south-east, the *Carpathia*, by coincidence, stumbled upon the survivors, and the green flares helped guide them in.

At around 4.05 a.m., Rostron saw another flare, and realised that it was from a boat and not *Titanic* itself. Soon thereafter, *Carpathia* came to a stop as Boat No. 2 rowed towards their side.

Even as the ship drew near, the survivors adrift in the lifeboats could still see the lights of the mysterious ship on the horizon to the north. Although it had been visible throughout the night, it never did respond or acknowledge the Morse signals, rockets or wireless calls. In that way, it is ironic that the *Carpathia*, a ship that had been much more distant than the stranger on the horizon, arrived much sooner. (Boat No. 8, following Captain Smith's orders to row towards this distant ship, attempted to do so for most of the night, but with the south-setting current in the area, never made much progress towards it.)

Arriving alongside *Carpathia* at 4.10 a.m., Boxhall saw some of the rescue ship's crew members peering down from the Bridge wing and asked them to shut down their engines and take them aboard, saying, 'I have only one sailor.' First Class passenger Mahala Douglas suddenly shouted out, '*Titanic* has gone down with everyone on board!', at which point Boxhall told her to shut up. Later, Mrs Douglas felt that he was 'perfectly right' in doing so.

It was breaking dawn as the first survivors climbed ladders into *Carpathia*'s gangways, while those who could not do so were lifted in bosun's chairs or canvas bags. The sky glowed

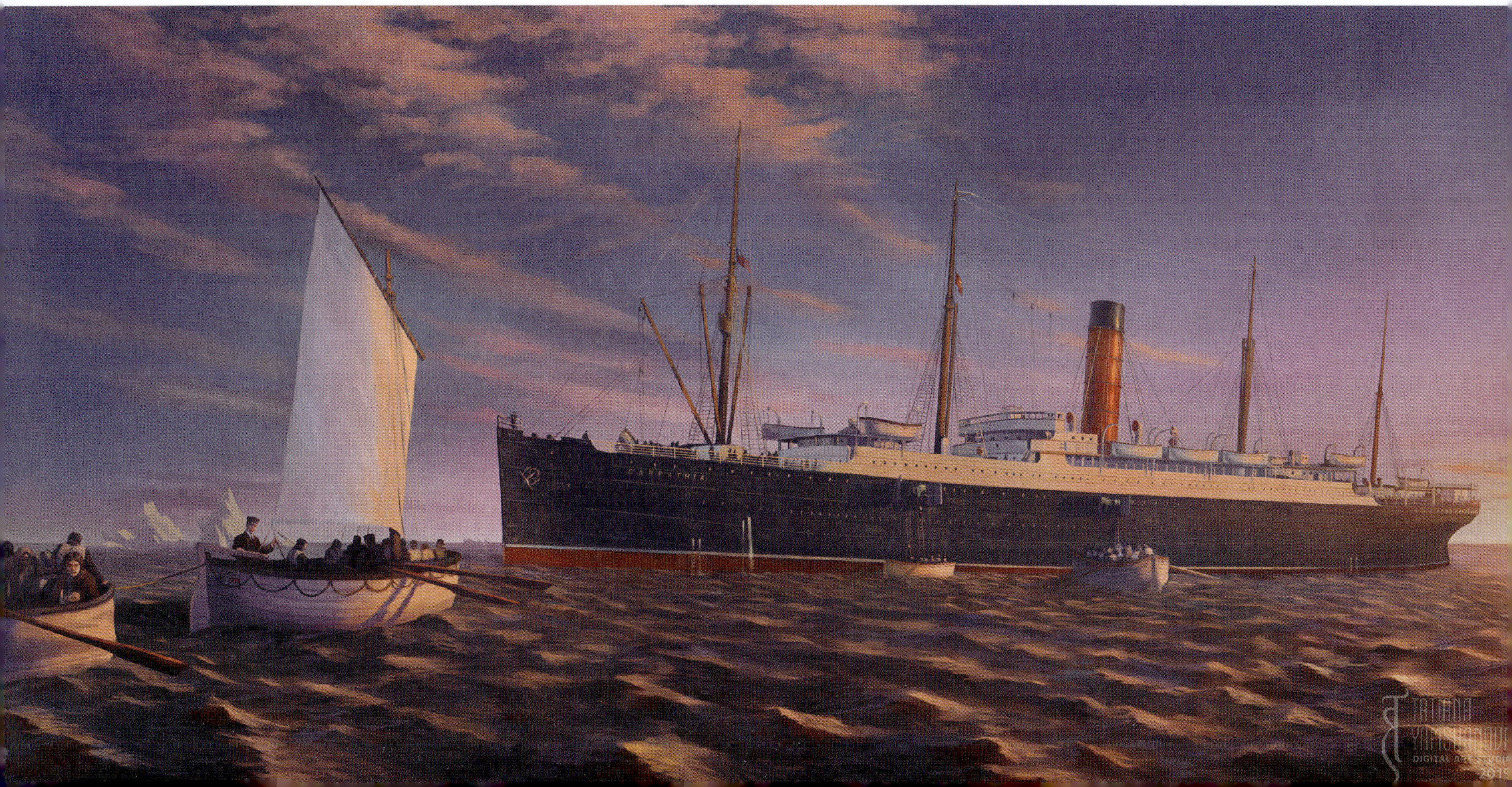

The Cunarder *Carpathia*, the first ship to arrive at the scene. On the left, Fifth Officer Lowe in Boat No. 14 tows Collapsible D toward rescue. Mrs Harris, her arm in a sling from a fall during the voyage, can be seen looking over the side. (Tatiana Yamshanova, commissioned by Will Kindler)

with orange, yellow and red hues, and there were numerous icebergs and growlers surrounding them. If it had not been for the terrible tragedy that had just played out, it would have been an inspiring view.

After Boxhall climbed aboard, he met with Captain Rostron. Although he surely knew the answer, Rostron asked where the *Titanic* was. The Fourth Officer replied, 'Gone! She sank at 2.20 a.m.' After a few moments of silence, Boxhall exclaimed that the ship 'was hoodoo'd from the beginning ...'

Rostron gently took Boxhall by the arm, saying, 'Never mind that, my son. Tell me, were all her boats got away safely?'

Boxhall replied that he believed so, but that it was difficult to see. He discussed how the collision happened at 11.40 p.m., and the 'boats were launched from 12.45 onwards'. He expressed dismay that many of the boats were half full when lowered. Rostron asked, 'Were many people left on board when she sank?' At that point, the young man, who had remained calm and steady throughout the sinking, broke down, his voice cracking with emotion. 'My God sir, they've gone down with her. They couldn't live in this icy cold water.' It was the first confirmation of the enormity of the disaster to the outside world.

The ramifications of the sinking were enormous, directly leading to significant safety regulation changes, and also spawned numerous myths and legends. Even as the true story has faded more and more into obscurity over time, the name *Titanic* and the legend of the ship and sinking has been resilient, with a large percentage of the world's population being at least somewhat familiar with the basic details of the story. *Titanic*, although its period of existence was short, has profoundly impacted popular culture. In some ways, this is a good thing, as it has prevented the memory of those aboard and those impacted by the tragedy from vanishing.

FOUR

# *OLYMPIC*

## THE NEGLECTED SISTER

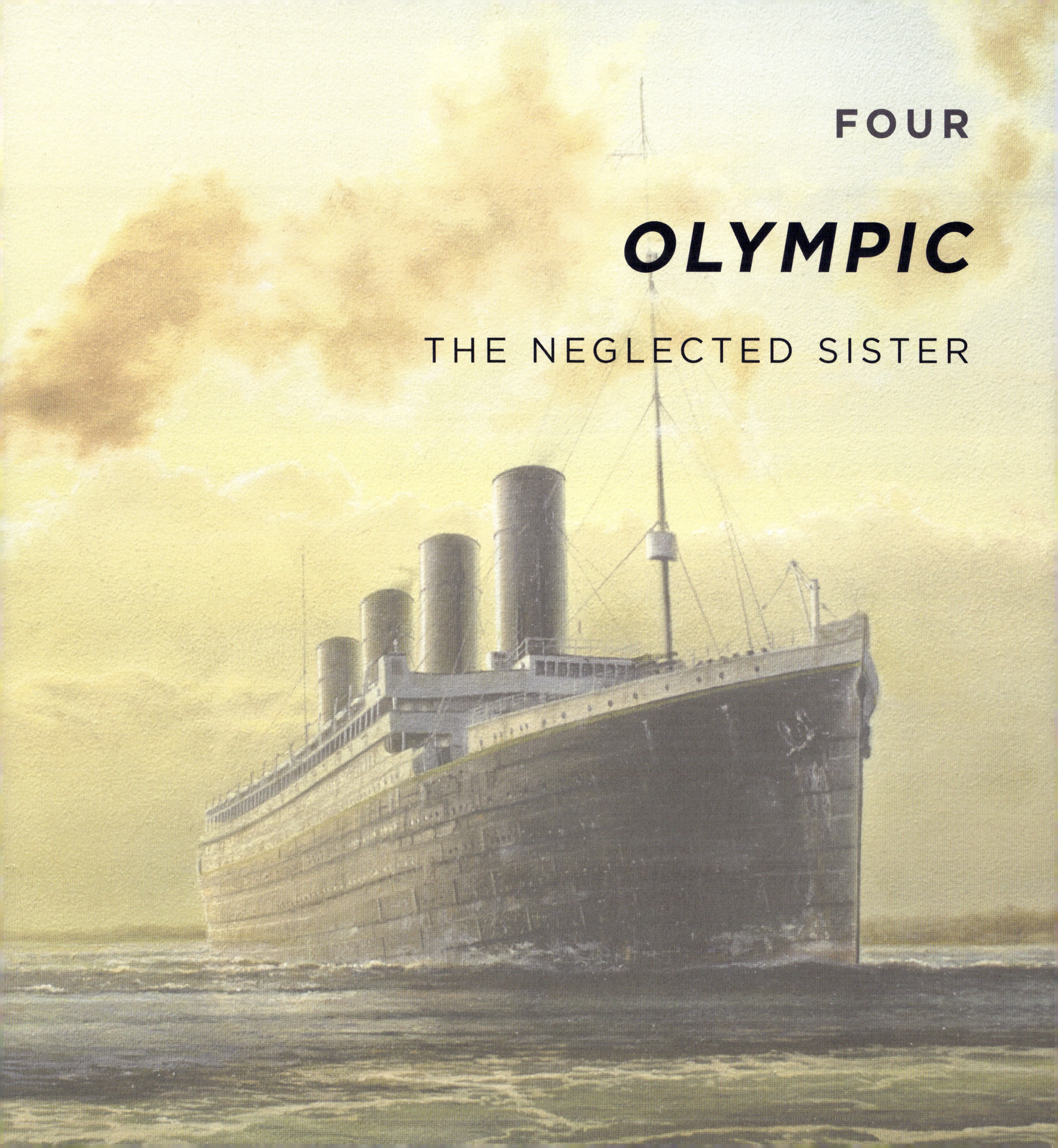

If you were to survey those with a casual interest in maritime history, nearly all would know *Titanic*'s name, and the majority would at least know something about its sinking. A smaller percentage of those same individuals would have at least heard of the *Britannic* and its tragic loss in the First World War. However, it is likely that fewer people still would be familiar with or know much about the RMS *Olympic*. This is in spite of the fact that the ship was the eponym of the *Olympic*-class liners, and the fact that she had a long and storied career, remaining in service for nearly twenty-four years.

While volumes have been written about *Olympic* for more serious enthusiasts, the purpose of this chapter is to give a general overview of its highly successful career. Through this, we hope that readers will gain some appreciation for the durability and reliability of the one ship among the trio of *Olympic*-class liners that actually experienced a full and successful career. Dubbed 'Old Reliable' during the First World War, this moniker stuck and aptly describes *Olympic*'s life. The fact that she was such a successful ship, and was eventually retired and scrapped – the fate of nearly all long-lived ships – is perhaps why she has faded into obscurity, while her sister ships, both of which met early and tragic fates, are more well known.

*Olympic*'s maiden voyage began on 14 June 1911, with Captain E.J. Smith commanding the ship. Shipbuilder Thomas Andrews and a 'Guarantee Group' from Harland & Wolff were also on board, monitoring the performance and looking for possible areas of improvement. Arriving in New York successfully and to welcoming crowds of onlookers, *Olympic* was billed as the most luxurious and safest vessel afloat. It attracted a large amount of international press and attention from the public.

The ship was opened up to the public after reaching port, receiving over 8,000 visitors. Over 10,000 spectators watched when it departed on its first return crossing from North America to Europe. *Olympic* quickly became a favourite among travellers, with many going out of their way to cross exclusively on the ship.

As discussed previously, on 20 September 1911, while outbound at the start of her fifth voyage, *Olympic* suffered a collision with the Admiralty cruiser HMS *Hawke*. In a testament to the high quality of her construction and the materials used to build her, *Olympic* stood up to the impact and was never in danger of sinking.

However, the damage took weeks to fix, necessitating a return to Harland & Wolff for repairs. Ironically, these repairs led to a delay in *Titanic*'s completion and maiden voyage. This delay has often been examined, with historians debating whether *Titanic* would still have rendezvoused with an iceberg if the maiden voyage had not been postponed. This is a question to which no one has a definitive answer.

*Olympic* was repaired as quickly as possible and put back in service. During January 1912, she fought her way through a tremendous gale when sailing towards New York. Captain Smith later recalled it as the worst storm he had seen in his long and storied career. The heavy seas tore rails from the Forecastle, ripped the heavy No. 1 Hatch cover from its mountings, and loosened steam and anchor windlasses. It wasn't the last time that *Olympic* would encounter severe storms during its career, but goes to show how durably built the vessel truly was.

On 24 February 1912, *Olympic* experienced another setback when it threw a propeller blade, and had to return to Harland & Wolff for repairs yet again. A replacement propeller was pulled from the still incomplete *Titanic*, and *Olympic* was quickly put back into service. She operated very successfully during subsequent voyages.

*Olympic* was approximately 500 miles away from *Titanic* on 14 April 1912, when her sister collided with the iceberg. Receiving the distress calls, Captain Herbert Haddock rushed to assist, but was ultimately too far away to help, with the *Carpathia* rescuing all of the survivors before *Olympic* had a chance to arrive. *Olympic*'s crew offered to take aboard the survivors, but White Star Chairman J. Bruce Ismay declined, concerned that the appearance of the virtually identical *Olympic* would cause survivors undue distress.

In late April 1912, *Olympic*, which at the time had the same lifeboat provisions as her sister, was hurriedly equipped with additional collapsible lifeboats by the White Star Line. However, a large group of firemen among the crew were sceptical about the safety and seaworthiness of these hastily provided boats, and went on strike prior to the ship leaving port. When the White Star Line brought aboard non-union crew members to replace them, a large group of sailors also decided to go on strike, leading to the scheduled voyage being cancelled.

All fifty-four sailors involved were arrested on the charge of mutiny once ashore; they were tried and found guilty, but were discharged and released due to the extraordinary circumstances surrounding the strike. The White Star Line relented and allowed them to return to work, fearful of even more bad publicity when the company had already received a huge blow from the *Titanic* disaster.

*Olympic* resumed its transatlantic service, and then in October 1912 returned to Belfast, in order to be overhauled and have additional safety features retrofitted, based on the lessons of the loss of *Titanic*. Some of these changes included the fitting of additional lifeboat davits and lifeboats to accommodate all of the people who could travel aboard; the addition of a complete double skin along the middle sections of the ship near the Boiler and Engine Rooms, adding to the cellular double bottom she and *Titanic* had been designed with; five of the watertight bulkheads were extended up to B Deck; and, finally, an additional watertight bulkhead was installed in the Electric Engine Room, dividing the compartment in two.

Besides these changes, extra cabins were added to B Deck, including the addition of Parlour Suites similar to those on *Titanic*. A Café Parisien, similar to *Titanic*'s, was also added, and additional bathrooms were added to some First Class cabins. White Star played up the new safety features when *Olympic* returned to service. The ship had a continued period of success following this, despite being usurped as the world's largest passenger liner when Hamburg-Amerika's *Imperator* entered service in June 1913.

When Britain entered the First World War on 4 August 1914, *Olympic* was immediately impacted. Westbound on a crossing from Southampton to New York when war was declared, *Olympic*'s crew received a warning from HMS *Essex* to look out for German vessels lurking outside of the neutral United States' coastal waters. The crew of *Olympic* extinguished nearly all of its lights, and the ship created a momentary panic when it slipped past Fire Island into New York under cover of darkness; many witnesses could not distinguish which large passenger liner it was.

*Olympic*'s return destination was altered from Southampton to Liverpool, and the ship quickly departed for its next voyage without passengers or cargo. Before leaving, the white superstructure of the ship was painted grey, and her buff-coloured funnels were painted dark grey. All twenty-nine of the ship's boilers were lit, and with assistance from the Gulf Stream, it reached 24 knots during the return trip.

The White Star Line wanted to maintain *Olympic*'s passenger service between Britain and America, but quickly found that, while the ship's westbound voyages to New York were crowded with passengers fleeing the war in Europe, very few passengers were booking passage on the return eastbound voyages from New York to Liverpool. On the 21 October 1914 eastbound crossing, *Olympic* had just 153 passengers aboard: 35 in First Class, 18 in Second Class, and 100 in Third Class.

During this voyage, on 27 October, *Olympic*'s crew rescued survivors from the battleship HMS *Audacious*, which had struck a mine and was sinking. Responding to the distress call, *Olympic*'s lifeboats and crew rowed through rough seas, picked up crewmen from the stricken vessel and safely landed them aboard *Olympic*. Worried about how the loss of *Audacious* would impact British morale, the Admiralty attempted to cover up the sinking. Because of this, *Olympic* was not allowed to proceed or disembark its passengers until 2 November.

With the lack of bookings causing a large drop in revenue, and with the increasing threat from U-boats and mines, White Star decided to pull *Olympic* from service following this voyage. It would remain laid up in Belfast for ten months.

On 1 September 1915, *Olympic* was formally requisitioned by the British Admiralty as a troop ship. It would now be known as the HMT (Hired Military Transport) *Olympic*. After receiving maintenance work due to having languished unused for so long, having her funnels painted a darker shade of grey, furnishings removed and interiors converted, and defensive deck guns installed fore and aft, she was ready for this service.

*Olympic* departed Liverpool on her first trooping voyage to the Mediterranean on 24 September, with 6,000 soldiers aboard. They would find eventual deployment in the Gallipoli Campaign.

Captain Bertram Hayes found himself under scrutiny when the *Olympic*, under his orders, stopped and rescued survivors from the French steamer *Provincia* on 1 October 1915. The ship had been sunk by U-boat *U-33*, and *Olympic*'s crew rescued thirty-four survivors. The British Admiralty believed that Hayes had put the ship in mortal danger, having stopped in the U-boat-infested waters and potentially leaving his ship open to attack. However, France held a different view, bestowing Hayes with the Gold Medal of Honour for his actions.

The *Olympic* would complete several more successful trooping voyages until the Gallipoli Campaign ended in January 1916. The British Admiralty discussed options for how best to utilise *Olympic* at that time, and considered laying the ship up again. Eventually, they transferred her from the Mediterranean service to the Liverpool–Halifax transatlantic route.

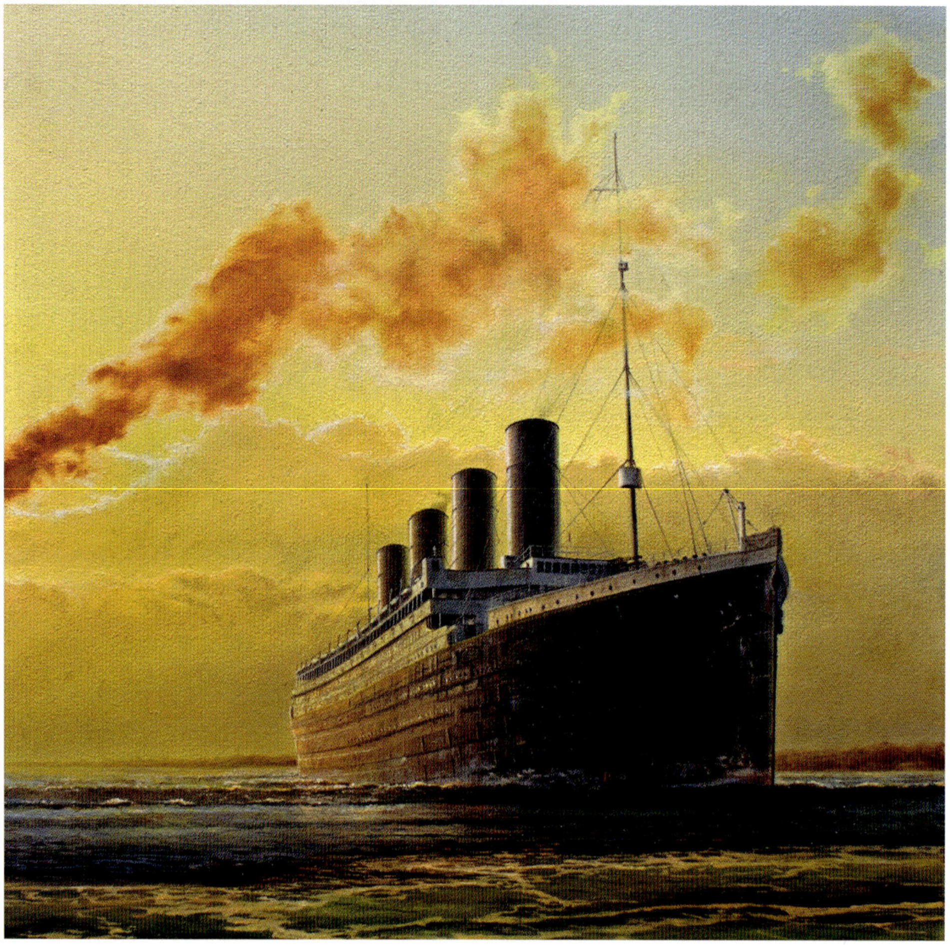

A splendid portrait of *Olympic* in late 1915 or early 1916, showing her greyed-out funnels and superstructure. (David Olivera)

The Canadian Government had been looking to charter vessels to transport large numbers of Canadian troops overseas. This kept *Olympic* in service, and the ship departed for her first voyage on this route on 22 March 1916. She would continue to make crossings for the rest of the spring and summer of 1916, without encountering any enemy action. *Olympic*'s speed helped make her a difficult target, but not an invulnerable one.

In early 1917, *Olympic* underwent a refit and was fitted with new 6in deck guns: two on the Forecastle, two in the Forward Well Deck and two aft on the Poop Deck. The ship was also painted in the first of two dazzle camouflage patterns that it would sport during the war. Large geometric shapes and designs were emblazoned on *Olympic*'s hull, in several shades of grey, blue and olive. Dazzle camouflage was designed to disguise the profile, heading and shape of the vessel when spotted, making targeting by U-boats or other vessels difficult.

Once the United States entered the war in 1917, *Olympic* began transporting American troops overseas, taking them aboard from New York. On 12 May 1918, *Olympic* was carrying American troops en route to France. She had recently been repainted in a second dazzle camouflage pattern, consisting of more distinctive geometric shapes and lines, coloured black, blue and blue-grey. Suddenly a U-boat, *U-103*, was spotted on the surface. *Olympic*'s deck gunners opened fire immediately, causing the submarine to make an emergency crash dive. After steering a parallel course, the *Olympic* turned and rammed the sub, piercing its pressure hull. The crew of the U-boat surfaced, scuttled it and abandoned ship.

It was later discovered that the U-boat's crew had been preparing to fire torpedoes but had not been able to flood the torpedo tubes before the collision. Thankfully, the only damage that *Olympic* sustained in the encounter was a bent stem and dented hull plates, which would be repaired in a post-war refit. This would stand as the only known incident in the war in which a merchant vessel sank an enemy submarine, but it had almost become a tragedy.

On 4 September 1918, *Olympic* again had a close shave with a U-boat. The crew of *U-53* had spotted the *Olympic* approaching, and manoeuvred into attack position, firing two torpedoes. When no explosion resulted, it was assumed that both torpedoes had missed. In fact, although nobody on *Olympic* felt it at the time, the second torpedo had hit the ship, but luckily failed to detonate.

Despite this, the torpedo had penetrated the outer skin of the hull between the third and fourth funnels and caused the inner skin to take on water. The double hull added during the 1912 refit had prevented catastrophic flooding, although it is possible that if the torpedo had detonated, it could have sunk the ship. Remarkably, the fact that *Olympic* had come within a hair's breadth of sinking was only discovered once the damage from the impact was found during a post-war refit in February 1919.

By the war's end, *Olympic* had proven herself to be a workhorse, truly earning her nickname of 'Old Reliable' by safely making thirty-six trooping voyages. After the Armistice was signed, *Olympic* was employed in repatriating thousands of Canadian troops back home. These voyages put a lot of wear and tear on the ship, particularly since maintenance was restricted to essentials with emphasis on quick turnaround times. While exact numbers vary, the most reliable analyses indicate that during its service as HMT *Olympic*, the ship carried over 200,000 troops, more than any of its contemporaries, including *Mauretania* and *Aquitania*.

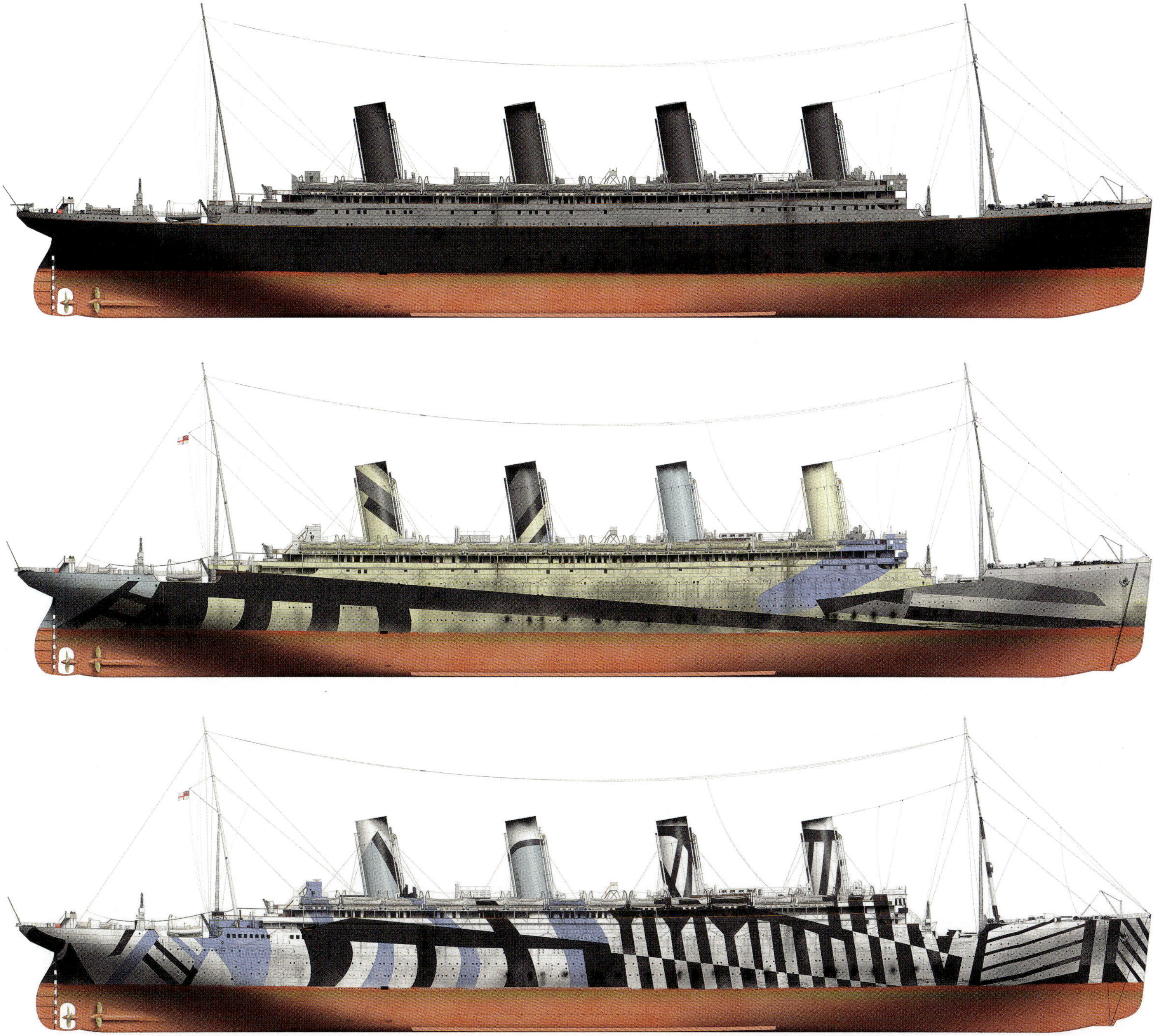

The First World War appearance of *Olympic* (top to bottom) in 1916 with the superstructure painted grey and the funnels dark grey; its first dazzle camouflage pattern in 1917; and its second dazzle camouflage pattern in 1918. (Michael Brady)

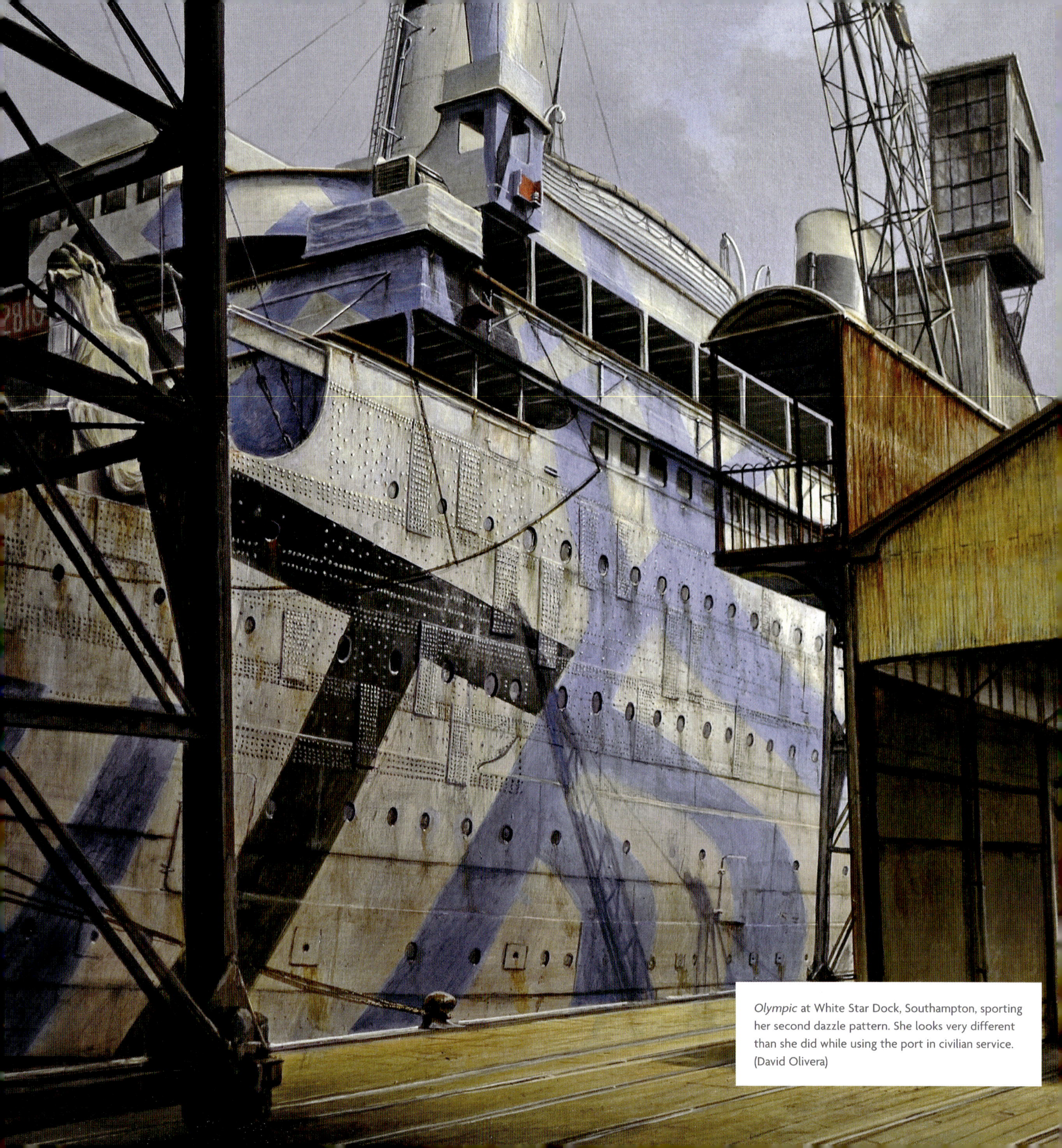

*Olympic* at White Star Dock, Southampton, sporting her second dazzle pattern. She looks very different than she did while using the port in civilian service. (David Olivera)

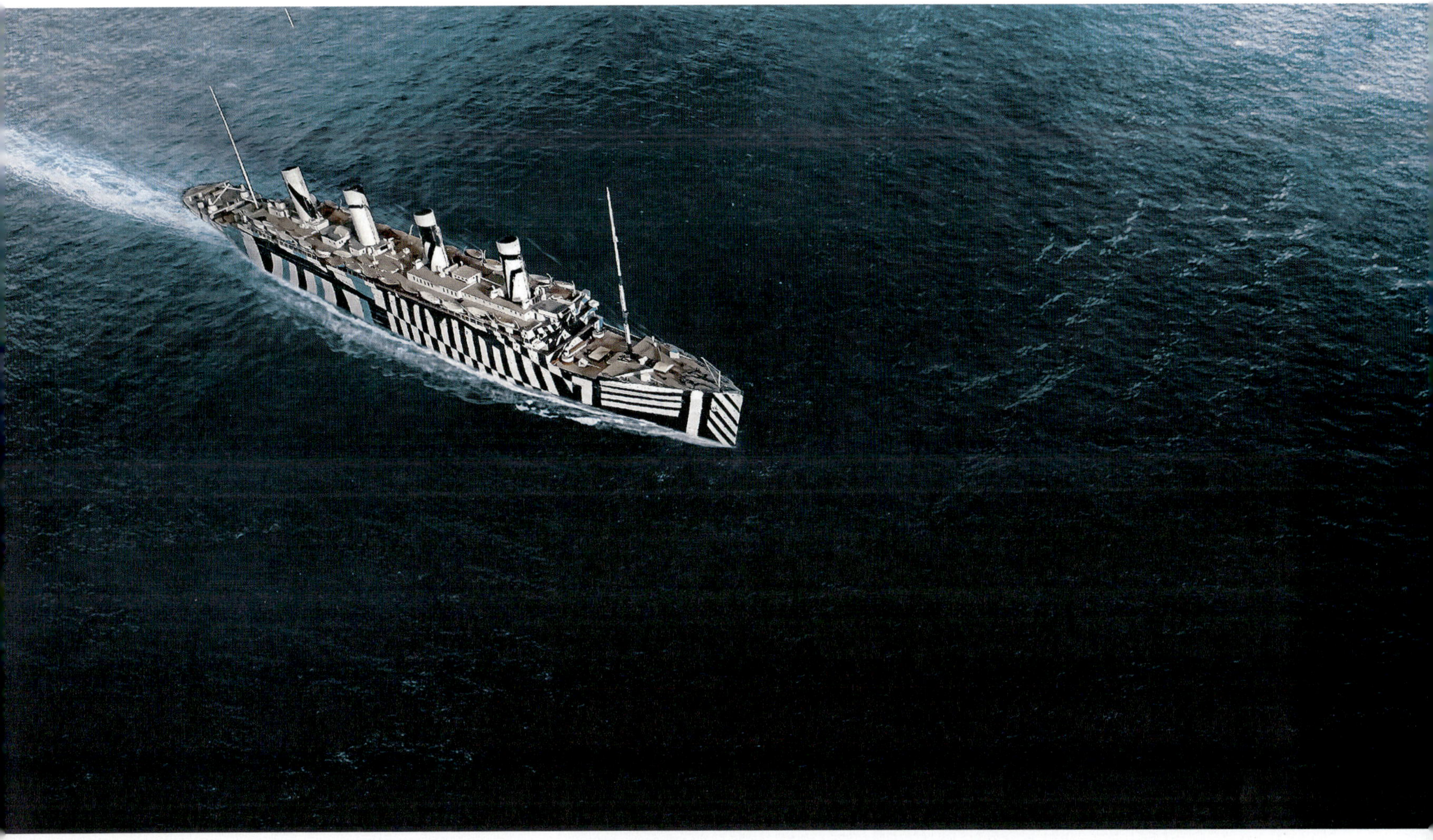

A splendid aerial rendering of what *Olympic* looked like at sea in late 1918 or early 1919. (Ala Kermiche)

In August 1919, *Olympic* returned to Belfast to be refitted and returned to civilian service. Given the rough usage of the ship, extensive work needed to be done. All of the interior spaces would need to be restored and refurbished, and the White Star Line used this opportunity to modernise them as well, to suit the evolving tastes of passengers. The furnaces were converted to burn oil and the coal bunkers were also altered to carry the liquid fuel, in what was perhaps the biggest change to the ship.

On 25 June 1920, *Olympic* made a triumphant return to civilian service after months of hard work at the shipyard. The ship carried 2,249 passengers on what some billed as her 'second maiden voyage'. As a result of the refits, *Olympic* had gained gross tonnage, allowing her to regain the claim as the largest British-built liner afloat, even though *Aquitania* still exceeded her in length.

Throughout the 1920s, *Olympic* fulfilled the true potential that had always been present in the *Olympic*-class liners. The ship was extremely popular and successful and was the choice of many celebrities when they travelled, including Prince Edward and Charlie Chaplin, among many others. The White Star Line also obtained the passenger liners *Homeric* and *Majestic* – which had been the former German liners *Columbus* and *Bismarck*, respectively, and had been handed over to the British Government as war reparations by Germany following the war. The two vessels joined *Olympic* as successful running mates on the North Atlantic passenger route, and helped fill the void left by the losses of *Titanic* and *Britannic*.

During this period of success, there was also an accident that caused significant damage to *Olympic*. On 22 March 1924, while *Olympic* was backing away from its berth, she collided with the liner *Fort St George*, which had crossed behind it. While the smaller liner, which was at fault, sustained significant damage and had to cancel its voyage, it was initially believed that *Olympic* had sustained only minor damage. However, it was soon discovered that *Olympic*'s stern frame was damaged enough that it needed to be replaced, necessitating another trip back to the shipyard.

After the run of success in the early 1920s, changes to immigration laws in the United States led to the need for some significant operational revisions. In 1924, when these laws restricted the number of immigrants allowed to enter the United States to just 160,000 per year, it caused a major reduction in bookings from such passengers. This had long been a significant portion of the customer base for transatlantic liners, and companies like White Star were now forced to pivot and cater to the tourist trade in order to fill the financial void that was left behind.

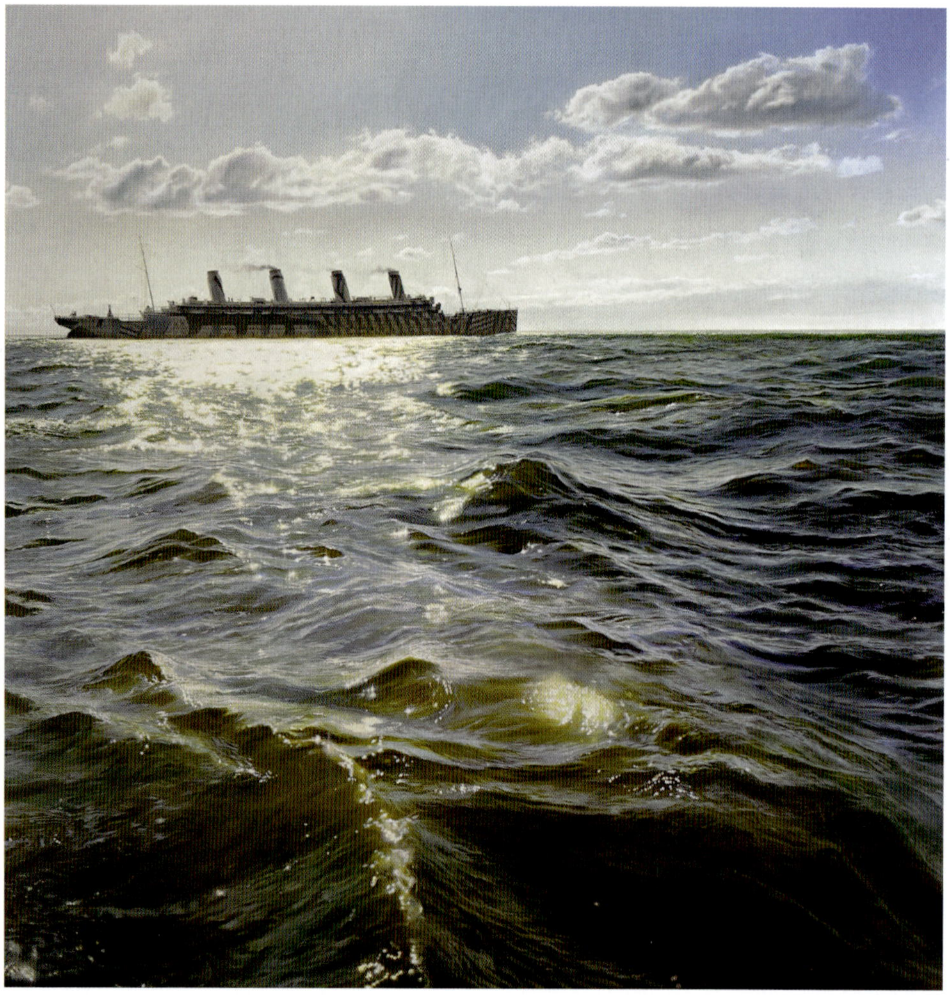

'The Ship that Brought us Home', a portrait showing *Olympic* engaged in repatriating troops who were heading home from the most catastrophic war the world had ever experienced. (David Olivera)

As a result, during her third major refit in December 1927 and January 1928, *Olympic* was converted to carry Tourist Third Cabin passengers, as well as First, Second and Third Class ones. This was an attempt to provide low-cost accommodation for tourists, instead of purely catering to immigrants in Third Class. A new Tourist Third Cabin Dining Saloon, Lounge and Smoke Room were added, as well as their own cabins.

This space was created by reductions in the number of Second Class staterooms and repurposing of some of the Second Class public spaces, as well as some of the original Third Class public spaces, such as the Third Class Smoking Room and Lounge underneath the Poop Deck, which was reassigned to the Tourist Third Cabin.

A Tourist Third Cabin enclosed Promenade was added by carving out a portion of the enclosed Second Class Promenade aft on C Deck. A new Third Class Smoking Room was constructed forward on the port side of D Deck to replace the one that was reassigned. Over time, Tourist Third Cabin and Second Class would be merged, forming the new Tourist Class in 1931. During this refit, it was noted how well *Olympic* was ageing, despite years of heavy use.

Throughout the 1920s, *Olympic* continued to experience success. New features such as a dance floor, updated decor, additional First Class bathroom facilities, as well as additional suites with private facilities on B Deck, were added, all in an attempt to keep up with competitors and to keep the look and feel of the ship fresh. Despite her age, *Olympic* experienced some of her greatest successes in 1929, carrying the highest average number of passengers per crossing since 1925.

Like all other passenger line companies, the White Star Line was badly impacted by the Great Depression in the 1930s. The number of passengers crossing the Atlantic per year was drastically reduced during this time period, being roughly halved. The popular *Olympic* was not immune, dropping from an average of 1,000 passengers per crossing up until 1930 to less than half of that figure by 1932.

The situation was so dire that during her lengthy refit at the end of 1932, there had been rumours that *Olympic* was to be withdrawn from service. In fact, the ship was found to be in quite good condition for her age, and was placed back in service in March 1933, being described by White Star as 'looking like new'. Attempts were made to update the interiors to appeal to more modern tastes, including painting the oak panelling and rails of the Grand Staircase in a dress of 'green and gold', which was considered 'unforgiveable' by some.

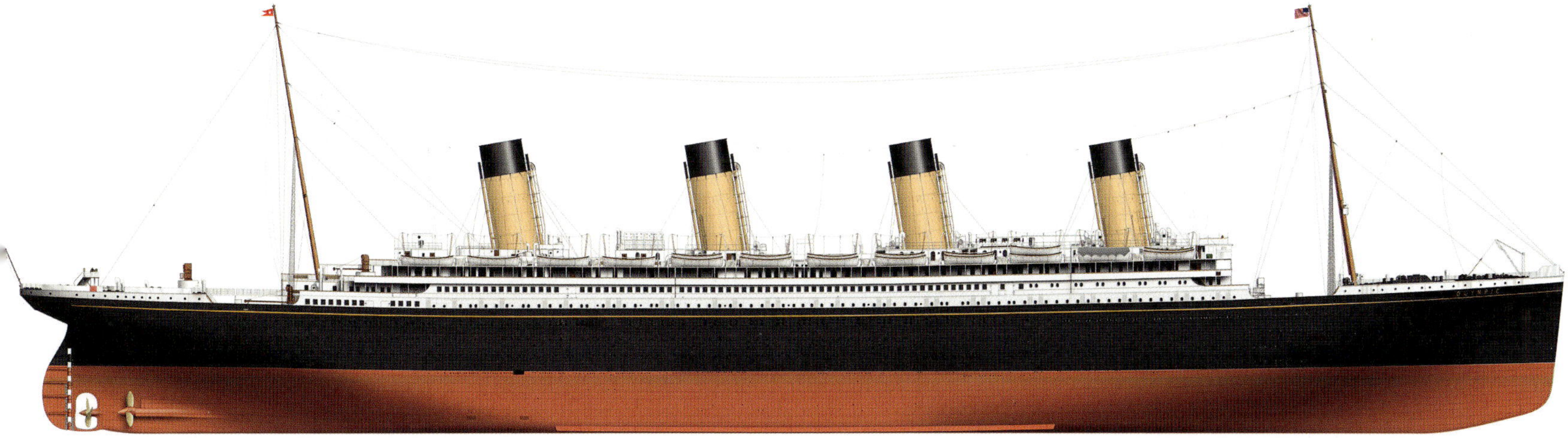

Sadly, during 1933 and 1934, *Olympic* suffered a net operating loss for the first time. As it turned out, 1933 would be the worst year that *Olympic* had ever experienced when it came to average passenger counts. In fact, White Star as a whole had lost money every year from 1930 to 1933. Economic conditions and increased competition were frequently cited reasons.

Cunard then had a massive unnamed passenger liner under construction at the John Brown & Company shipyard in Scotland. This ship would eventually become the RMS *Queen Mary*. However, due to financial difficulties during the Great Depression, the new ship had sat unfinished and Cunard had been unable to secure a loan to finance its completion.

The British Government ultimately agreed to the loan, under the condition that the struggling Cunard Line merge with White Star. This agreement provided enough money to complete the *Queen Mary*, and also to eventually build a running mate, the RMS *Queen Elizabeth*. This two-ship tandem meant that, when completed, the two state-of-the-art vessels would handle the majority of Cunard–White Star's North Atlantic passenger service and made it inevitable that their fleet of older liners would begin to be retired.

A particularly sad incident occurred on 15 May 1934, just five days after the Cunard–White Star merger became official. *Olympic*, inbound towards New York in fog, collided with and sank the Nantucket Lightship, *LV-117*. While nobody aboard the *Olympic* was hurt, seven of the eleven lightship crew members were killed. In hindsight, it has been viewed as a symbolic death knell of sorts to *Olympic*'s career.

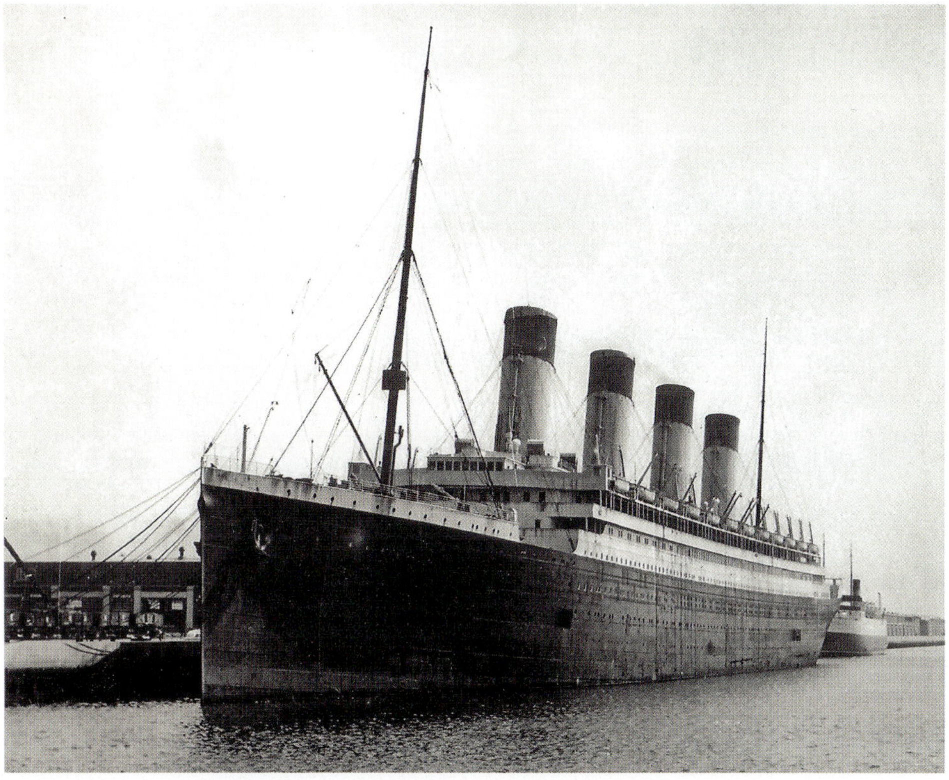

*Top*: A splendid starboard profile of *Olympic*, *c*. 1930, shows how the ship's appearance had changed in nearly twenty years. The yellow, sheer stripe had been lowered in 1922 to help match the appearance of her new running mates, *Homeric* and *Majestic*. (Michael Brady)

*Bottom*: This August 1935 photo shows the *Olympic* lying forlornly at Berth 108. *Mauretania*, her former rival but final companion, had already departed for the breakers. *Olympic* herself would go up for auction within days, and eventually head for the breakers herself. (The National Archives, Authors' Collection)

With her service now considered redundant, *Olympic* was withdrawn from the transatlantic service and departed New York for the final time on 5 April 1935. There were multiple attempts to save *Olympic*, including Cunard–White Star considering her use as a summer cruise vessel, and even once the ship was put up for sale there were discussions of making it a floating hotel off France. None of these ideas came to fruition.

*Olympic* would be laid up alongside her former competitor *Mauretania* for five months; both great ships and former rivals were destined for the scrapyard. After being sold to Sir John Jarvis, who was a Member of Parliament, *Olympic* was to be partially demolished in Jarrow, England, in order to provide work for the region.

She departed Southampton for the last time on 11 October 1935. The once-mighty ship would be slowly dismantled over the next two years, with many of her fittings being stripped out and sold. Many of these still exist to this day.

While the end of *Olympic*'s career was a sad one, and arguably premature – at the time of her scrapping, *Olympic*'s Chief Engineer stated, 'The engines are as sound as they ever were' – there is no denying that it was both long and successful. Succeeding where her two fellow *Olympic*-class sisters, *Titanic* and *Britannic*, had failed, *Olympic* truly proved herself 'Old Reliable', plying the Atlantic successfully in both peace and war, calm and storms, all while becoming a favourite of the cultural elite and common traveller alike.

By the time she was removed from service, *Olympic* had safely completed 257 round trips across the Atlantic, transporting over 430,000 passengers on commercial voyages. This, combined with her successful military record, is all the proof that should ever be needed that the *Olympic*-class vessels were well constructed, durable and of a successful design. Sadly, while *Olympic* experienced great success, her story has been overshadowed and somewhat forgotten by the general public over time, eclipsed by the tragic losses of her better-known sister ships. *Olympic*'s record is one that truly deserves to be celebrated and preserved, rather than forgotten.

FIVE

# BRITANNIC

## TRAGIC TWIN

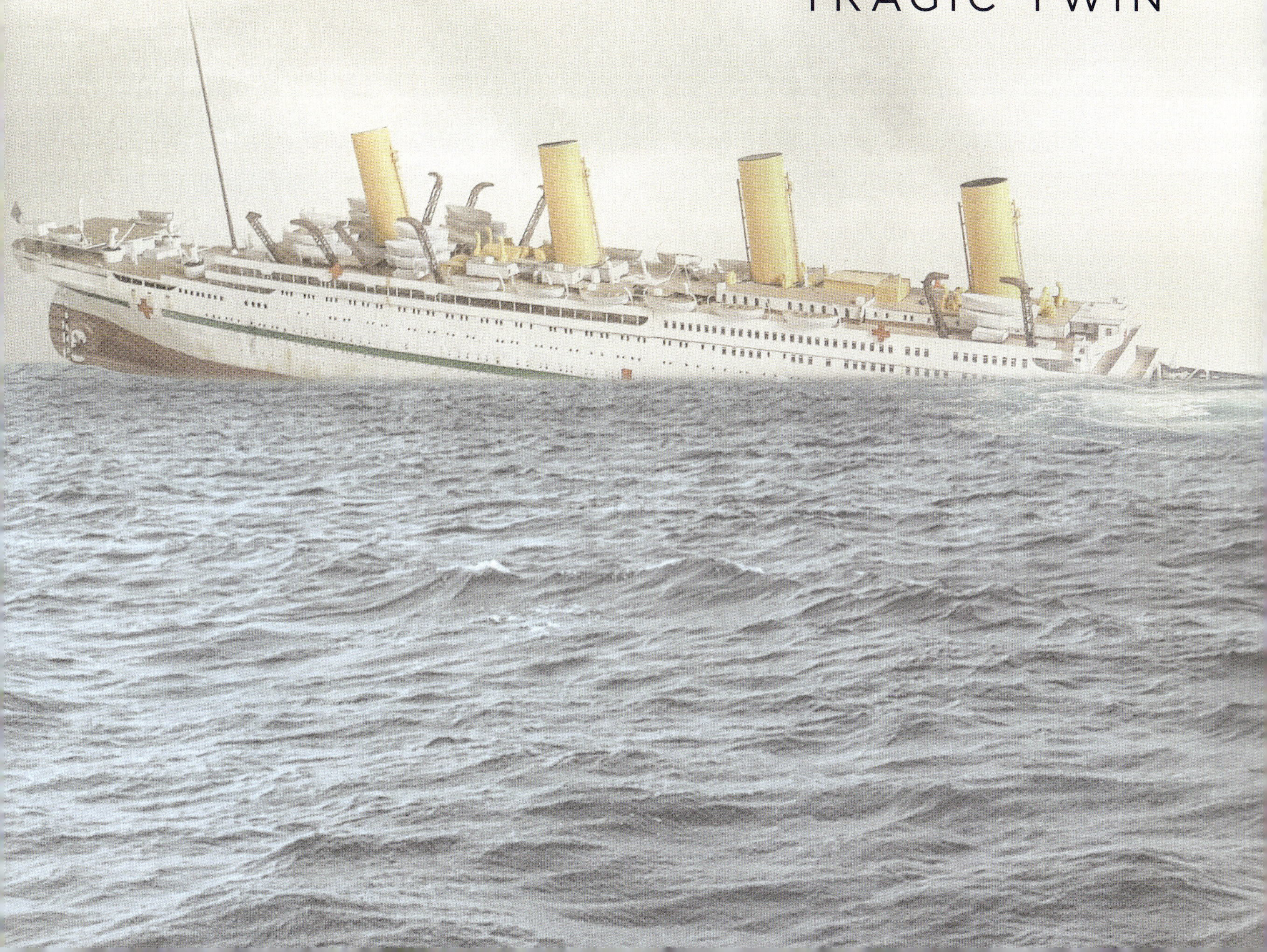

Envisioned as the third vessel among the trio of *Olympic*-class liners, *Britannic*'s keel was laid down in November 1911. Conceptually, the White Star Line had planned a number of improvements in the accommodation, decor and design of *Britannic* over her sisters.

Besides having the benefit of hindsight and feedback from passengers and crew on what worked well and what did not in the designs of *Olympic* and *Titanic*, there was also the matter of competition from other steamship companies. Innovations had continued at lightning speed, and by 1914, many of the features of the *Olympic*-class ships were being surpassed, both in terms of luxury and comfort. For example, White Star continued to be pushed by the Cunard Line, particularly with *Aquitania* slated to enter service by summer 1914. *Aquitania* would surpass *Olympic* in both size and luxury, a fact that White Star was well aware of and had been monitoring closely.

Additionally, there was competition from the Hamburg-Amerika Line, in the form of both *Imperator*, which entered service in 1913, and *Vaterland*, due to enter service in May 1914. Both of these vessels also surpassed the *Olympic*-class liners in terms of size and amenities.

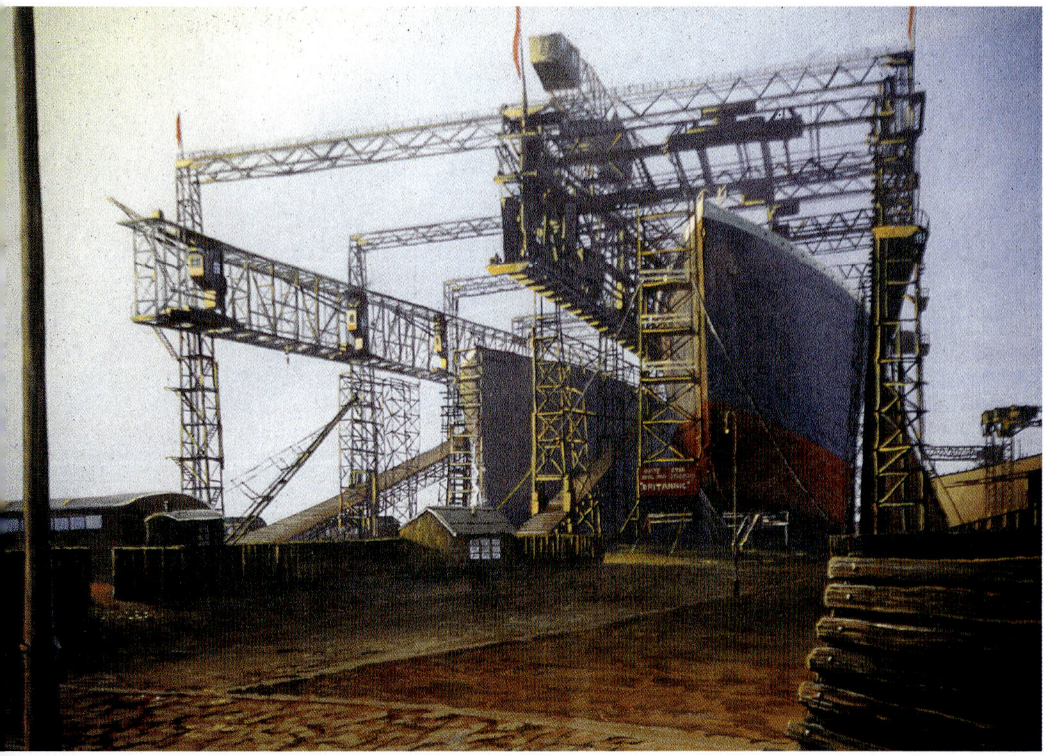

*Britannic* stands proudly on Harland & Wolff's Slip No. 2, where *Olympic* had been built, just before her launch in February 1914. Beside her on Slip No. 3 is the *Statendam*, later renamed *Justicia*. (David Olivera)

With this in mind, multiple refinements and additional accommodation were planned that would allow *Britannic* to both surpass the beauty and comfort of its sister ships and compete with the onslaught of new competitors on a more level playing field. For example, the ship's Swimming Bath, which was decorated in a very spartan manner on *Olympic* and *Titanic*, would be more luxuriously appointed, more in keeping with the pools on *Imperator* and *Vaterland*. The hull beams and structural members of the ship, which were bare and exposed in this area on her sisters, would be covered behind luxurious panelling and decorations.

The Grand Staircase, already a focal point aboard *Britannic*'s sisters, would also be improved with the addition of a multi-deck pipe organ and more sumptuous carved panels and balustrades, along with a different tile pattern on the deck. The ship's A Deck was dedicated entirely to First Class passengers, with the inclusion of a Salon, two Verandah Cafés, a Smoking Room and a Reading Room. Additional deluxe Parlour Suites were added to B Deck, and nearly every First Class cabin would have its own en-suite bath facilities, a rarity at the time.

While there were too many design changes to list here, some of the more significant improvements included: a children's playroom; a dome being added to the central portion of the First Class Smoking Room, which was to be decorated differently than its sisters; the addition of a Shade Deck atop a new deckhouse fitted at the stern; a ladies' Barbers Shop, a first on an Atlantic liner; an additional lift; a larger First Class *à la carte* Restaurant than either *Olympic* or *Titanic*; and a larger Second Class Promenade, aft of the Second Class staircase and Smoking Room, owing to the addition of a roof over the Aft Well Deck.

Harland & Wolff was originally supposed to deliver *Britannic* to the White Star Line in order for it to be ready for service at the beginning of 1914. However, due to the safety improvements that were needed as a result of the *Titanic* disaster, *Britannic* was not even launched until 26 February 1914, let alone ready for service. The design changes made to the then under construction *Britannic* were largely in line with the post-disaster changes that were made to *Olympic*.

Some of the safety changes included adding a double hull along the majority of her length amidships, including the Engine and Boiler Rooms, whereas *Titanic* only had a double bottom, which did not extend up the side of the ship's hull;

raising six of the fifteen watertight bulkheads up to B Deck and strengthening them for the increased height; and adding another watertight bulkhead which essentially divided the Electric Engine Room into two separate compartments. Modifications were made to the ship's hatchways and entrances, in order to allow the ship to remain afloat even with six watertight compartments flooded, and a more powerful turbine engine was provided. Provisions were made to fit eight large crane-like gantry davits to the deck, each operated electrically, and every set was capable of launching six lifeboats at a time. The gantry davits were capable of reaching lifeboats on the opposite side of the ship, allowing them to be launched even if a list developed in an emergency situation. *Britannic* would be capable of carrying forty-eight lifeboats, with a capacity in excess of the ship's capacity of passengers and crew.

## THE *OLYMPIC*-CLASS: SIMILAR YET UNIQUE

*Olympic*, *Titanic* and *Britannic* were sister ships, drawn from a common design concept, yet there were clear differences between the three. It should be noted that, contrary to what is often said, all three ships were identical in overall length at 882ft 9in, and the frame numbering and spacing was identical for each of the three ships. *Britannic*, however, was slightly wider than the first two ships, at 94ft instead of *Olympic* and *Titanic*'s 92ft 6in. It is often incorrectly said that this 18in increase was to accommodate the inner watertight skin to protect her from an event like that suffered by *Titanic*. However, this decision was made before mid-October 1911, before her keel was ever laid and well before *Titanic*'s disastrous maiden voyage.

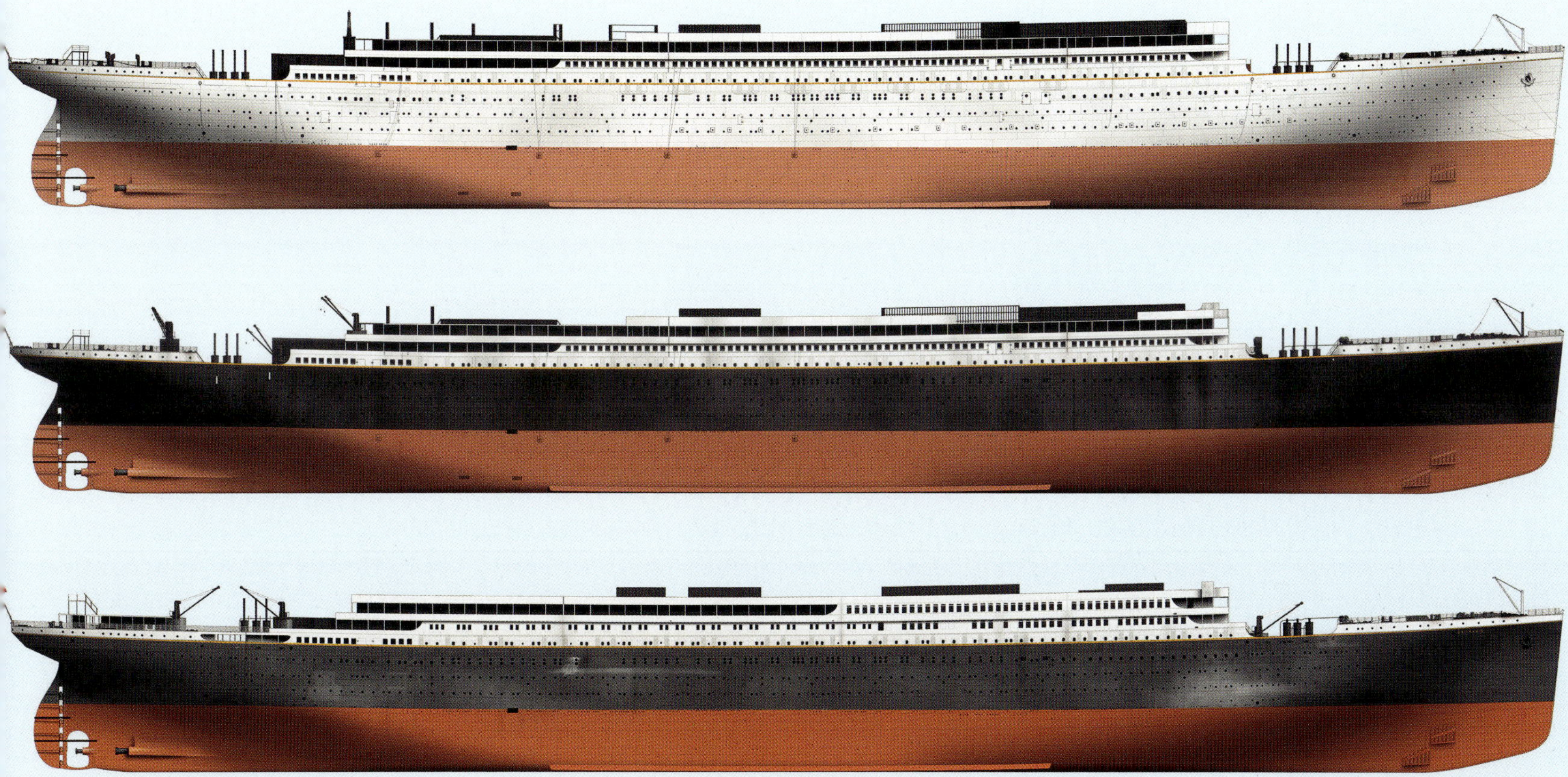

This portrait shows the subtle differences between each member of the *Olympic*-class at the time of their launch. *Olympic* was painted 'virgin white' when launched, *Titanic* was in her more traditional colouring, and *Britannic* was different entirely, as her forward Promenade was already enclosed at the time of her launch. (Michael Brady)

There were other differences. *Titanic* had seen a number of significant improvements over *Olympic*: the extra B Deck staterooms and Parlour Suites, the enlarged Restaurant and new Restaurant Reception Room, the new Café Parisien, the enlarged Reception Room and the enclosed forward Promenade. Many of these upgrades were later retrofitted into *Olympic*, though her B Deck Promenade remained in that guise, and only during her 1928–29 refit was the forward end of this space enclosed for more suites like *Titanic* had. As a result, her A Deck Promenade was never enclosed as *Titanic*'s was.

By comparison, *Britannic*'s A Deck Promenade was enclosed from the very beginning, albeit with windows of a different pattern and size than *Titanic*'s. Aft on B Deck, there was no Café Parisien, but the Restaurant was made the full width of the ship instead. A fully equipped hospital suite was added on the port side of C Deck, while a children's playroom was added on the port side of the No. 2 funnel casing, opposite the Gymnasium.

Although *Olympic*, *Titanic* and *Britannic* were sister ships, clear and critical differences made each one a distinct vessel. If she had ever entered civilian service, *Britannic* would doubtless have been the finest and most comfortable of the three ships for prospective passengers.

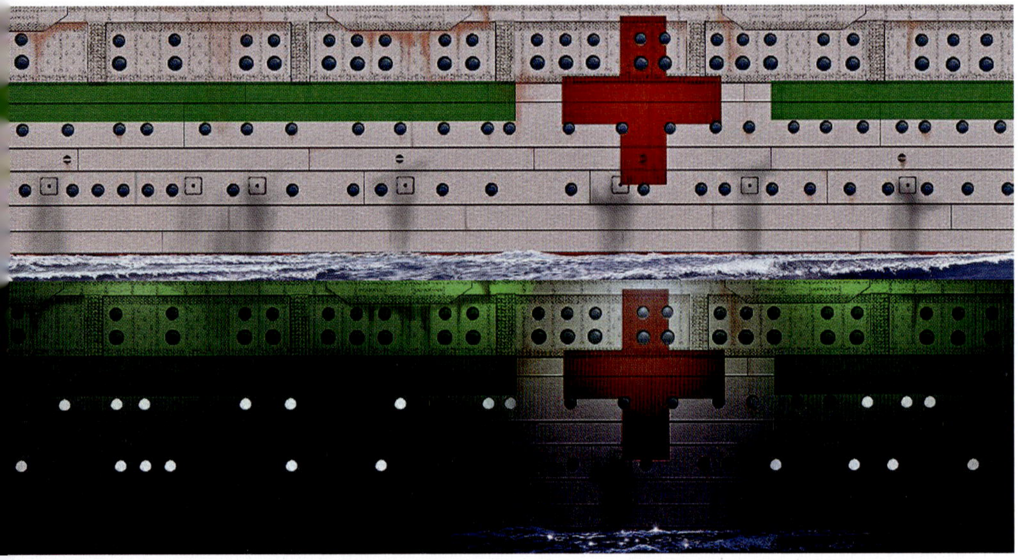

A close-up view of how *Britannic*'s side would have looked in her hospital ship livery both during the day and at night. (William Barney)

With the outbreak of the First World War in July 1914, before *Britannic* could enter transatlantic service, all shipyards with British Admiralty contracts were given priority access to available raw materials and resources. Due to this, *Britannic*'s construction was significantly impeded. However, by May 1915, *Britannic* had undergone mooring trials of her engines. It was estimated that construction had proceeded to the point where the ship, if work was limited solely to that required for making the vessel seaworthy and accommodating a crew, could be ready for service in ten to twelve weeks, with that time period being reduced if the British Government gave it priority.

By June 1915, the Admiralty had decided to utilise some of the large passenger liners that it had requisitioned, such as *Aquitania* and *Mauretania*, as troop transports in the Gallipoli Campaign. Both ships had been laid up since earlier in the war, when their large size and massive coal consumption had led to them being branded as unsuitable for use as armed merchant cruisers.

With the campaign in the Dardanelles failing, the need to ship as many troops to the Mediterranean slackened, but as massive numbers of casualties continued to mount, it soon became evident that there was a much greater need for shipping wounded troops back home to England. Additional hospital ships, with a larger carrying capacity, were needed. As a result, in August 1915, both *Mauretania* and *Aquitania* were converted into hospital ships.

In November 1915, the British Admiralty formally requisitioned *Britannic*, in a move that had seemed inevitable given the state of the war. Even with *Mauretania* and *Aquitania* serving as hospital ships, another large-capacity passenger vessel was needed to carry the growing number of casualties back from the front. *Britannic* was the largest British liner ever built up to that point, in gross tonnage, but was not yet complete. With the Admiralty giving the ship precedence, it was estimated that it would take four weeks to get it ready for service. *Britannic*'s hull, which had never been completely painted in its intended White Star Line livery, was rusty and dirty from sitting unused in the water at the shipyard.

Harland & Wolff soon focused their efforts on completing *Britannic*. The corrosion was cleaned off and the hull was painted white, with a green stripe around it. Green lights were installed that would illuminate the green band and openly identify it as a hospital ship at night.

Three large red crosses were also painted at intervals along the side of the hull for daylight identification. These

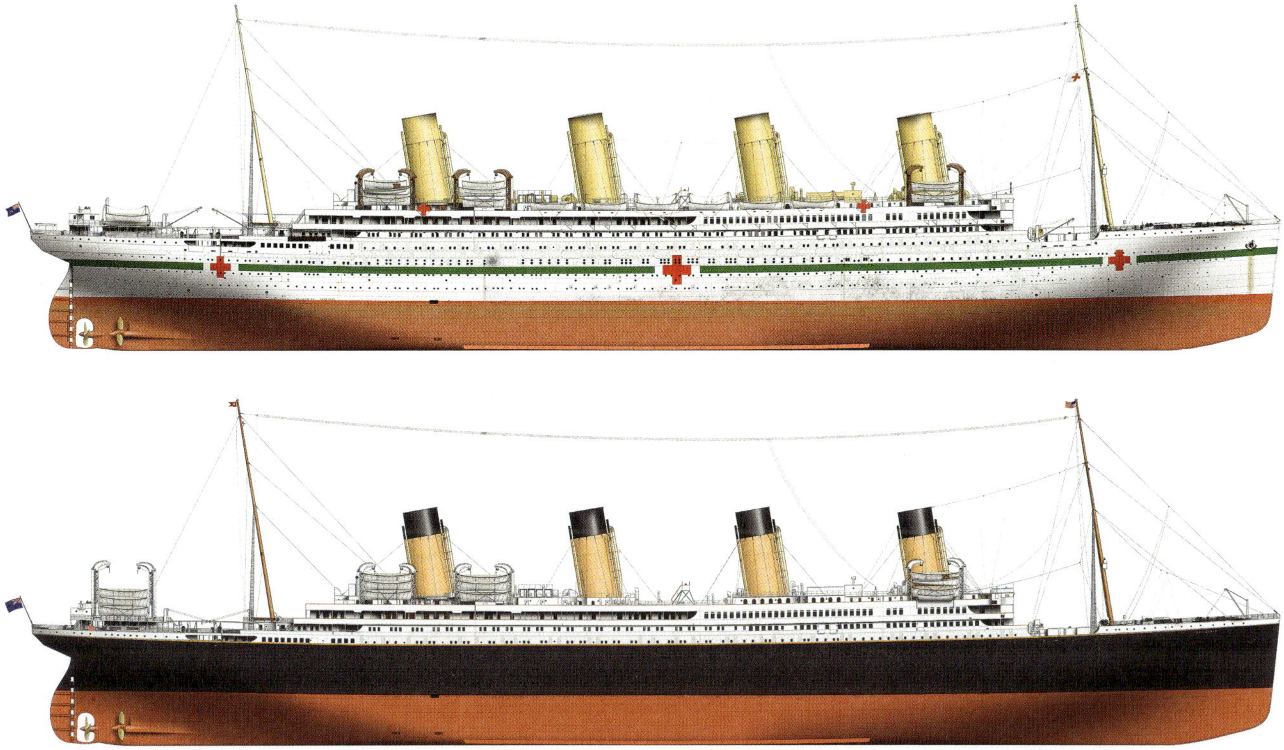

*Britannic*, in her actual guise as a hospital ship (top), and as she was intended for civilian service on the North Atlantic (bottom). (Michael Brady)

would be illuminated by a spotlight at night. Two additional red crosses, smaller than the ones painted on the hull, were attached to the Promenade Deck on the port and starboard sides and would be illuminated red at night. These would clearly identify the ship as a hospital ship to any lurking U-boats or enemy vessels, protecting her from potential attack. The funnels, masts and some structures on the deck – such as the weather covers over the domes – were painted buff yellow.

The interiors of the ship, which were never completely furnished, were hurriedly completed. There is debate over how many of the intended luxurious fittings were already installed at the point when the ship was requisitioned, but it is clear from surviving examples that many of the fittings, if they ever were installed, were removed prior to the ship being pressed into service. *Britannic*'s public rooms, including the First Class Dining Room and Reception Room, were converted into operating theatres. Other spaces, such as those on the Promenade Deck and some of the other upper decks, were converted into recovery wards for patients. The cabins on B Deck were used to house doctors.

Just five of the eight planned sets of giant gantry davits were installed before *Britannic*'s maiden voyage, with smaller Welin-type davits being utilised to finish the job more quickly.

The ship's completed lifeboat arrangements consisted of forty-one wooden lifeboats, fourteen Engelhardt collapsible boats and a series of floats placed around the boat deck, each capable of holding twenty-five persons. The need for this many lifeboats is understandable, since *Britannic*, when the conversion to a hospital ship was completed, would have the capacity to carry 4,473 people, including 675 crew members, 489 medical staff and 3,309 casualties. When all was said and done, Harland & Wolff would complete work on the *Britannic* three days earlier than their original four-week estimate.

On 13 December 1915, Captain Charles A. Bartlett was placed in command of the vessel, and HMHS (His Majesty's Hospital Ship) *Britannic* embarked on its maiden voyage on 23 December. Between arriving in the port of Mudros at Lemnos on New Year's Eve 1915, and departing on 3 January 1916, *Britannic* would take on around 3,300 wounded soldiers from smaller hospital ships and hospitals on shore.

In January 1916, *Britannic* safely completed two voyages carrying casualties back from the front. After arriving back in port at Southampton on 9 February, *Britannic* would be laid up at the pier until the middle of March. With the end of the Gallipoli Campaign, the number of wounded troops needing to be transported through the Mediterranean slackened, and there was no immediate need for *Britannic*'s carrying capacity. *Britannic* would complete one more voyage beginning on 20 March, and by 11 April, the medical personnel aboard were all transferred to other ships.

The vessel would sit idle in an anchorage off Cowes for weeks, before eventually being moved to Southampton and again back to Cowes. On 21 May 1916, *Britannic* was discharged from military service and was soon moved back to Belfast, where work was to commence on converting the ship back into a passenger liner. The White Star Line received £76,000 from the British Government for this purpose. The Harland & Wolff yard workers and craftsmen were engaged in this task, which proceeded at a slow pace, until August 1916, when action in the Mediterranean once again heated up, particularly along the front in Salonica.

On 28 August, the British Admiralty requisitioned the *Britannic* once again. How far the conversion back into a passenger liner had proceeded at that point is not well documented. Regardless, work was halted, and the workers began the task of converting the vessel back into a hospital ship for the second time. Captain Bartlett was placed in command of the ship yet again.

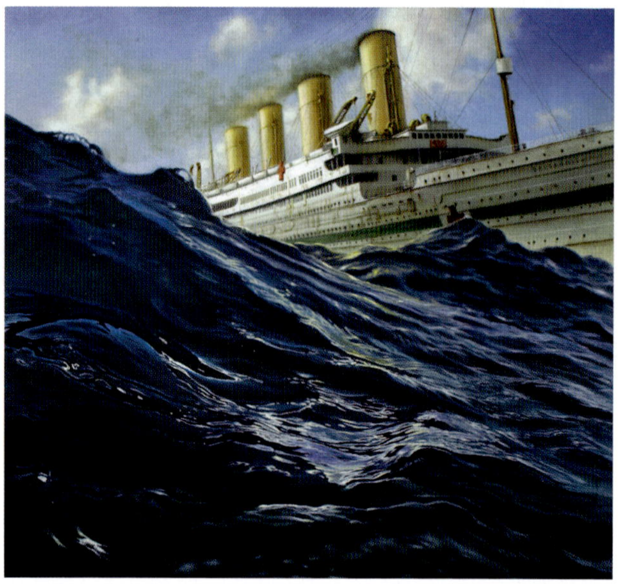

*Right: Britannic* steams through high seas early in her career as a hospital ship. Her original ID number, G608, is clearly visible below the Bridge. (David Olivera)

*Opposite:* An artistic view showing *Britannic* during her second stint as a hospital ship, after August 1916, bearing the new ID number G618. (Ala Kermiche)

HMHS *Britannic* departed on its next voyage on 24 September, with an entirely new medical staff, since all the previous crew who served in that capacity had been transferred elsewhere when the ship was released from service.

*Britannic* departed Southampton on its sixth voyage to the Mediterranean on 12 November 1916, bound for the island of Lemnos, where it would take on wounded Allied soldiers and personnel. Arriving in Naples on 17 November, *Britannic* took on coal and water. There were no passengers aboard the ship, unlike on previous outbound voyages, which allowed the medical staff to have free rein aboard – a freedom that they greatly enjoyed. Typically, policies dictated that passengers and medical staff remain separated, which prevented the latter from being allowed into certain public spaces on the ship.

Due to stormy weather conditions, Captain Bartlett wisely decided to postpone the departure from Naples until the situation improved. The storm persisted until the afternoon on 18 November. Finally, *Britannic* resumed its voyage, although the heavy seas made things extremely uncomfortable for many aboard.

By the next morning, the weather had cleared up and the ship's speed increased. The medical staff hurriedly made preparations for the wounded that they anticipated receiving. Even though it was against regulations to do so in a war zone, many of the portholes on the ship were left open, in order to air out the rooms and patient wards.

*Britannic* reached the Kea Channel, off the Greek island of Kea, on 21 November. The weather was sunny and clear, the deep-blue waters were calm and the mood aboard was relaxed. Many were enjoying the beautiful morning on deck, or were just settling down for breakfast.

Suddenly, at 8.12 a.m., an explosion shook the ship. Crew members and nurses leapt up from the tables immediately to see what had happened. Near the bow, many aboard believed that the ship had been torpedoed. Others further aft felt a slight jolt and believed the ship had collided with a smaller vessel.

On the Bridge, the serious nature of the incident was immediately apparent to Captain Bartlett and Chief Officer Robert Hume. The explosion had occurred somewhere between Cargo Hold Nos. 2 and 3, and water quickly began filling the first four forward watertight compartments. Bartlett closed the watertight doors and ordered a distress signal to be sent.

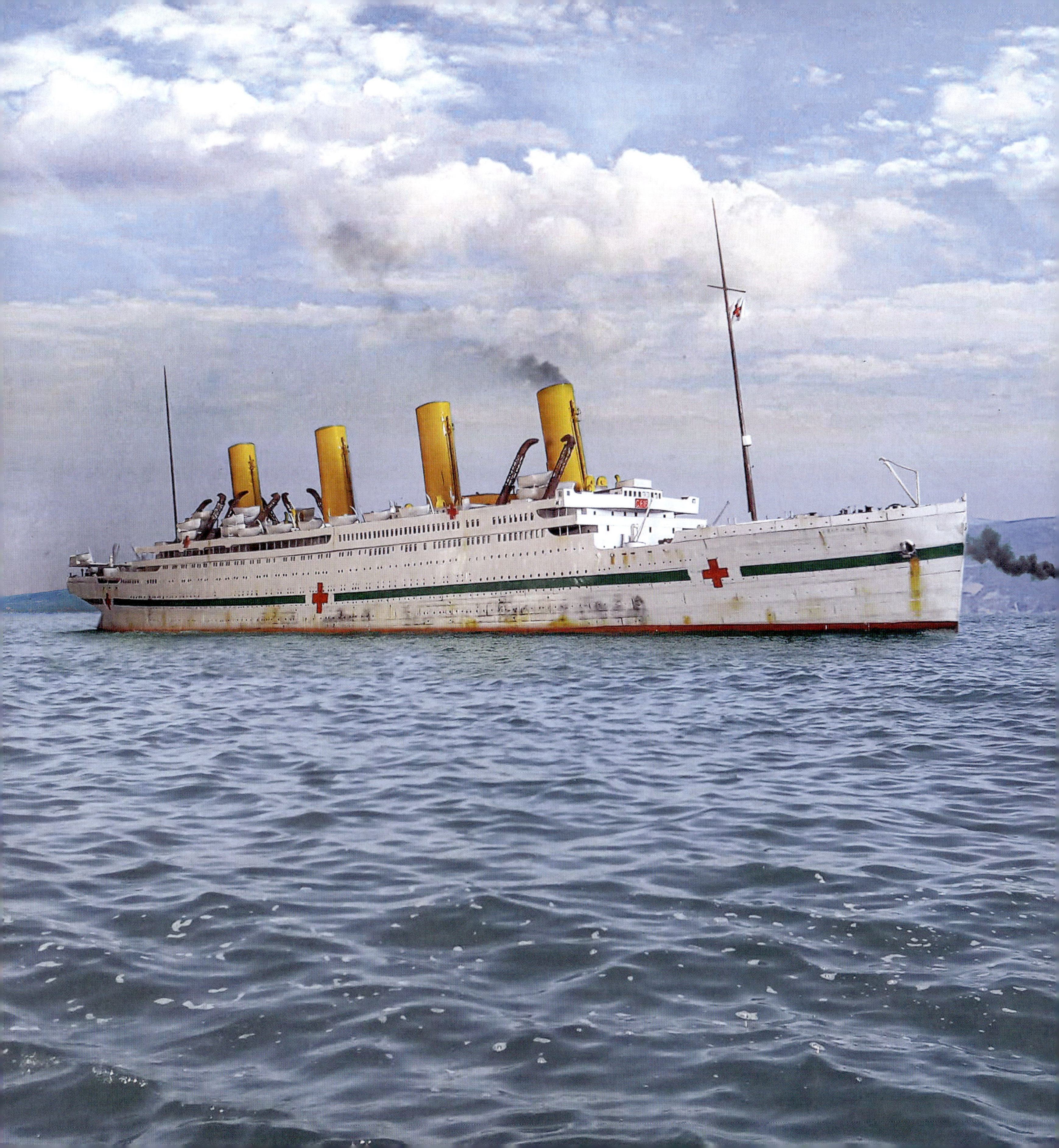

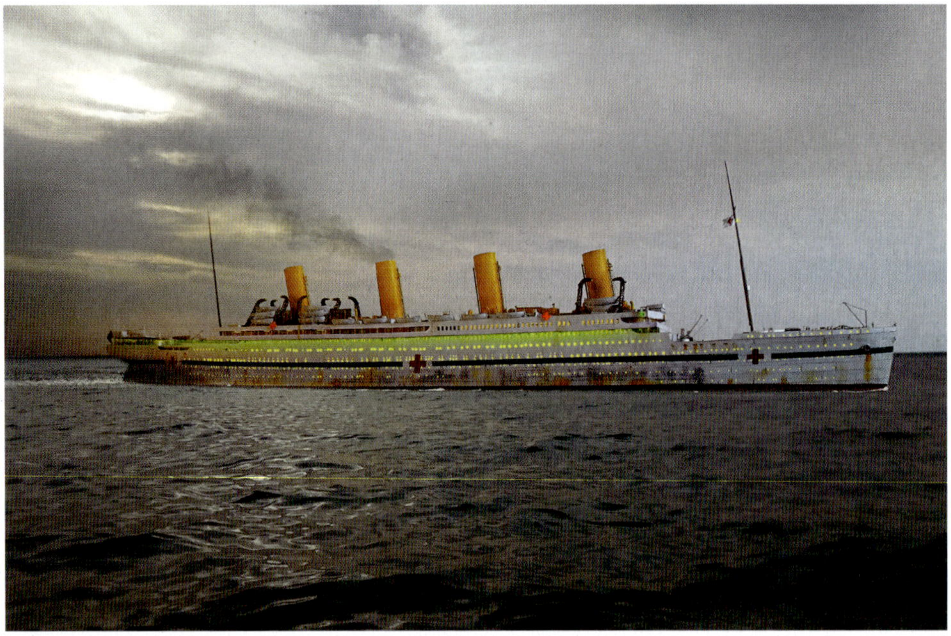

*Above:* A fine artistic view of *Britannic* in the evening, illuminated for easy recognition as a neutral hospital vessel. (Ala Kermiche)

*Below:* An artistic depiction of what *Britannic*'s last sunset might have looked like on the evening of Monday, 20 November 1916. (William Barney)

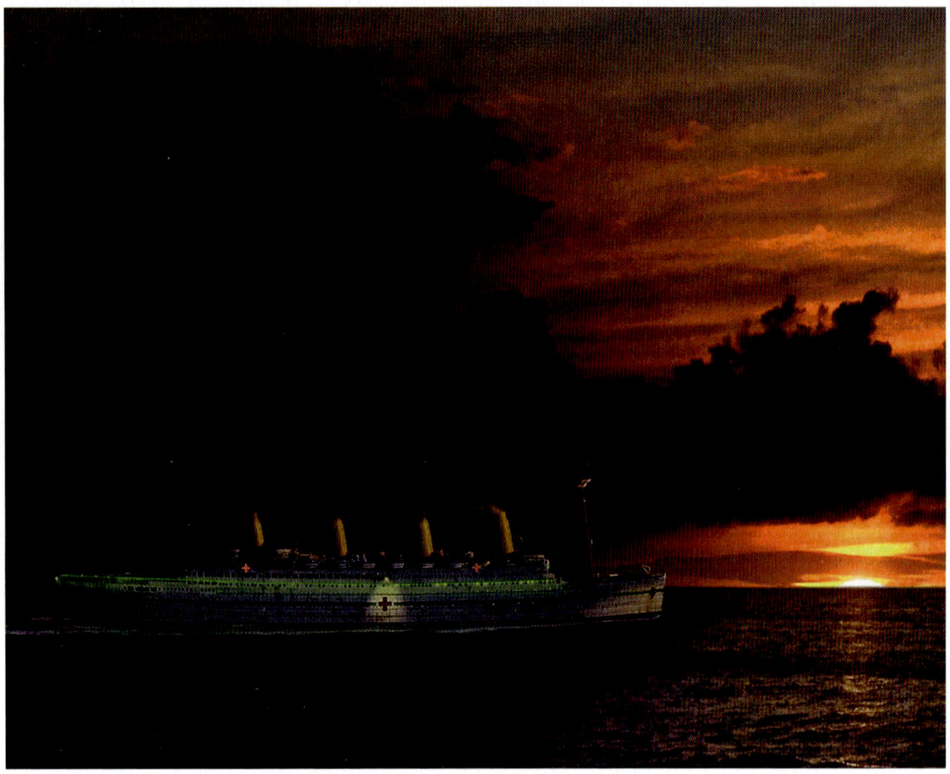

While he ordered the lifeboats to be prepared, Bartlett did not believe that the ship was mortally wounded. Given all of the safety features that had been added to the ship's design following the loss of *Titanic*, this is perhaps understandable. Despite Bartlett's initial optimism, to those below deck it was apparent that the situation was dire. Not only were Cargo Hold Nos. 2 and 3 flooding, but Cargo Hold No. 1 and Boiler Room Nos. 5 and 6 were taking on water as well. The Forepeak would later flood as the bow settled, which made a total of six flooded watertight compartments, the maximum number that the ship was designed to stay afloat with. The ship began to take on a significant list to starboard.

On the Bridge, Captain Bartlett, with the shoreline of Kea visible just 2 miles away, turned *Britannic* toward the island. With the engines running at full speed, he hoped to prevent his ship from sinking by beaching it in the shallow waters offshore. Because the ship was still underway and he believed that they could safely make it to shore, Bartlett did not order any of the lifeboats lowered away.

However, within fifteen minutes of the explosion, *Britannic*'s bow was down significantly by the head and she was listing heavily to starboard. Water soon reached the lowest lines of portholes on E Deck, which normally stood 20ft above the waterline. Because many of these portholes had been opened to ventilate the ship, this provided another entry point for water, which only hastened the flooding.

Besides the open portholes, there is evidence that some of the ship's watertight doors below deck were open. Normally, when travelling through a war zone, crewmen were ordered to keep the watertight doors closed as a precaution. Unfortunately, some of these doors were opened due to the 8.00 a.m. change in watch, so that the firemen and other crewmen below deck could freely move between the boiler rooms and their quarters. The doors had not yet been closed when the explosion occurred, about twelve minutes after the shift change. There is speculation that the explosion warped *Britannic*'s hull, preventing some of these doors from closing.

On deck, members of the crew began assembling at their assigned lifeboat stations. With most of those aboard aware of the seriousness of the situation by this stage, some began to get unruly. Poor communication led to misunderstandings, and some of the crew were apparently unaware that the Captain had not yet given orders for the lifeboats to lower away. Stewards and some members of the deck crew started panicking.

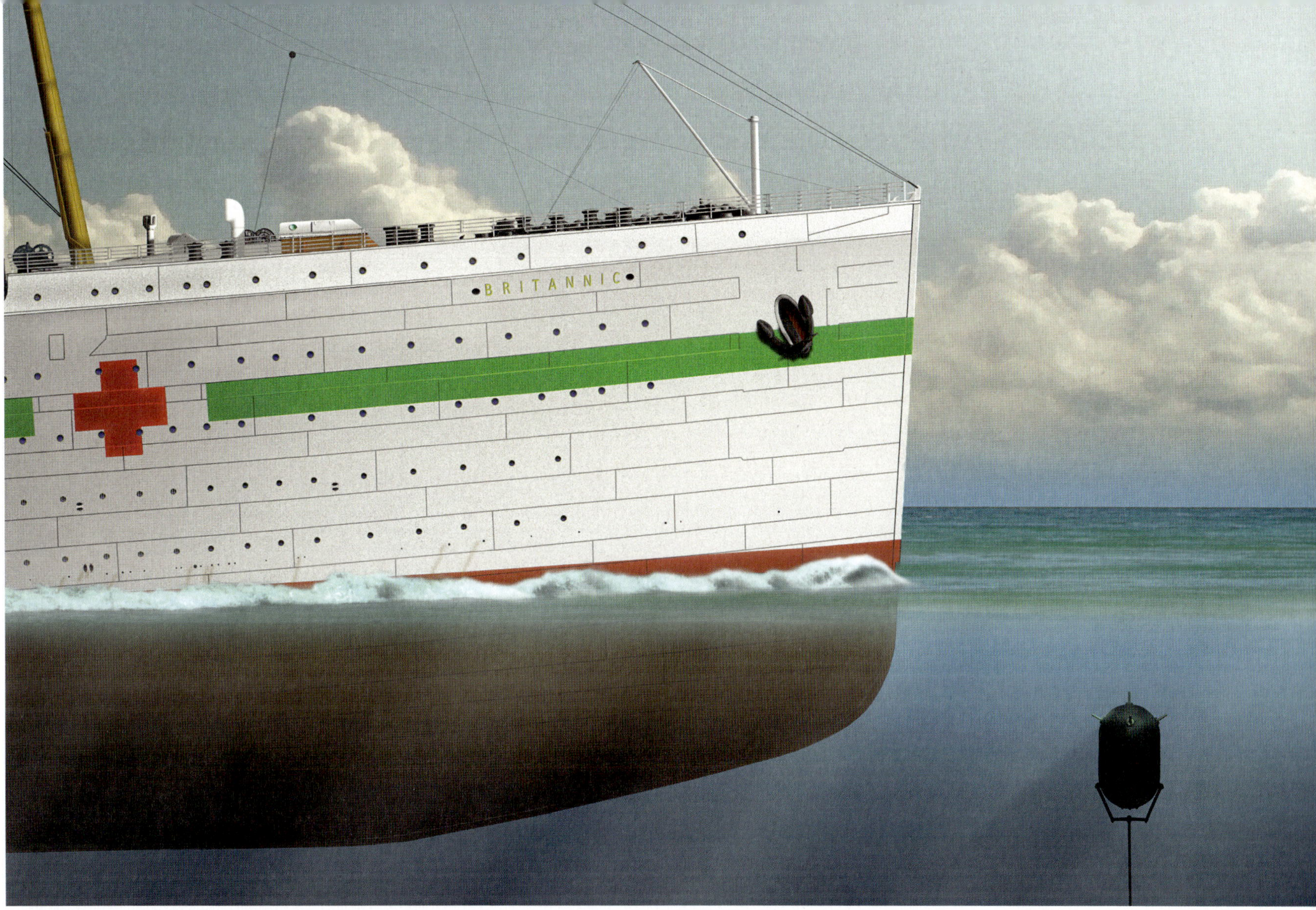


In one instance, one of *Britannic*'s officers began lowering away some of the lifeboats near his assigned station, but suddenly realised that the ship's propellers were still churning and halted their descent when they were just 6ft above the water. He then awaited orders from the Bridge, while the occupants of the lifeboats dangled in the air, cursing and yelling, apparently convinced that they were being left stranded and attached to a dying ship.

On the Shade Deck, located aft, a group of a dozen or so firemen commandeered a lifeboat and lowered it away, despite being the only occupants aboard. Some aboard *Britannic* began jumping overboard and Assistant Commander Harry Dyke ordered the firemen in the commandeered lifeboat to rescue the swimmers.

At 8.30 a.m., a lifeboat from the port lifeboat station assigned to Third Officer Francis Laws was lowered away. Laws was unaware of this, since those aboard the boat used the automatic release gear. A horrific and gory scene then played out, as the lifeboat drifted aft and was drawn directly into *Britannic*'s massive port propeller. The lifeboat was quickly reduced to kindling, and the occupants of the boats were chopped to bits.

Despite what had just happened, Boat No. 4, also on the port side, was lowered away. Like the lifeboat before it, it was also drawn aft into the turning propeller. As those aboard realised their oncoming peril, they began jumping overboard in an attempt to save themselves. Despite this, many were eviscerated. The blue waters of the Mediterranean turned red. One of the lucky survivors from this boat was Nurse Violet Jessop, who had also been aboard *Olympic* when she collided with HMS *Hawke* in 1911 and had survived the *Titanic* disaster in 1912!

Seconds to disaster: *Britannic* approaches the mine that would seal her fate. (William Barney)

## VIOLET JESSOP – SURVIVOR

Violet Jessop originally found work as a stewardess on the *Orinoco* for the Royal Mail Line, and she transferred to the White Star Line in 1910.

At White Star, she started on the *Majestic* in 1910. From there, she moved to a number of other White Star ships and transferred to the brand-new *Olympic* in June 1911. Consequently, she was on board when *Olympic* was struck in the stern by HMS *Hawke* on 20 September 1911. No one was seriously hurt, and though the ship was damaged, *Olympic* was able to limp back to Southampton for repairs. Miss Jessop left no account of this collision, though her biography does go into detail of her time on *Olympic*. She left the ship on 30 March 1912.

On 10 April, she signed on to *Titanic* at Southampton. Late in the evening of 14 April 1912, she heard a 'Crash! Then a low, rending crunching ripping sound, as *Titanic* shivered a trifle, and the sound of her engines gently ceased.' She and her roommate, thought to be Stewardess Elizabeth Leather, were roused by a steward and started waking their passengers and fitting them with lifebelts.

After a while, she and her roommate made their way up to the Boat Deck and escaped the sinking ship in Boat No. 16. As she entered the lifeboat, an officer – possibly Sixth Officer Moody – gave her someone's baby to look after. She watched *Titanic* sink into the sea and recalled:

> One of the huge funnels toppled off like a cardboard model ... The ship seemed to right herself with a broken back ... She settled for a few minutes, but one more deck of lighted ports disappeared. Then she went down by the head with a thundering roar of underwater explosions, our proud ship, the beautiful *Titanic* gone to her doom.

Lifeboat No. 16 reached the rescue ship around 6.45 a.m. At that time, Miss Jessop boarded her, along with the baby she had been holding the entire time. The baby was taken by an unknown woman survivor on *Carpathia*.

Miss Jessop was taken back to England on the ship *Lapland* with the rest of *Titanic*'s crew, and did not testify at either the American or British inquiry. By 5 June, she was back on board *Olympic*, continuing her job as a stewardess.

In April 1915, she went into training as a nurse for the VAD (Voluntary Aid Detachment) for the British Red Cross. She was then assigned to the third of the *Olympic*-class ships, *Britannic*. She only served one cruise, and for only nine days, before *Britannic* was sunk by the naval mine in the Aegean Sea, just off Kea.

On 21 November, she was at breakfast, and stated, 'Suddenly, there was a low deafening roar. *Britannic* gave a shiver, a long drawn out shudder from stem to stern, shaking the crockery on the tables, breaking things until it subsided as she slowly continued on her way.'

Miss Jessop returned below to help her 'special charge' to dress and put on her own heavy coat, and they then headed up to the Boat Deck. An officer rushed her to her assigned boat, No. 4, forward on the port side. She noticed the churning water at the stern of the ship, due to the propellers that were still spinning, as *Britannic*'s engines had not been stopped.

While Boat No. 4 was being lowered, *Britannic* was listing to starboard, and the lifeboat hooked itself on an open porthole, which tilted the boat precariously. Once clear, they continued lowering, scraping along the side of the ship going down and landing with a 'terrible impact' in the water, among a number of other boats. As these boats were held against the ship and being pulled toward the propellers, the men in the boats started jumping overboard to avoid being pulled into the props.

Never having learned to swim, Miss Jessop was very afraid of drowning, and hesitated about going into the water as almost everyone else had already done. She finally had no choice and jumped but was pulled underwater by her heavy coat. She later said about getting to the surface:

> I felt myself rising and my head came into violent contact with something solid, something that prevented me from reaching the surface. Then again, there was another terrific crash above me and something very solid struck the back of my head a resounding blow, but happily on that part where my plentiful hair was thickest. My brain shook like a solid body in a bottle of liquid. It was an unpleasant feeling and at the third repetition I imagined the next time would be the last.

It appears that she may have been pulled into the spinning propeller.

She did make her way free from the blades, but when she surfaced, she saw 'an indescribable scene of slaughter'. One of *Britannic*'s motorboats saw her in the water, came over and picked her up. As she climbed in, she discovered that her leg 'was badly torn and deeply gashed'. Her coat had been ripped and her 'hair had come loose; attached to the very end of it was the dilapidated remnant of the cap I always wore'.

Other survivors who were pulled into the motorboat were also in bad shape. Some died and were dropped overboard. More people were picked up, and then the boat made its way to the nearby island of Kea.

As a nurse, and not feeling that she herself had been that badly hurt, Miss Jessop did what she could for some of the injured, before finally being told to help herself to fresh clothing and clean herself up. She complained:

> My whole body felt one big ache, starting from my head which I felt would burst at any moment. But I was afraid to mention this lest I should again have people rushing at me with aspirins. It never occurred to anyone, then or later, to investigate the pain I endured.

Years later, she said:

> I leapt into the water but was sucked under the ship's keel which struck my head. I escaped, but years later when I went to my doctor because of a lot of headaches, he discovered I had once sustained a fracture of the skull!

Miss Jessop left the sea for a number of years, but then went back to White Star and *Olympic* in June 1920. She continued working at sea for many years, for a number of shipping lines, until the beginning of the Second World War in 1939. She spent the next years on land, returning to sea in late 1948, before finally retiring in December 1950. She passed away in May 1971.

As an aside, Miss Jessop wasn't the only survivor of the three *Olympic*-class ships' accidents. Fireman John Priest had been on *Olympic* during the *Hawke* collision, and on *Britannic* during its sinking. Lookout Archie Jewell was not on *Olympic* during the *Hawke* collision but was an early escapee of the *Titanic* disaster, leaving the ship in the first lifeboat launched, No. 7. On *Britannic*, Jewell was also in one of the boats chewed up by the propeller blades, but like Jessop managed to be pulled into a rescuing boat. Sadly, he would be killed when SS *Donegal* was torpedoed and sunk by a U-boat in April 1917.

By 8.35 a.m., *Britannic* was down by the bow and listing to such a degree that Captain Bartlett realised they would never make it to Kea in time to beach the vessel. He ordered the engines stopped. This command came just in the nick of time, since a third lifeboat had been lowered and was then drifting aft towards the propeller, which stopped churning right before the boat reached it. The lifeboat actually bumped into the still propeller and the occupants had to push off of it to get free of the ship. Bartlett, still unaware that lifeboats had been lowering without his permission, finally gave the order to abandon ship.

Despite the early chaos and disorder, the remaining lifeboats were lowered away quickly and safely. The heavy starboard list did interfere with the lowering of some of the boats on the port side, particularly where the traditional Welin davits had to be used, versus the gantry davits. By 9.00 a.m., the list was so great that the davits could no longer be used. However, most of those aboard had already been evacuated safely.

In all, twenty-eight lifeboats and two of the ship's motor launches had been successfully lowered. The motor launches cruised about, picking up swimmers from the relatively warm water, which was approximately 70°F.

As *Britannic* settled, the water pressure inside and outside of the ship temporarily reached an equilibrium, slowing the rate of sinking. Monitoring the situation from the Bridge, and with the ship largely evacuated, Captain Bartlett decided to make one last-ditch attempt at saving his vessel by reaching the shallows off Kea. He ordered the ship ahead, and the engines restarted. Unfortunately, the ship's rudder was no longer responding, since so much of it, as well as the propellers, were now above the waterline. The ship simply could not get adequate forward thrust. Seeing that continued efforts to save the vessel were futile, Bartlett rang off the engines for the final time.

Bartlett had *Britannic*'s steam whistles blown, signalling all remaining crew to abandon ship. Bartlett was then washed from the ship as the Bridge submerged around him. He swam to a nearby lifeboat and was taken aboard. He immediately began organising the survivors and overseeing the rescue attempts.

As *Britannic* settled, its bow hit the sea floor while a significant portion of the ship was still above the surface. The waters in the Kea Channel were shallow, approximately 400ft deep, compared to *Britannic*'s 882ft length. Already listing heavily, the vessel rolled further over to starboard as it sank, with all four of its funnels coming unseated from their mountings before it submerged, collapsing into the water.

By 9.07 a.m., *Britannic* was gone. It would be the largest ship sunk during the First World War. All of the survivors were soon rescued by Greek fishermen, Royal Navy vessels or by making it to the nearby shore on their own. In all, it took just fifty-five minutes to sink, far less time than *Titanic*, even with her additional safety features.

Miraculously, just thirty people were killed in the sinking, most of these being chopped up in the port-side propeller. It was fortuitous that *Britannic* sank when she was not fully loaded with Allied casualties, or the death toll would surely have been much higher.

In the aftermath of the sinking, many were quick to accuse Germany of having torpedoed *Britannic*. However, evidence indicates that the sinking was unintentional. *Britannic* struck a mine that was laid by the German submarine *U-73*. Researchers have found solid evidence that *U-73* laid mines in the area that *Britannic* sailed through. During a 2003 expedition led by the late diver Carl Spencer, evidence of a minefield was spotted near *Britannic*'s wreck, using sonar scans. This included mine anchors and chains.

While some survivors did claim to have spotted torpedo wakes, no evidence has ever emerged to support the notion, either in the form of U-boat logbooks or operational records, that any submarine discharged torpedoes at a ship of *Britannic*'s size and description on the date in question. While some might suggest that the attack could have been scrubbed from the records, such was not the case with even the most controversial sinkings during the war. It was far more typical for U-boat commanders to document the attacks, then come up with excuses to justify them later, such as with claims of mistaken identity.

This startling recreation depicts *Britannic*'s bow and forward decks slipping beneath the surface on 21 November 1916. (Ala Kermiche)

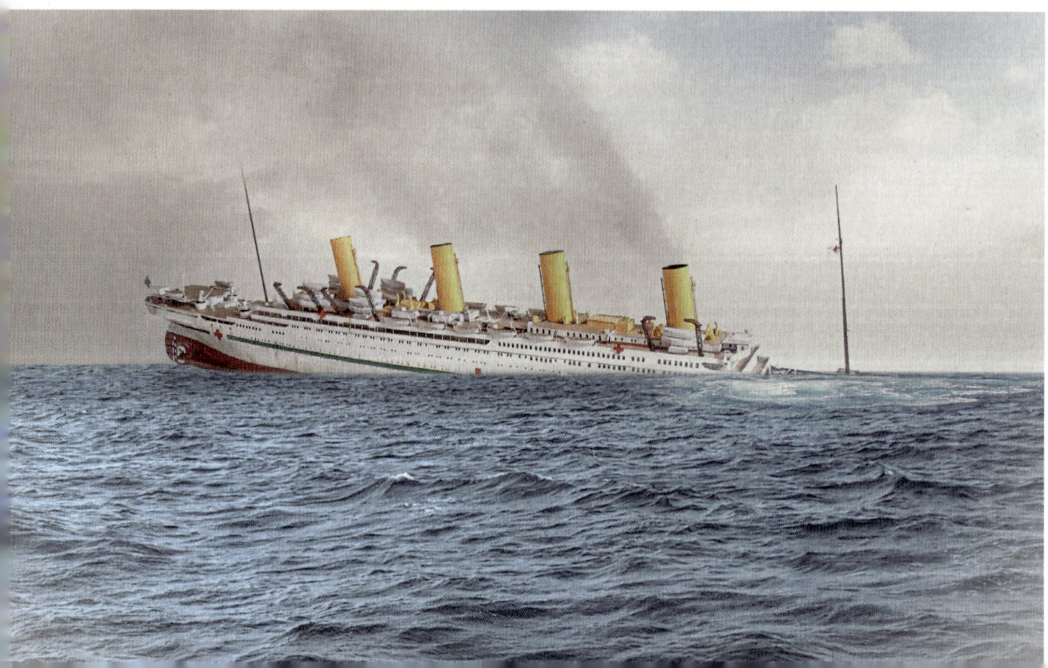

Tellingly, Captain Bartlett himself assumed that *Britannic* had struck a mine, since his ship was clearly marked as a hospital ship and there was no geyser of water accompanying the explosion, which was the telltale trait of a torpedo strike.

Regardless of the specifics of the sinking, *Britannic*'s loss was yet another significant blow to the White Star Line. They had now lost two of the three grand *Olympic*-class liners. Two of them, *Titanic* and *Britannic*, had never even completed a single transatlantic crossing carrying paying passengers.

Ironically, a large percentage of *Britannic*'s surviving crew, now without a ship, were later assigned to HMT *Justicia*. The story of this vessel serves as a coda of sorts to *Britannic*'s loss. *Justicia* began life in the Harland & Wolff shipyards as the *Statendam*, built for the Holland America Line. When the First World War erupted, the British Government negotiated for and acquired her use for the war effort. The ship was originally to be chartered to the Cunard Line, hence the name change, but Cunard simply did not have the necessary number of crewmen available to man her at the time. Instead, she was handed over to the White Star Line, which would be responsible for operating the troop transport.

From 19 to 20 July 1918, *Justicia* was torpedoed multiple times, and by at least two different U-boats. The ship – of a similar design but on a smaller scale to the *Olympic*-class liners, and also built by Harland & Wolff – was clearly well constructed. She took six torpedo hits before finally succumbing and being sent to the bottom. Once again, *Britannic*'s former crew members found themselves abandoning ship due to actions taken by U-boats. Thankfully, the majority, apparently a resilient bunch, would survive to fight another day.

When all is said and done, the story of the *Britannic* is a tragedy, albeit on a much smaller scale than that of *Titanic*, because of the smaller number of lives lost. However, in a completely separate type of tragedy, *Britannic* herself was denied the chance to enjoy what surely would have been a robust and celebrated post-war career.

SIX

# THE WRECKS OF *TITANIC* AND *BRITANNIC*

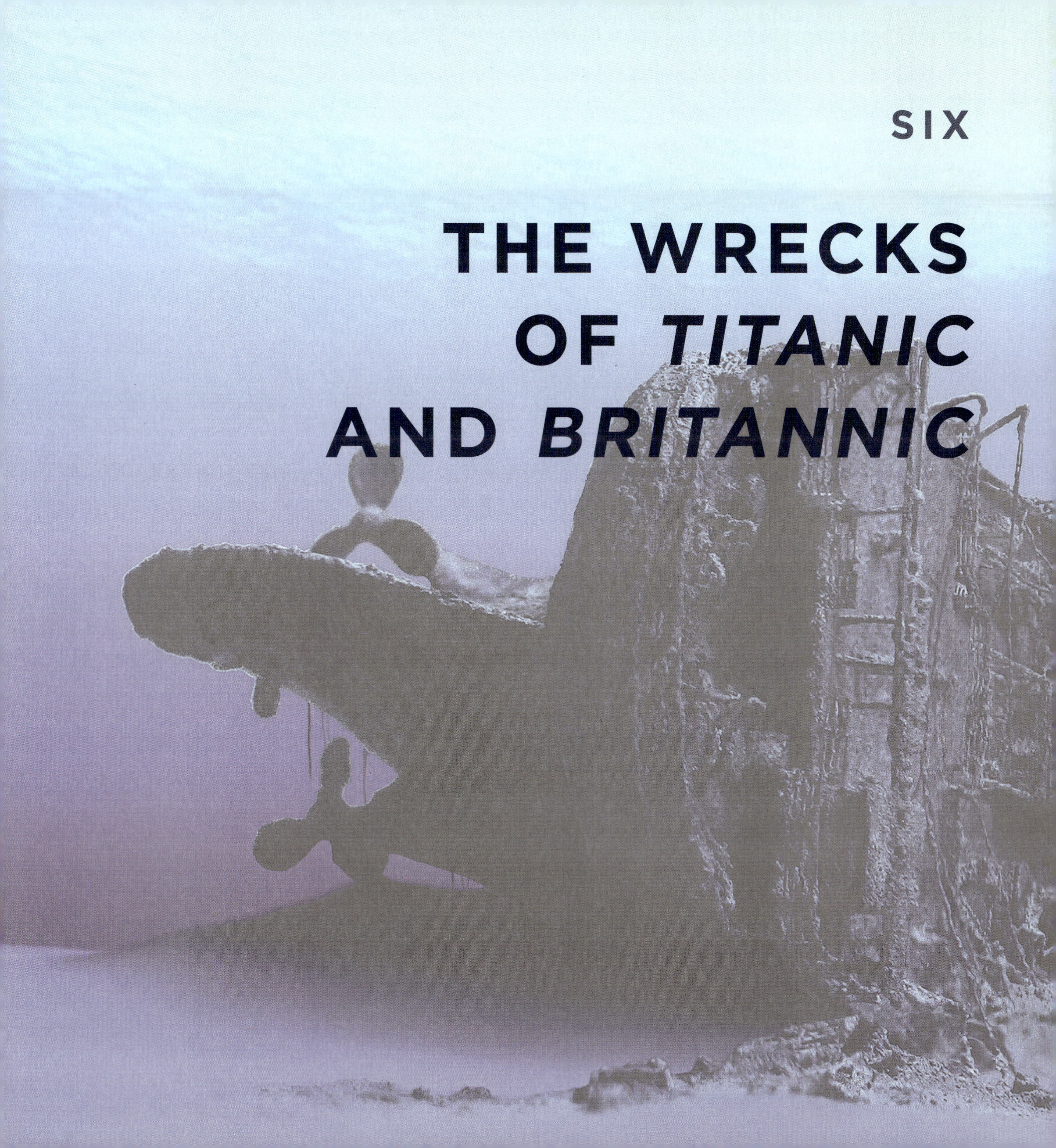

It is remarkable, when one thinks of the labour, the effort and the expense that was lavished on *Olympic*, *Titanic* and *Britannic*, that only one of the three ships ever completed a transatlantic voyage carrying paying passengers. But while *Titanic* and *Britannic* died prematurely and tragically, whereas *Olympic* had a long and happy career, in a sense *Titanic* and *Britannic* have lived far longer than their successful sister.

Decades after *Olympic*'s steel was cut up and recycled and her lavish interiors torn out of the ship and strewn all over the world, we can still visit the wrecks of *Titanic* and *Britannic* today. These wrecks can give us some feel for what both of these ships were like before they were lost, and what happened to them as they sank, or afterwards as they decomposed on the ocean floor.

The two wrecks could hardly have ended up in more different conditions on the sea floor. Whereas *Titanic* heaved her stern high out of the water and tore herself to shreds as she left the surface, *Britannic* sank at an oblique angle, saving her from the risk of breaking up the way *Titanic* did, and ended up on her side in much shallower water. The depth and temperatures of the water in which they sank have also greatly affected their natural state of deterioration. We will now briefly tour each of these wrecks through words and art.

## THE WRECK OF *TITANIC*

*Titanic* broke into two primary segments as she sank from the surface, just abaft of amidships. The bow section, from the tip of the prow to the area just forward of the No. 3 funnel, left the surface almost entirely intact and remained largely that way as it descended through over 12,000ft of water into the crushing, black abyss. Tank tests and computer simulations show that the bow would have largely retained the course that it was on when it broke clear from the stern, and would have fallen in what is referred to as a 'fall-and-stall' descent pattern – diving steeply and picking up speed as it dove, then stalling out, recovering its nose-down attitude until nearly on an even keel, before diving again and repeating this pattern a number of times as it went.

When the bow arrived at the sea floor, it was in a somewhat 'nose-down' attitude, and its prow knifed down and forward into the sediment, burying itself up to the anchors forward. The areas behind the Well Deck were not supported because of the break, and the vast weight of the wreck, along with its momentum, broke the ship's keel near the aft end of the Well Deck. As the remainder of the bow slammed down flat upon the sediment, the upper decks amidships 'pancaked' down precipitously, crushing the Lounge, the Reading and Writing Room, the Dining Saloon and everything else in those areas into a mass of compressed steel.

The quick stop tore the cover for the No. 1 Hatch off and flung it forward of the wreck, and this became the northernmost piece of wreckage. The side shell plating in the area where the ship's back broke at the aft end of the Well Deck was placed under enormous strain and ultimately tore open on the starboard side, looking as if it had been blown open by an internal explosion and exposing spaces like the Mail Room.

Incredibly, part of the iceberg damage is still visible, namely that which opened up Boiler Room Nos. 5 and 6 and actually sealed the ship's fate. The damage is right where eyewitnesses like Fireman Fred Barrett said it would be, and although the damage further forward is hidden under the mud, this damage is openly exposed to a viewer's eye.

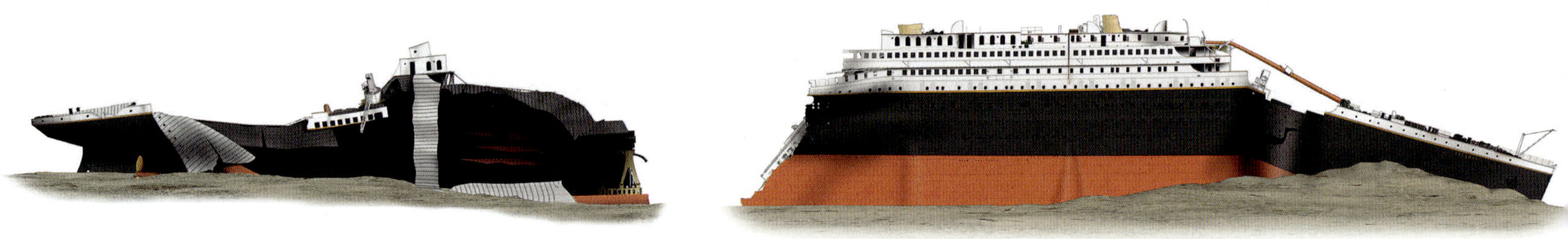

A starboard profile of what the bow and stern section of *Titanic*'s wreck looked like on the morning of Monday, 15 April 1912. (Michael Brady)

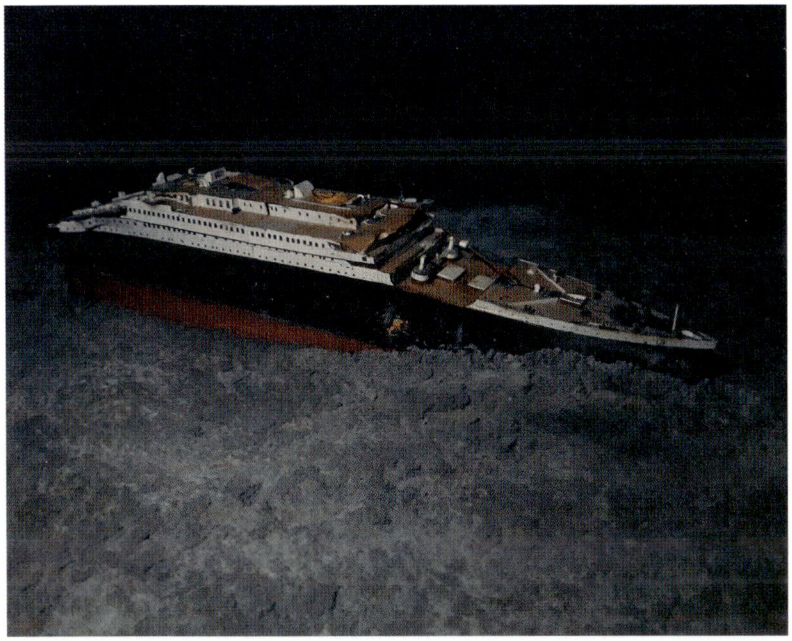

*Above left:* This splendid artistic view shows what *Titanic* probably looked like on the morning after the disaster. (Ala Kermiche)

*Below left:* This view shows what the break at the aft end of *Titanic*'s bow section looks like when seen from above, looking forward. Again, the boilers are visible, though firmly seated in their beds, and the Lounge roof and Compass Platform can all easily be identified on the model. (Ala Kermiche)

*Below right:* An artistic view looking at *Titanic*'s port side shows a submarine shining its lights along the starboard hull plates of the wreck, near where she broke in half. The boilers in Boiler Room No. 2 are clearly exposed, and the upper decks are pancaked down, crushing them out of existence. (Ala Kermiche)

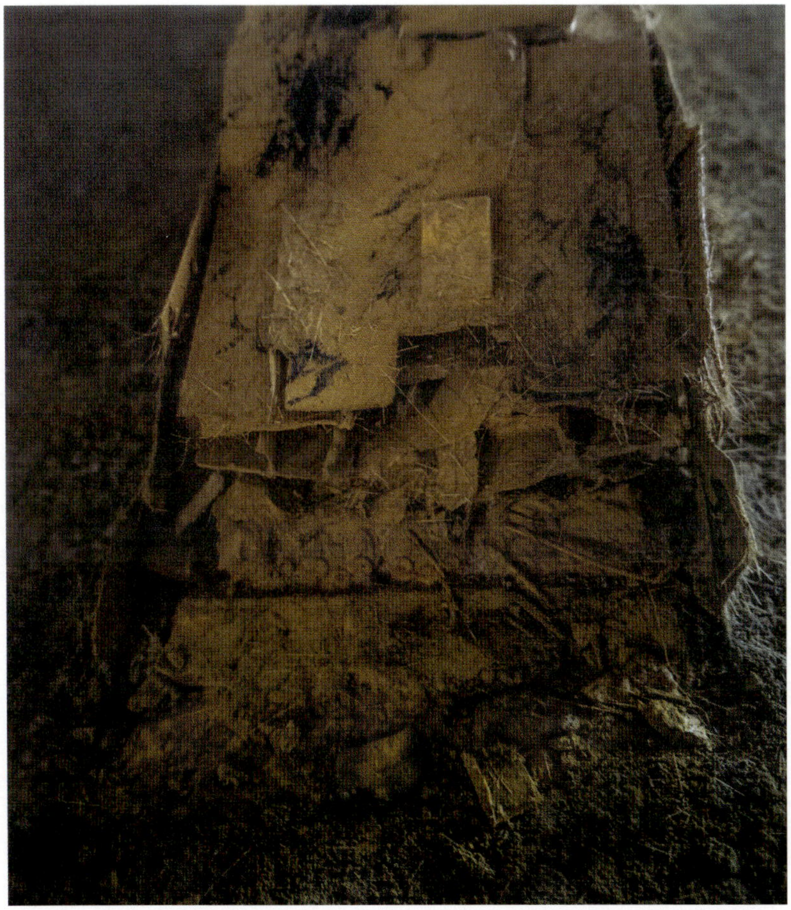

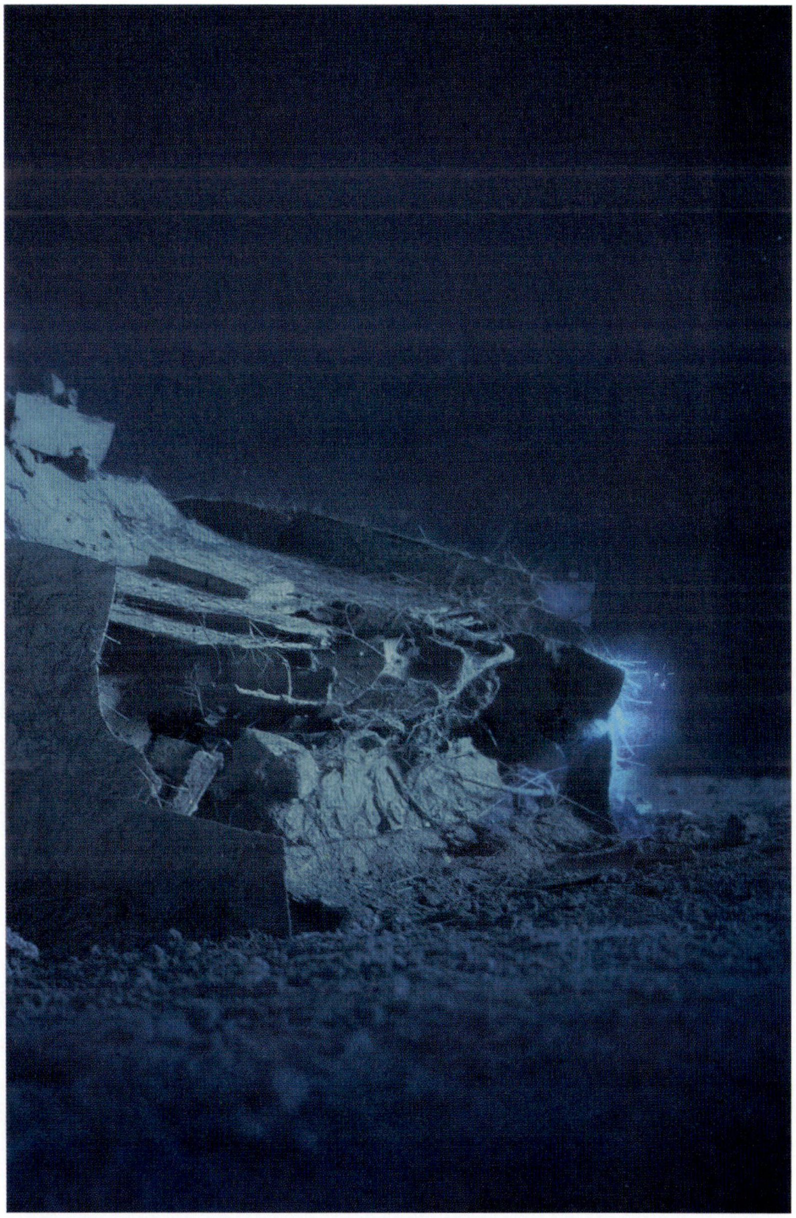

The stern section of *Titanic* experienced a far more horrific fate. This mess really began at the surface as the two main portions of the ship were separating. Two large sections of superstructure and upper decks – known generally as the two 'tower' sections – at the forward end of the stern, began to come apart. Two sections of double bottom that held the two main halves together also tore off, snapping off the front cylinders of each reciprocating engine in the process. As the stern reared back up, it also spun in an anticlockwise direction, indicating that something was still connecting the two halves of the ship and creating an otherwise inexplicable rotation. By the time the stern sank completely beneath the surface, it was in a nearly vertical, nose-down orientation. As water rushed past its open, shredded face, great damage was caused to its forward spaces and the Poop Deck was peeled back on itself.

Eventually, the stern began to flatten out and began a very wide clockwise circular spin as it descended. As pressure built up, what few spaces had not flooded likely imploded. By the time it reached the sea floor, not only was it spinning like the hands across the face of a watch, but it also seemed to have inverted into a stern-down position to at least some degree.

When it came into contact with the sea floor, much of the starboard side shell plating was peeled off. The wing propellers were shoved up into an unnatural position so that the props themselves are still visible even though the stern is deeply buried. Sadly, the controversial centre propeller, which first-hand documentation shows was of a three-bladed design, remains invisible beneath the mud.

An artistic side view of a submersible and ROV exploring the forward superstructure of *Titanic*'s wreck. (Ala Kermiche)

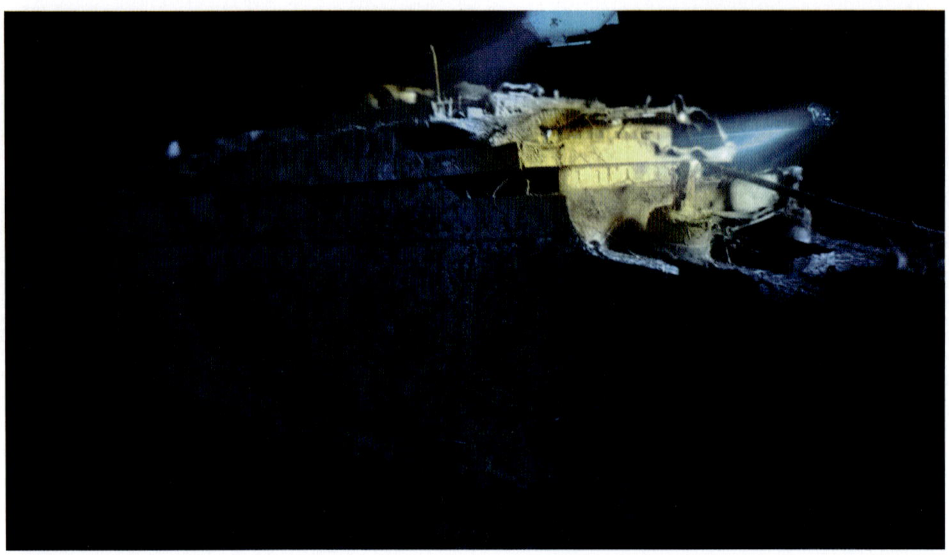

By the time the wreck had settled to the sea floor, the stern section looked almost like a bomb had gone off inside of it as it sank. Whereas the bow section retained much of its dignity, the stern is an ugly mess, torn open so deeply that the after three-quarters of the two reciprocating engines stand exposed like sentinels at its face.

Sadly, despite years of study at the wreck site, many questions about the way *Titanic* broke apart and sank remain unanswerable through forensic analysis alone. Numerous analyses based on the forensic evidence alone have been performed, and none of them agree with each other.

Damage patterns sustained by portions of the wreck once it left the surface can cause significant confusion when trying to piece together what happened before *Titanic* actually sank. Strict non-disclosure agreements and archival footage that is kept all but completely out of the public's eye further cloak the ship's demise and current condition in mystery.

*Titanic*'s wreck was discovered on 1 September 1985 by a joint US–French expedition headed by Dr Robert Ballard and Jean-Louis Michel. Since then, numerous expeditions to the wreck site have been mounted, and virtually every square inch of the wreck has been photographed. Thousands of artefacts have been recovered from the site – from teacups and passengers' luggage to a Welin davit and even the 17-ton piece of *Titanic*'s starboard hull plating that became known as the 'Big Piece'. Many of these objects have been carefully preserved and put on stationary or travelling exhibitions around the world.

Much has been learned about what *Titanic* was like in 1912 through these expeditions. Some have sent remotely operated vehicles (ROVs) deep inside the ship, capturing fantastic imagery of the remains of the First Class Reception Room and the forward end of the Dining Saloon, individual passengers' cabins and even exploring as far down as the Turkish Baths on the starboard side of F Deck, last seen by Steward Wheat at about 12.45 a.m. on the night of the disaster as water was trickling down the stairs from E Deck above.

From the open gangway door and the inner gates being opened on the port side of D Deck, we have learned that the men Lightoller sent below to open the door to help load boats from closer to the water likely succeeded in their task, even though the plan itself did not succeed. We've learned that the pattern on the lift gates was slightly different from that found on *Olympic*. We know that the clock on the mantle of the Strauses' suite still stands in place after falling 2½ miles from above, and countless additional details that would otherwise have been lost to history have now been recorded for posterity.

It has also become clear that *Titanic* is returning to the elements and being consumed by nature. In the more than thirty-five years since she was discovered, nearly half of the time that she originally spent on the sea floor before being found, it is plain to see where her form is degrading and weakening.

The windows along the forward A Deck Promenade, installed so late before the maiden voyage, are falling out. Certain portions of the ship that were once visible or accessed by ROVs are no longer reachable. The more lightly built superstructure seems to be taking the brunt of the decay, with some sections of the Boat Deck collapsing down into A Deck. It is likely that the actual hull of the ship, built much more robustly, will take longer to collapse, but it is only a matter of time.

Salvage efforts once raised so much controversy that the *Titanic* community was torn apart over the matter, with accusations of 'grave-robbing'. Over the years, however, to a large degree, the quarrelling on this subject has abated to a dull roar. The things that we have learned from the recovered items and the way that the displays of these items have brought *Titanic* to people around the world has touched many individuals who would never be able to visit the wreck in person. Furthermore, no photograph taken at the wreck site can stand in for being in the actual physical presence of these items, and their loss if left on the sea floor is assured.

The bankruptcy proceedings surrounding the legal owner of the wreck site and legal battles over who owns the artefacts continue to make headlines. Yet nothing can hide the fact that, one day, *Titanic*'s wreck will be lost forever. Many feel that it is best to preserve what we can, while we can, and attempt to keep her memory alive as a tribute to those who perished aboard her.

## THE WRECK OF *BRITANNIC*

Whereas *Titanic* was a brand-new liner engaged in passenger service when she sank, her younger sister *Britannic* was born into the horrors of the First World War and sank in a sea that she was never intended to sail. In spite of the improvements to her design after *Titanic*'s loss, she sank nearly three times faster, thanks to the enormous size of the damage caused by the mine she struck.

Yet *Britannic*, in spite of being robbed of the civilian service she was supposed to have enjoyed, is a fantastic wreck. She lies on her starboard side in around 400ft of water, with some sunlight still outlining her enormous form.

To this day, she remains the largest passenger liner wreck on the ocean floor. Whereas *Titanic* became a shattered, fragmented ghost of her former self as she plummeted into the sea floor, *Britannic* sits in one primary piece and even retains her full original width.

The worst damage occurred to her forward sections, where her weight came down on the bow as she was sinking, but was not yet fully submerged, snapping her keel and tearing open her decks and shell plating below the Forward Well Deck.

*Below:* This near-perfect profile view depicts how *Titanic*'s bow section appears in more recent years. Compare its condition with the profile view of the ship the morning after the disaster. (Ala Kermiche)

*Bottom:* This view shows the depth of *Britannic*'s wreck in the Kea Channel. A surface vessel gives some sense of scale. (William Barney)

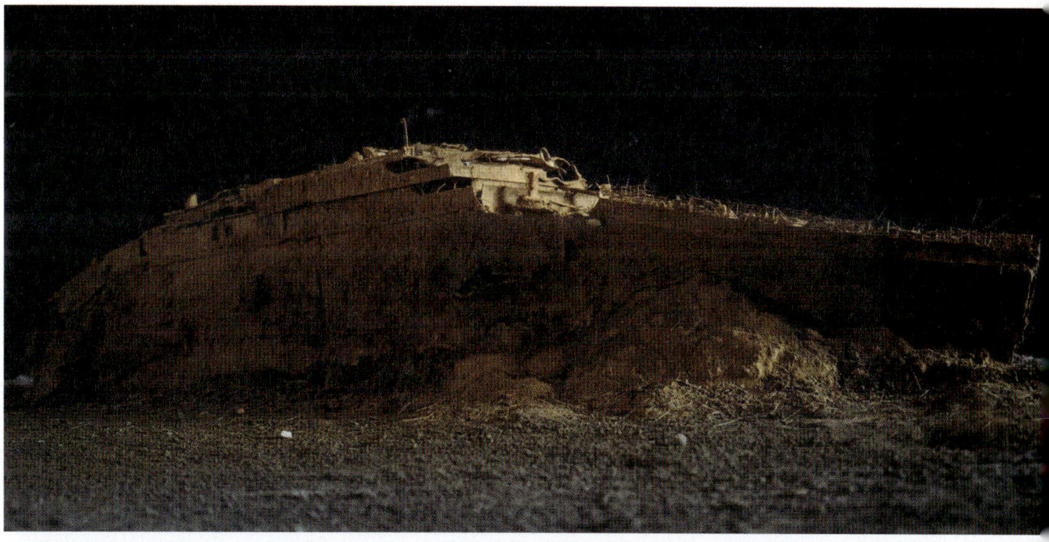

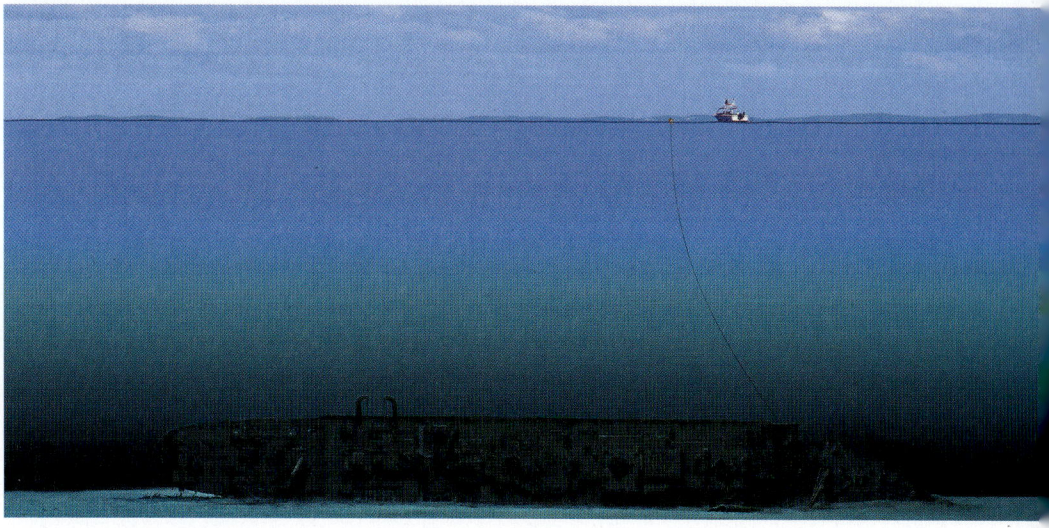

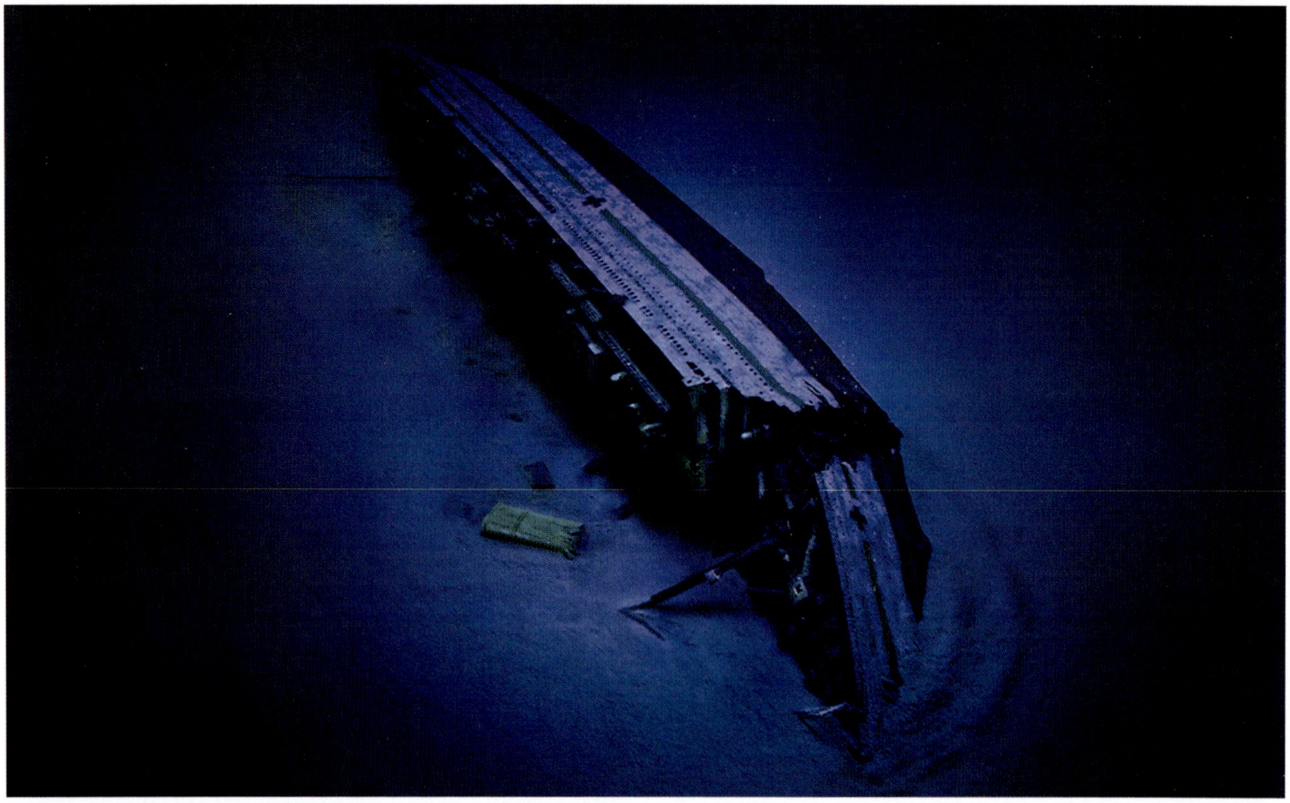

*Right:* How *Britannic* was likely to have appeared on the morning of 22 November 1916. Note the forward funnel on the seabed, so close to its original location on the ship. (William Barney)

*Below:* All of *Britannic*'s propellers are visible on the wreck today, and the rudder remains turned slightly to port. Note the enormous gantry davits reaching for the surface from the after port Boat Deck. (William Barney)

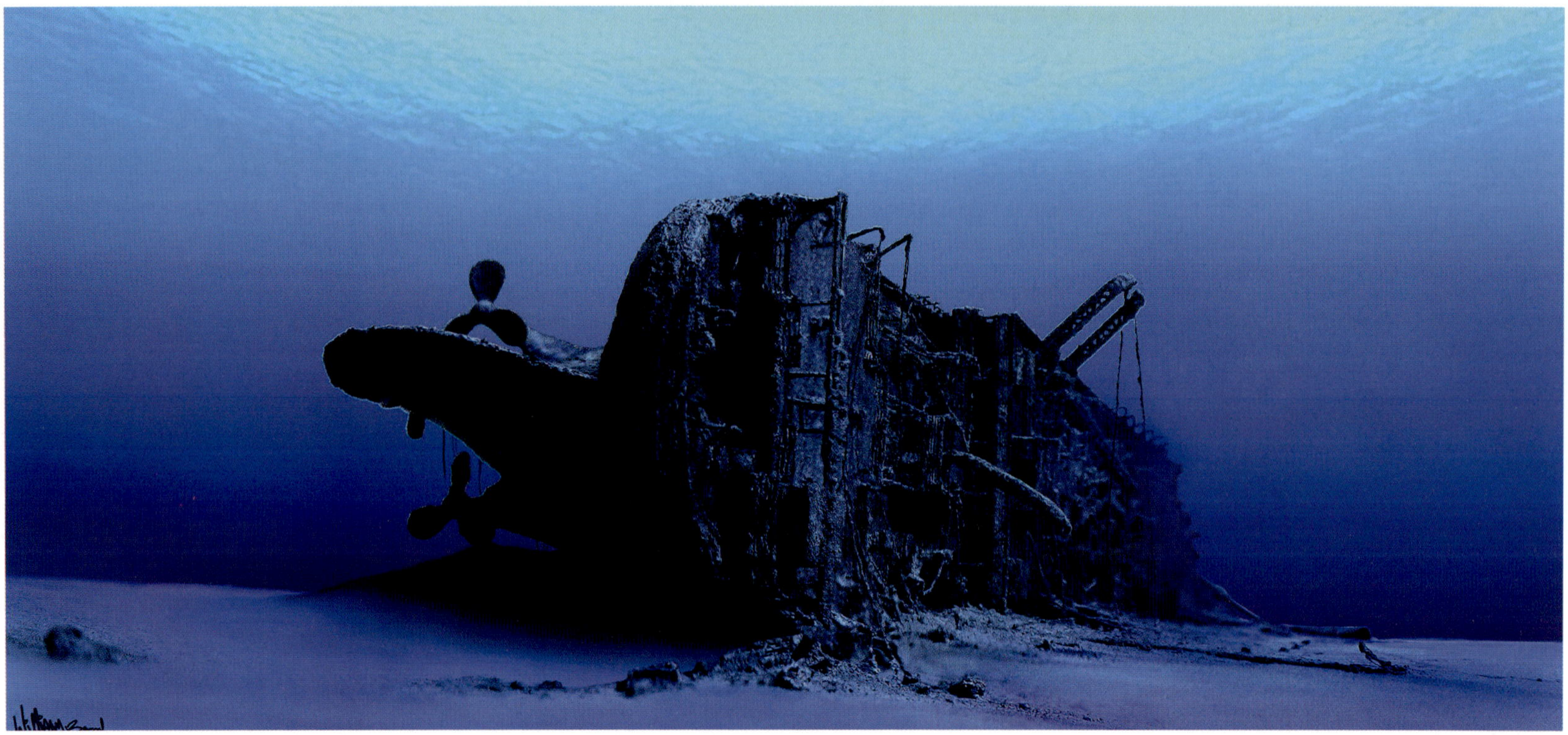

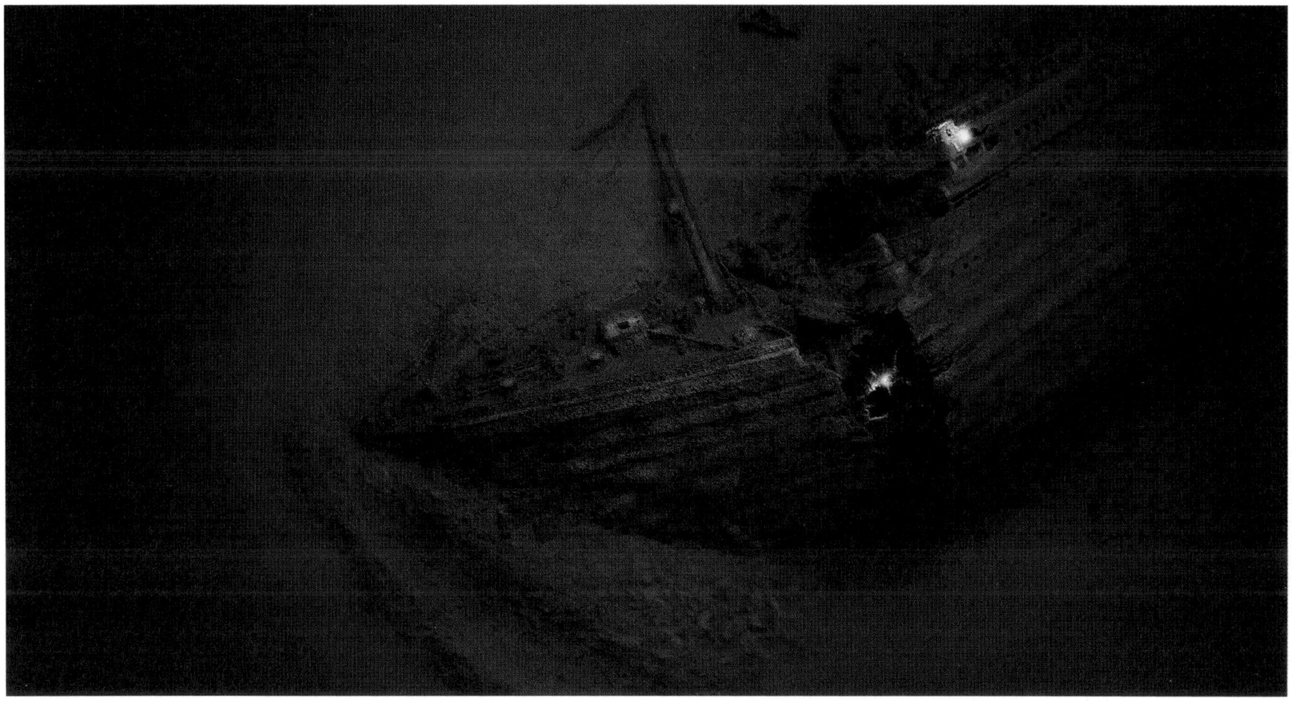

Divers penetrate deep within the wreck of *Britannic* in this artistic view. In reality, divers have identified open watertight doors within the wreck that either contributed to or caused the ship's rigorous system of watertight subdivision to be overcome after the mine exploded. (William Barney)

*Britannic* was originally found by French oceanographer Jacques Cousteau in November 1975 and explored by an expedition he mounted in September and October of the following year. Further expeditions to the site have been carried out over the years, and divers have managed to penetrate deep within the ship's remains, capturing footage that, for example, proves that the watertight door between the Firemen's Tunnel and Boiler Room No. 6 did not close after the mine exploded and the watertight door between Boiler Room Nos. 5 and 6 also remained open. It has also been proven that *Britannic* struck a mine, rather than being torpedoed, as the remains of the mine responsible for the disaster have been found.

All three propellers remain in place on *Britannic*. The port wing propeller, responsible for all of the casualties during the disaster, still looks like it is reaching for the sky above. The centre prop is of a single-cast, four-bladed design. It is more or less a match for *Olympic*'s original design, and the design *Olympic* again adopted after briefly sporting a three-bladed one similar to *Titanic*'s following her 1912–13 refit.

It seems that by the time *Britannic*'s propellers were ordered, experience with *Olympic* had proven that the three-bladed centre propeller did not have as many advantages as originally hoped, and *Britannic* was thus fitted with a four-bladed one from the start. The starboard prop is accessible beneath the massive rudder, which remains turned slightly to port.

Divers have penetrated *Britannic*'s boiler rooms and this has shown clearly that the boilers in that compartment remain seated firmly in their beds, in spite of the fact that the ship rests on her side. Contrary to the pathetic state of *Titanic*'s wreck, suggesting bad manufacturing and cost-cutting to the more conspiracy minded, the way *Britannic* survived her violent demise really is a testament to the fact that Harland & Wolff built the *Olympic*-class ships very well. *Titanic* was simply, to all intents and purposes, put through the proverbial shredder.

Sadly, in spite of the many things we have learned from numerous expeditions to *Britannic*'s wreck, two divers have also perished: the first to die was diver Carl Spencer, who passed away in 2009 while returning to the surface after his dive. Then, in September 2019, a second diver, Tim Saville, passed away while at about 115m depth, diving the wreck. Not much information has been released about the more recent incident, but the deaths prove the lesson that *Titanic* and *Britannic* keep giving us – a lesson that rings as true today as it did in 1912 and 1916 – in spite of the best-laid plans and precautions, at times technology fails us. And when it does, the results can be incredibly costly.

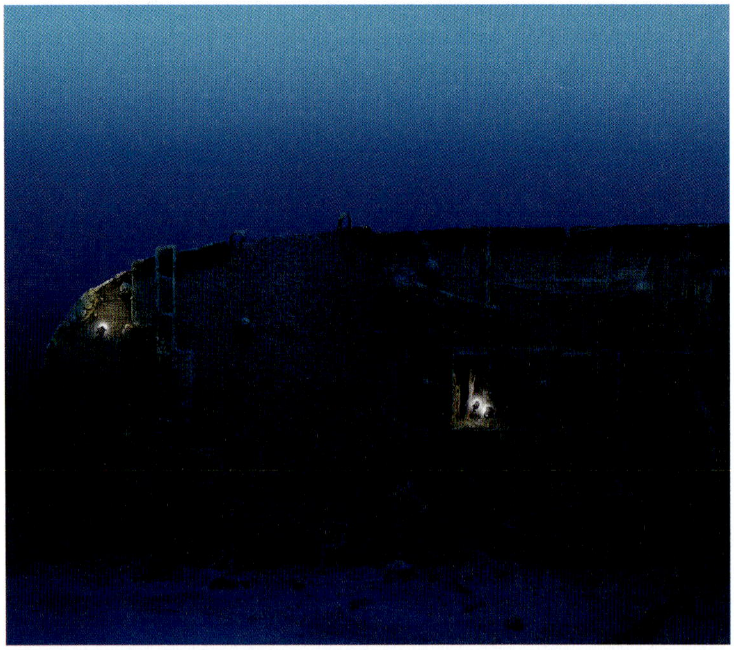

*Right:* In this incredible view of *Britannic*'s stern, divers explore the ship's remains. (William Barney)

*Below: Britannic*'s wreck on the sea floor today. This view shows the cant of her broken bow, crushed and twisted out of position as it rammed the sea floor, and shows the length of her Boat Deck. (William Barney)

* * *

*Titanic* and *Britannic* were never intended to be wrecks littering the floor of the sea. They were intended to conquer the sea. Sadly, those dreams never came true for either ship. Yet, through their loss, and through ongoing investigation of their wrecks, we can learn about an era entirely different from our own, an era that is becoming increasingly difficult for us to understand in the tumultuous world that we live in today. This is particularly true of *Titanic*, which passed into eternity before the horrors of the First World War, with all of the resulting reverberations that we feel right into our own time. As we continue to learn about the *Olympic*-class liners, and as we endeavour to bring them to life once more, we preserve the memory of an era – one that otherwise might be all but forgotten today. Better yet, we preserve the memories of those who lived during those times, and we pass on their stories with the utmost respect and care for future generations to learn from.

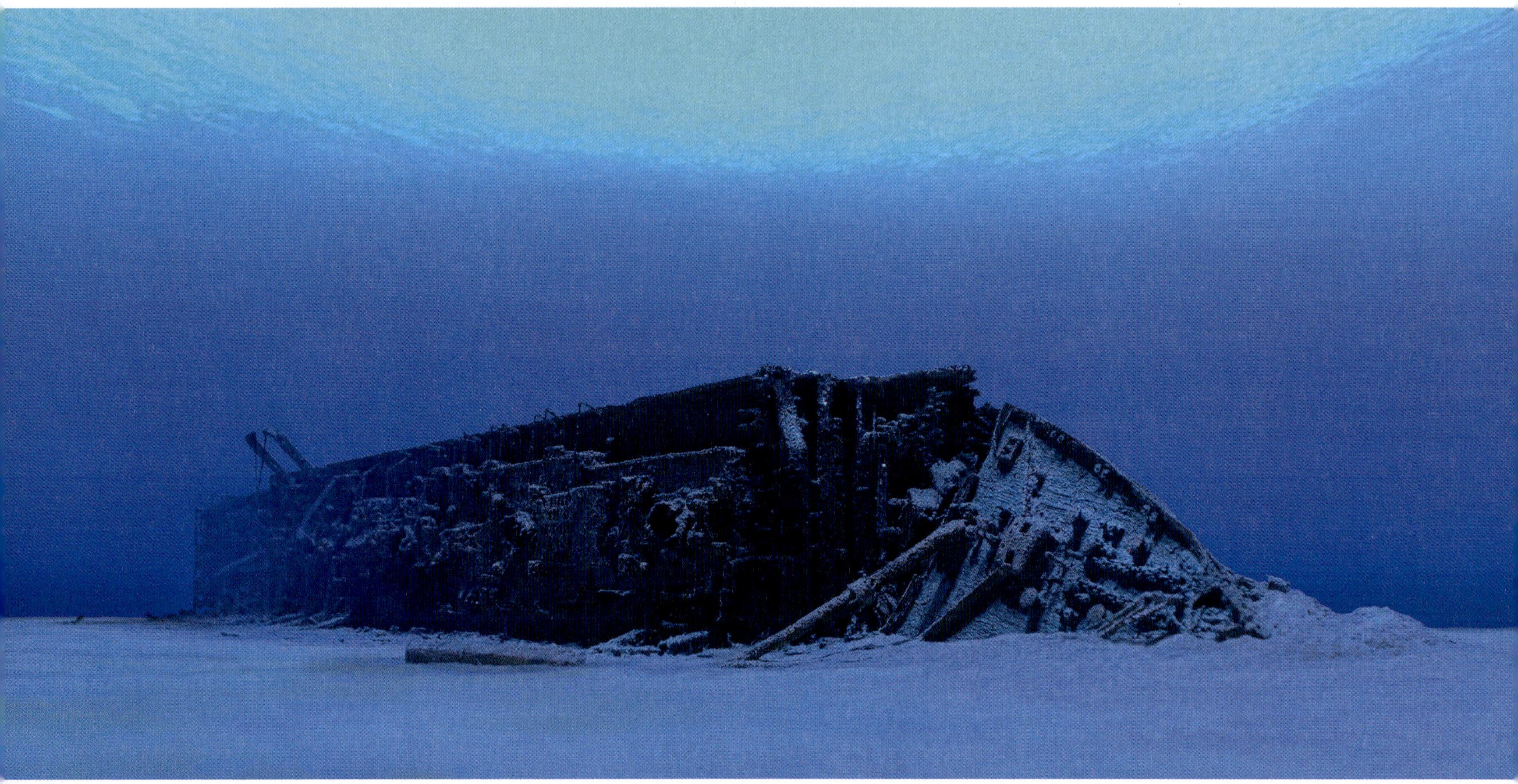

SEVEN

# THE ARTISTS BEHIND THE IMAGERY

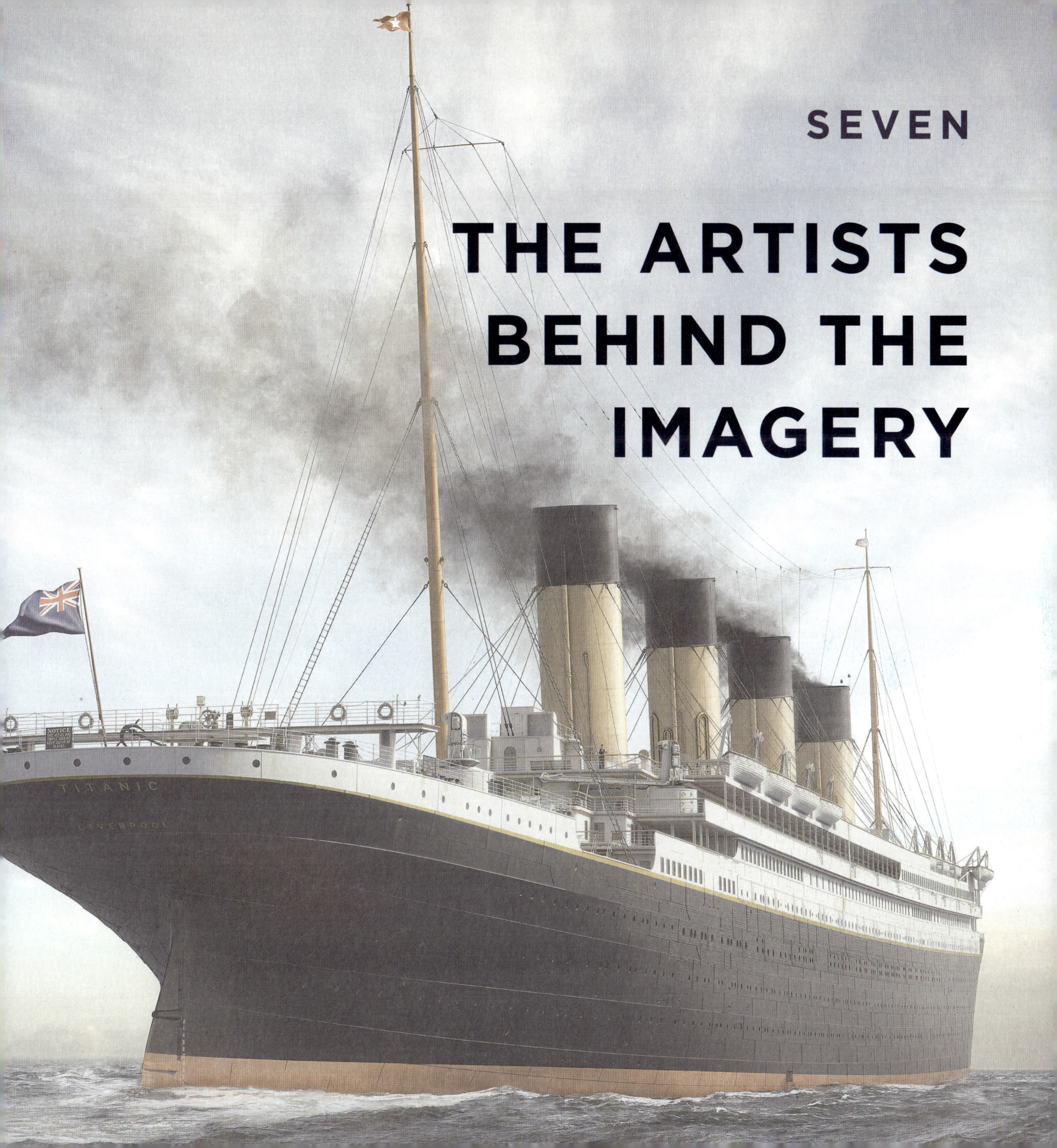

This book would never have been written without the participation of an incredibly talented group of artists from around the world, working in every conceivable medium available. Their talent and their enthusiasm for the *Olympic*-class ships knows no bounds, and we wanted to take the opportunity to briefly introduce each of them to you, and to the world in general.

We hope that you will take the time to become acquainted with their works, and to support their talents. We also hope that, together, we can continue to encourage enthusiasm for preserving the knowledge of *Titanic* and her sisters as we move further and further from their era.

## WILLIAM BARNEY

**Nationality:** American

**Website or social media contact:**
www.williambarney.wixsite.com/williambarney/home

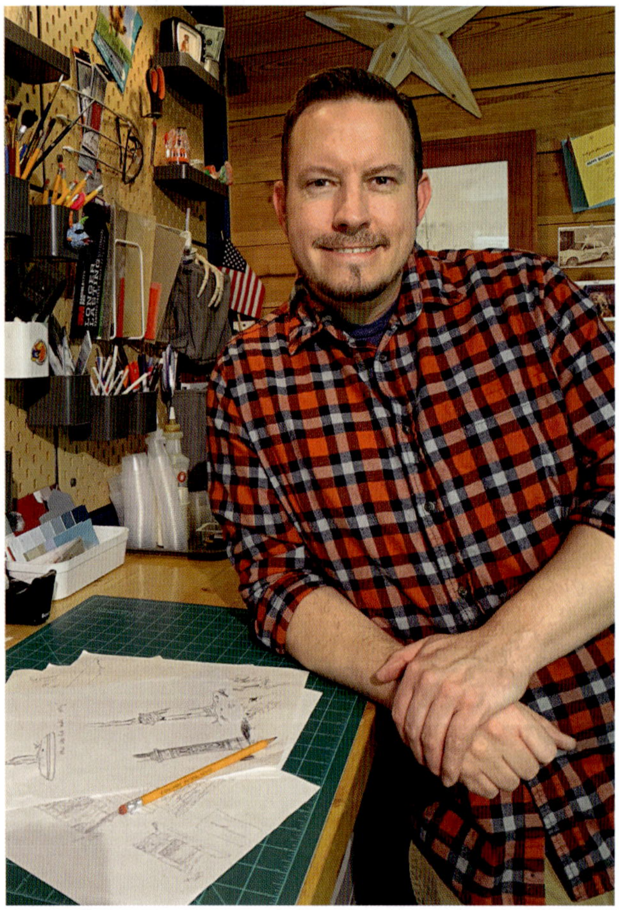

William Barney

When news broke that *Titanic* had been discovered on 1 September 1985, William Barney was enthralled, yet it was the artwork of Ken Marschall that inspired William to grab a piece of paper and pencil when he was between the ages of 6 and 8. Of Marschall's *Titanic* sinking artwork, William says, 'They were so beautiful and realistic that I felt like I was there!'

Over the years, as each new expedition goes out to the wreck and comes back with new images and data – such as the RMS *Titanic*, Inc. expeditions of the 1990s, or those made by James Cameron and his team – William has felt a new thrill of discovery. He is also fascinated by *Britannic*, and works very hard to bring her back to life through his art, terming her work as a hospital ship as 'heroic'.

William works in a variety of media, from pencil or pen sketches to paintings, to building models and then using photography and Photoshop to bring them to life. He also does from-scratch digital artwork depicting *Titanic* and the wreck of *Britannic* and says, 'When I build a model or do paintings by hand, I love the old-school methods of showing artistic talent.'

Most of William's favourite ships are the older Atlantic liners with funnels, which are so different from modern cruise ships. He says:

> For over a decade, I have especially tried to raise awareness about *Britannic*, bringing her out of *Titanic*'s shadow and into the spotlight. I can see people starting to become more and more interested in *Britannic*, doing models of her, doing artwork, and more. I love to see *Britannic* remembered, rather than forgotten.

## JAN BOUDESTEIN

**Nationality:** Netherlands

**Website or social media contact:**
www.deviantart.com/jboudestein

As a small child, Jan went on holiday with his parents, and they took an overnight ferry trip. The experience ignited a love of passenger ships and liners. He soon learned about the tragedy of *Titanic* through a children's book, and a few years later, when he was only 12, he heard the news that *Titanic*'s wreck had been discovered. When he read Ballard's book, *The Discovery of the Titanic*, 'the tragic and fascinating story of the great liner really started to grab my attention'.

He was particularly enthralled by the artwork of Ken Marschall, as seen both in Ballard's book and later in Don Lynch and Ken Marschall's book, *Titanic: An Illustrated History*. Seeing these paintings inspired Jan to begin drawing and painting *Titanic*. He paints in a variety of media but prefers acrylics, oil on canvas or oil on panel. The painting process itself is his main focus. However, if the finished painting is to be refined in the service of historical or scientific accuracy, he photographs the art and makes minor improvements in Photoshop.

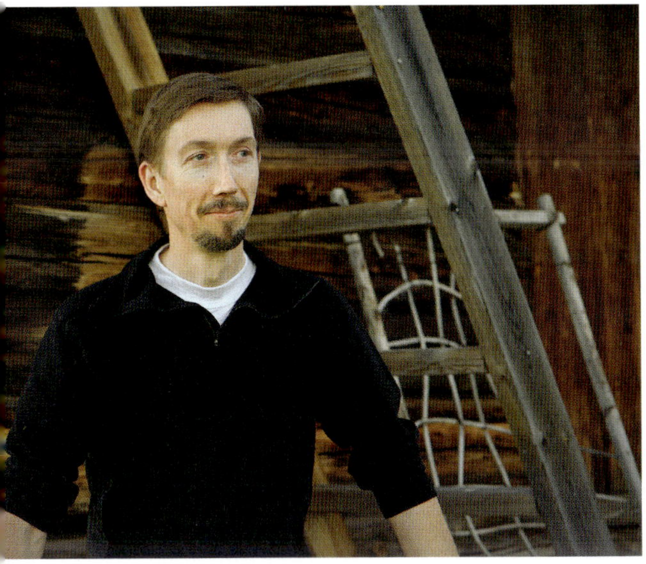

Jan finds something magical about the golden age of ocean travel, and is fascinated with the romance of the transatlantic liners. He loves the ex-Cunard liner *Queen Elizabeth 2* which he once had the privilege of sailing on and which he misses greatly. Yet, he is particularly drawn to the sheer beauty of the *Olympic*-class liners, and to *Titanic* in particular. He says:

> *Titanic* doubly fascinates me. She embodies all of what fascinates me about the golden age of ocean liners, but she is also a symbol of so many other things, including the end of that era, and of pointing the world forward to a new one. Her tragedy shattered the illusion of her age, and brought an end to the seemingly unbridled confidence, progress, and hubris felt then.

Although Jan paints for his own personal enjoyment, he very much hopes that others take pleasure in the pieces of art that he creates and shares.

## MICHAEL C. BRADY

**Nationality:** Australian

**Website or social media contact:**
www.oceanlinerdesigns.com
**Instagram:** @linerdesigns

Michael's father emigrated to Australia on the P&O liner *Strathnaver* in 1959, and Michael was exposed to *Titanic* with the release of the 1997 film *Titanic*. After that, he was given a steady stream of *Titanic* documentaries to watch, and found himself fascinated by the artwork of Ken Marschall, 'whose dedication to detail and accuracy produced artwork often described as photo-realistic'.

At the age of 12, he joined the website juniorgeneral.org, which specialises in the depiction of historic military figures and equipment in side-profile views, which can be used in table-top games. He recalls, 'My Mother would listen with intent as I recounted facts and figures about the *Titanic* ad infinitum, and she has always encouraged me to pursue my passions and interests.'

Michael now creates orthographic depictions of many liners, intended to represent ships at various stages in their careers, both transatlantic liners and post-war immigrant ships like those his father came to Australia on. 'I do not bill these as "plans",' he says, 'but rather "illustrations", since they fall somewhere in between art and design.' He adds, 'My aim from the outset was to draw ship profiles at such a scale and magnitude that viewing them in person at full size would, for a moment, create the same sense of excitement that one might have felt when seeing the real thing at a wharf.'

Of the *Olympic*-class, Michael says, 'The story of these three ships is so vast in its scale and depth that it is possible to just always keep learning, and considering fresh perspectives that I would not personally have thought of.'

He has a deep, special connection with *Titanic* in particular, because it was her story that drew him to learn about liners in general. Yet he admits, '*Olympic*'s career is perhaps my favourite of any ship in history.'

Because of his ground-breaking work in drawing attention to many lessser known liners, he hopes that his greatest legacy will be helping to preserve their memory, in addition to those of *Titanic* and her sisters.

*Left:* Jan Boudestein

*Below:* Michael C. Brady

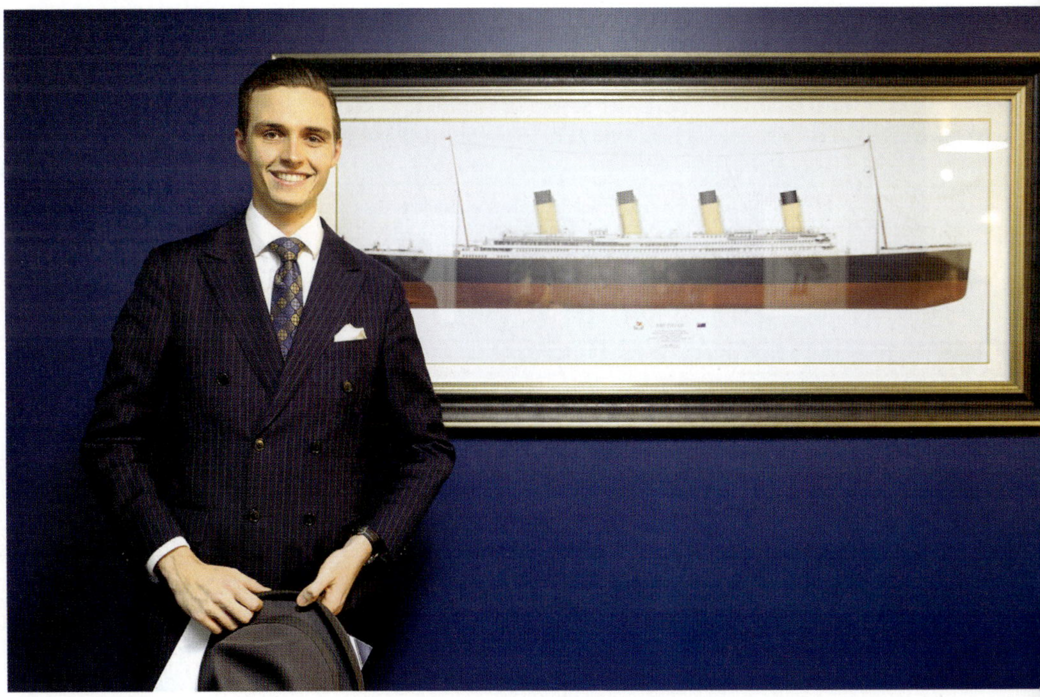

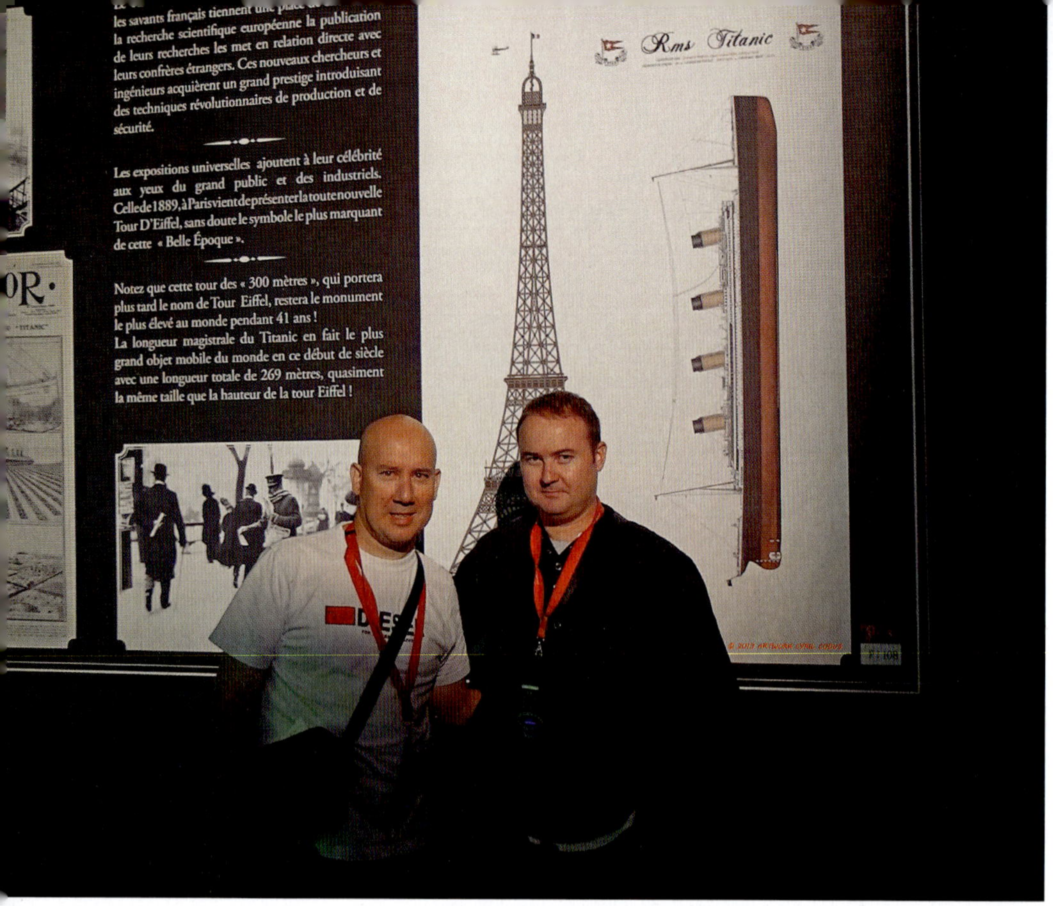


Cyril and Lionel Codus

## CYRIL AND LIONEL CODUS

**Nationality:** French

**Website or social media contact:**
www.rms-titanic-artwork.com
www.adesdubd.skyrock.mobi

Brothers Cyril and Lionel Codus have been intensely researching liners for years. Cyril loved *Titanic* from a very young age, fascinated with the concept that this beautiful new ocean liner sank on its maiden voyage. He began drawing plans of the ships from an early age. At the age of 14, he came across a scale model of *Titanic* and 'bought it immediately'.

When Lionel was 15 years old, he remembers Cyril sharing with him his copy of the Robert Ballard book *The Discovery of the Titanic*, featuring the artwork of Ken Marschall. Lionel was hooked as well, and was inspired to begin learning as much as he could about *Titanic* and her sisters.

'She was the most luxurious floating vessel of her time, and the largest ever built,' Lionel says. 'She was at the forefront of technology for her time, and she will forever remain a majestic liner.'

Cyril has done several plans of these liners: the RMS and HMT versions of *Olympic*, RMS *Titanic*, and both the intended RMS and the HMHS versions of *Britannic*, not to mention one of the tender *Nomadic*.

Lionel Codus does acrylic paintings on canvas, but he also works on detailed illustrations with the software program Illustrator, creating warships, private ships, cruise ships, planes and all sorts of vehicles. His work is intended to further historical research for himself and others who are attempting to learn specific details about these ships.

Lionel loves the *Olympic*-class ships, whereas Cyril has a special attachment to the Cunard liner *Lusitania*. Cyril is particularly fascinated by the fact that long before the *Lusitania* was designed, naval architects of the time were working on refining the lines of a ship built to travel at very high speeds. 'The *Lusitania* won the Blue Riband several times,' Cyril says, 'and the ship was a very large and luxurious one for her time.'

It is unique to see two siblings working together so closely on a shared passion for learning about and illustrating ocean liners, but they have much they would still like to do. Both would like to work on creating 3D projects of the *Lusitania* and *Titanic*, in order to help spread the information that they have learned and help to recreate the 'perfection and beauty' of *Titanic*.

## ELANG ERLANGGA

**Nationality:** Island of Bali, Republic of Indonesia

**Website or social media contact:**
**Facebook:** Elang Erlangga
**Instagram:** @dalemelangerlangga
**Email:** dalemelang@gmail.com

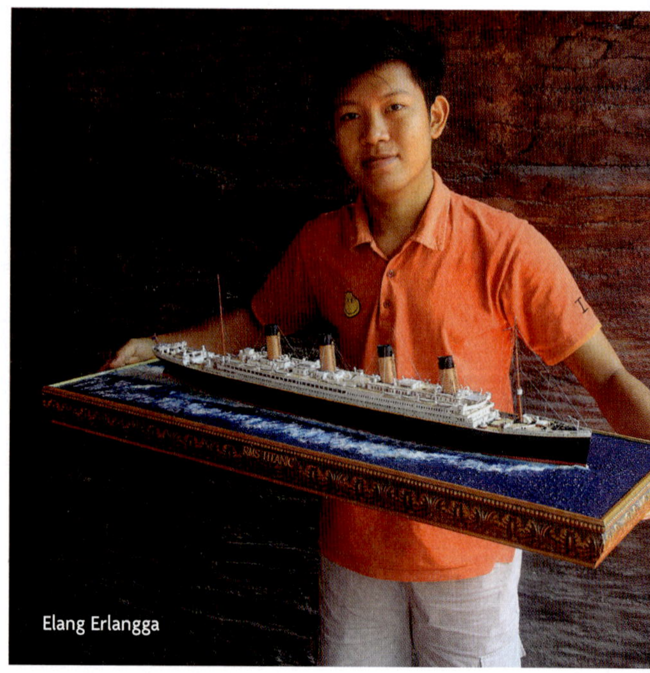

Elang Erlangga

As a millennial born in 1996, Elang has no direct connection to the Atlantic liners of old, yet when he saw the James Cameron film *Titanic* at the age of 9, in 2005, something clicked. 'At that moment I was struck by the beauty of the ship's design, and the story seemed surreal to me,' he says. Then he began to see artwork from artists such as Stuart Williamson and Ken Marschall, and by 2017 he was inspired to begin depicting the liners in his own original artwork.

Elang uses traditional painting techniques and tools. He is constantly researching the subject, in order to try to make the paintings as accurate as possible, looking into such details as the setting, the events taking place and the details of the ship itself, all before he puts pencil to paper or paint on canvas. He always tries to make sure that he doesn't ignore the historical facts for art's sake.

His list of completed pieces continues to grow, but he says, 'I still have many thoughts in my head that I want to depict in artistic form. I want to show many less-explored events in the liners' history.'

Of all the *Olympic*-class ships, Elang considers *Titanic* to be the most beautiful, in spite of her catastrophic end. However, he adds that he loves *Olympic* because of her illustrious career, and he has plans to explore and then portray many more specific events in her life. Indeed, he has come to view his artistic expression as more than a hobby. He says, 'I want to develop my skill, and commit myself to a professional career as a maritime artist. I hope that my work will inspire others and can be enjoyed for years to come.'

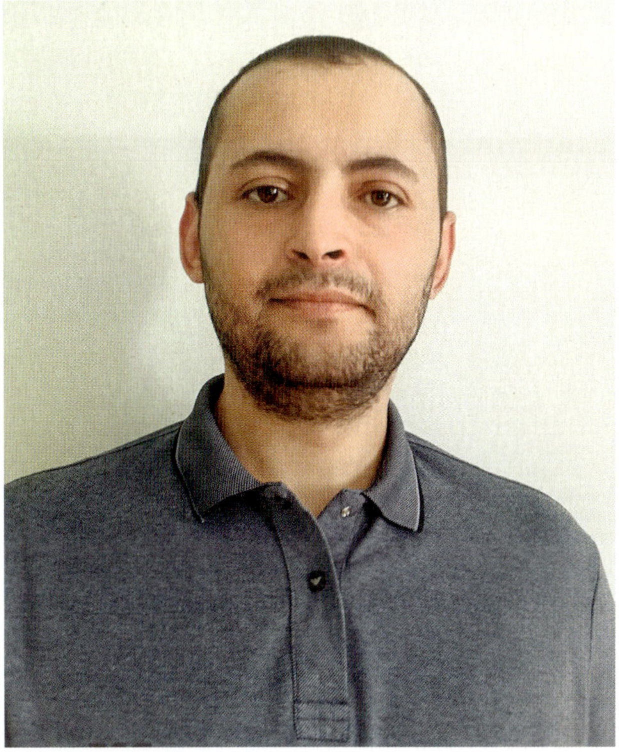

Ala Eddine Kermiche

## ALA EDDINE KERMICHE

**Nationality:** Algerian

**Website or social media contact:**
**Facebook:** Ala Kermiche (alakermiche7)
**Instagram:** @alakermiche7
**Email:** k.alaeddine964@gmail.com

Every time new photos of *Titanic* surface, historians around the world scramble to see what they can learn from them. Some while back, we were astonished to see a series of what seemed to be vintage photographs of *Titanic*'s fitting out appearing online – breathtaking, incredible views of the sort that we had always wanted to see. When we learned that they were actually artwork, not only were we stunned at their quality but we wanted to know more about the talent behind the works, and how the artist had achieved such realism.

We learned that the artist was Ala Kermiche, and he would begin by photographing models of the three *Olympic*-class liners. Then he would assess how much editing the photo would need 'in post'. For simple transformations, he might employ an Android app called Photo Editor, but for more complex edits, he turned to Adobe Photoshop to help him in making the photographs look like vintage photos.

Ala has a passion for the history of ocean liners, the Edwardian era and anything related to the sea. His particular focus of research on *Titanic* has to do with the time period between her launch and when she was complete. 'The photos are intended to shed some light on a phase of *Titanic* history when she was converted from an empty 25,000-ton giant hull into one of the finest British-built ships,' he says.

Although he has a special affinity for *Olympic* during her first year of service, saying that her appearance during that era rendered her the 'finest-proportioned' of ships, he is also a tremendous fan of the Cunard liner *Aquitania*. He says that, to him, subjectively, *Aquitania* was 'the best ship that was ever made or ever will be made'.

## THOMAS LYNSKEY

**Nationality:** American

**Website or social media contact:**
**Facebook:** HFX Studios
**YouTube Channel:** Part-time Explorer, soon to be rebranded as 'HFX Studios'

'When I was about 3 years old, my dad used to make plastic models,' Tom remembers. 'While he did that, he'd let me play with an old, smashed-up 1/350-scale *Titanic* model that he had built but accidentally broken. I had no idea what it was, but it was fun to play with.'

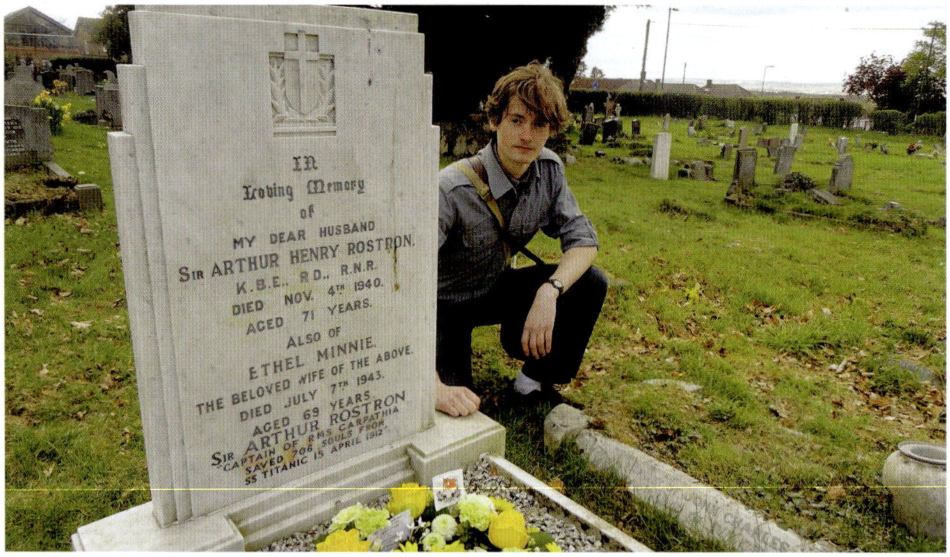


*Above:* Thomas Lynskey with the headstone of Arthur H. Rostron, the captain of RMS *Carpathia*

*Right:* Eugene Nesmeyanov

Eventually, Tom picked up a copy of Robert Ballard's *Exploring the Titanic*, and after watching the 1958 film *A Night to Remember*, he was totally hooked. In 1997, when the James Cameron film was released on his 6th birthday, he remembers, 'I suppose that sealed the deal.'

Tom was originally inspired by his dad to try to find interesting ways of depicting *Titanic* and other ships. He also found inspiration in the work of people like Bill Sauder, Parks Stephenson and Ken Marschall.

Tom now works almost completely in the digital medium, attempting to tell stories from history, especially maritime history, in an engaging and educational manner. He makes films, documentaries and direct VR experiences, and is enjoying working with a great team of artists and animators to bring these experiences to life. He says:

> I cannot explain what fascinates me so much about *Titanic* and her sisters. 1912 was such a foreign world to us – a world that has changed so much due to global conflicts and changes in life, both for better and worse. The era fascinates me, and that era is perfectly encapsulated in the story of *Titanic*. Nearly every aspect of it is represented on it.

In addition to his keen enthusiasm for *Titanic*, Tom has branched out to help tell the stories of other ships of the era: '*Britannic* holds a special place in my heart because of the effort my team and I put into rebuilding and animating her when I produced *Britannic: Patroness of the Mediterranean*.'

Every now and then, Tom hears from someone that his work on films or other projects has inspired their love of history. 'I don't need to teach them everything,' he says, 'and I'd be honoured if their studies surpass the knowledge I have on these subjects. I just want to have helped kindle that spark.'

## EUGENE (EVGENY, 'GENYA' OR 'JENYA') NESMEYANOV

**Nationality:** Russian

**Website or social media contact:**
**Facebook:** Eugene Nesmeyanov

Eugene's father was a retired navy officer and had a professional interest in the *Titanic* disaster. Around 1992 or 1993, Eugene began to read his first books on the subject from his father's home library. It was at about the same time that he was introduced to films like *SOS Titanic* (released to theatres in the UK and aired on television in a longer version in the United States in 1979). This film, he recalls, introduced him to the world of visual and cinematic representations of *Titanic*.

The first *Titanic* artwork that made a strong impression on him came from Fortunino Matania's 'Women and Children First', as well as Chris Mayger's 'Sinking of *Titanic*', which adorns the cover to the 1998 Russian edition of Walter Lord's *A Night to Remember*. Later on, he was introduced to the work of Ken Marschall and many other great *Titanic* artists.

Eugene's love of *Titanic* moved him to undertake extensive research on the matter, and he eventually became a published author. His excellent book, *The Titanic Expeditions: Diving to the Queen of the Deep 1985–2021*, is a must read. He says:

> *Titanic* provides us with an endless number of topics for discussion. Some of the problematic topics are debunked and their 'cases closed', but new ones invariably emerge in their place. Many of these mysteries will never be solved, and I think this is most intriguing and challenging.

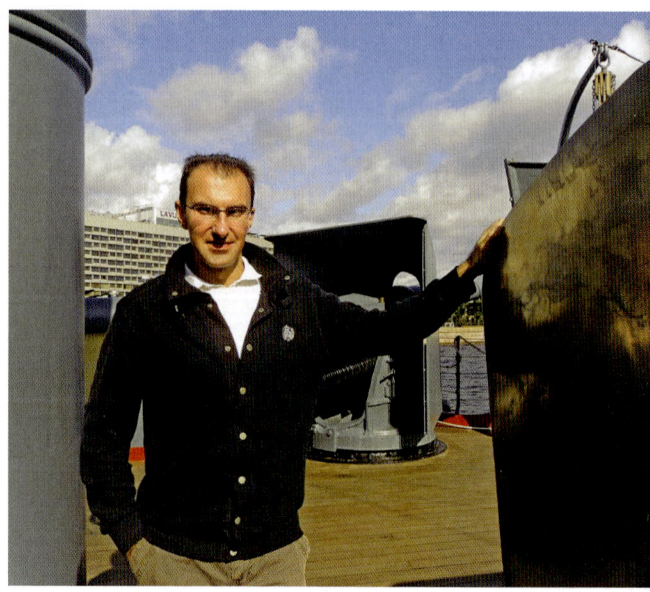

In addition to becoming an author on the subject, Eugene was moved to create a series of artwork featuring *Titanic* – which seems natural when one remembers that his mother is a professional artist and teaches both drawing and painting. Eugene refers to his own work as 'quite traditional' and of the 'classical Russian school of drawing and painting'.

'I've always been *Titanic*-centred, so to speak,' he says, pointing out that he restricts his focus primarily to *Titanic* because of time and budgetary constraints. He is quick to highlight the generous help he has had from his friends through the years to further his research on the subject. When asked if he hopes that his art might leave some sort of legacy, he admits that he had never thought about that, because he considers himself first an author and a researcher. But he adds, 'At the same time, I do try to use at least some of my artworks as illustrations in all of my books, along with the works of others.'

## ALEXANDRE DE ARAÚJO NUNES

**Nationality:** Portuguese

**Website or social media contact:**
**Facebook:** dearaujonunes
**Instagram:** @araujoalex1988

Alexandre was first attracted to *Titanic* and her sisters because of their decorations. He was fascinated with the idea of having so many different decorative styles combined, and all sent to sea in each individual ship, almost like a seagoing museum of European art and history. 'I really think the *Olympic*-class ships were a true work of art and an open book for us to learn from,' he says.

After seeing the work done by digital artists in bringing the ship's interiors back to life, including the work that went into the recreations for the game *Titanic: Honor and Glory*, Alexandre was inspired to take the next step and build one of *Titanic*'s staterooms in physical form. He began by doing extensive research, using photos of the *Olympic*-class ships. He did a lot of design work digitally in CorelDRAW, and then rendered all of these into a variety of materials, such as wood, urethane, plaster, iron and aluminium. He did many of the custom cuts in the wood carvings with a laser cutter.

The project began as something very personal. Then he began to take photographs of his stateroom and even showed what it might have looked like 'populated' by individuals in period garb, sharing these photos for *Titanic* enthusiasts to enjoy.

But it isn't all about new photos. 'I love old photography,' he says. 'Each photo I find allows me to continue improving the details.'

Alexandre is still fascinated by *Titanic* and her sisters, and points out that there is something new to learn every day about them. He does, however, admit to giving *Titanic* an edge over *Olympic*, in his mind, as she was the 'upgraded version' of her older sister and 'won the title with her great name'.

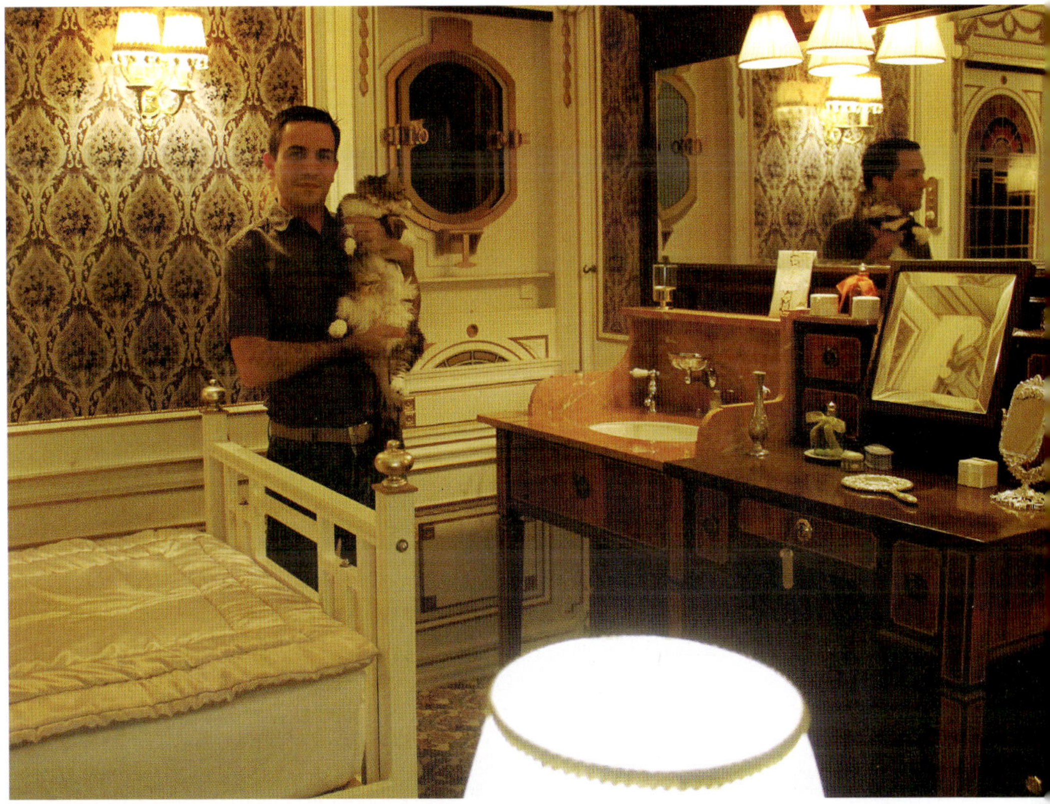

Alexandre de Araújo Nunes and his cat, Susy, in *Titanic*'s rebuilt Stateroom B-58 or C-63

At this point, the lovely stateroom recreation featured in this book remains incomplete, but the project forges forward. We wish him and the project all success and applaud the craftsmanship that has already gone into the project.

## DAVID E. OLIVERA

**Nationality:** American

**Website or social media contact:**
www.davideolivera.com
**Instagram:** @davidolivera

David Olivera recalls the Robert Ballard book *Exploring the Titanic*, which prominently featured the art of Ken Marschall. 'It opened my young mind to the bygone age of ocean liners,' he says, adding, 'Ken Marschall will always have a special archetypal artist mentor role in my life. He directly and indirectly inspires my artistic journey.'

David works mostly in oil paints on wood panels or canvas, but he also uses acrylic paints interchangeably. Each painting – including the one we asked him to do for this book, showing Francis Browne taking his famous photograph down *Titanic*'s port side on the morning of 10 April – begins with a mental vision. Then he merges the original concept with strong historical evidence, attempting to present a unique visual based firmly on historical accuracy.

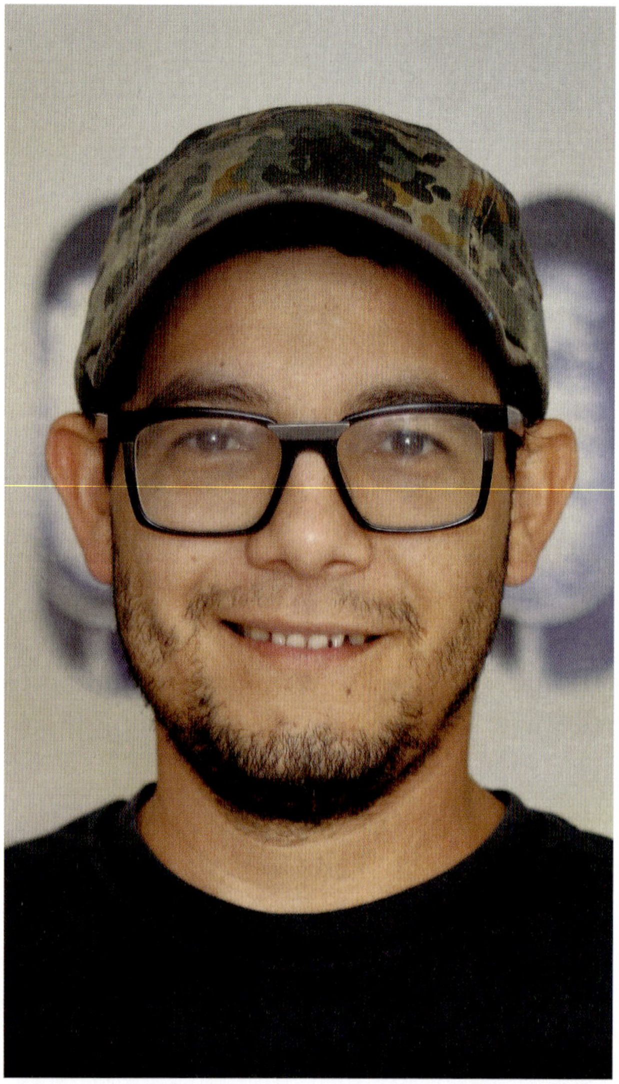

David E. Olivera

To him, *Olympic* is a clear favourite of the three sisters: 'She clearly lived an eventful life full of adventure, drama, thrills and tears. She managed the changes of times in stride. Ultimately, she left a legacy that represents the last of her technological kind and of her entire era.'

He speaks enthusiastically of how his research for his artwork has led him to travel all over the world and meet amazing people and gain access to things he would never have been able to experience otherwise. 'My artistic voyage has thus far rewarded me with wonderful life experiences and amazing realisations,' he reflects, but he acknowledges that his journey is just beginning: 'Being able to empathise and share knowledge in all forms, not limited to art, is definitely a clear message that I wish to underscore.' He adds, 'As Carl Sagan once said, "The visions we offer our children shape the future. It matters what those visions are."'

## VASILIJE RISTOVIĆ

**Nationality:** Serbian

**Website or social media contact:**
**Facebook:** RMS *Titanic* – 3D Illustrations by Vasilije Ristović (RMSTitanicVasilijeRistovic)
**ArtStation:** Vasilije Ristovic (waskogm)

'When I was a kid,' Vasilije remembers, 'my sister casually mentioned *Titanic* and the film *A Night to Remember*, which she had watched. Hearing that ship's dramatic name for the first time immediately captivated my attention, and from that point I just wanted to know more and more.' He eagerly remembers how the 'vivid and lifelike recreations' in the 1997 film and in Ken Marschall's art 'had a huge visual impact on me'.

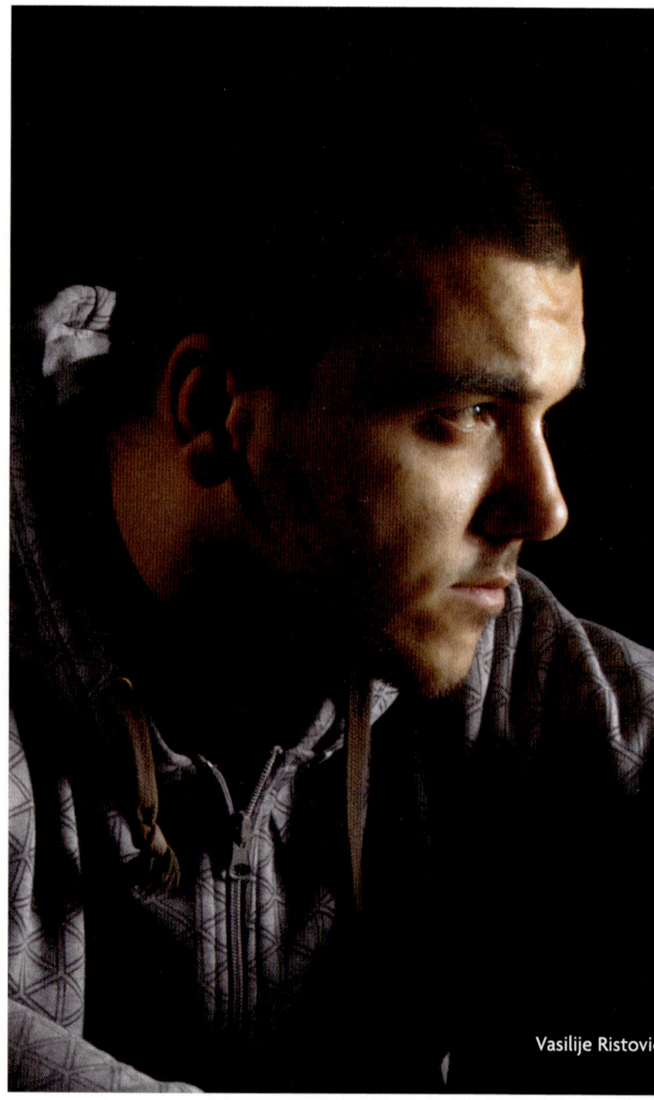

Vasilije Ristovi

At first, he began his CG project as a detailed, accurate study of the ship for his own personal enjoyment, but he soon began to realise that anyone could benefit from it. 'My main goal is the production of photorealistic still images,' working in such a way as to avoid the compromises that animation work often demands.

Regarding the ships themselves, Vasilije says, 'The elegance and aesthetic of the *Olympic*-class ships are one of a kind. *Olympic* was the prototype, and *Titanic* was the perfected one.' He points out that her demise 'forever changed how future ships would look and feel' and it really was 'the end of an era and of that way of thinking'.

*Titanic* remains his favourite of the *Olympic*-class, but he also has a special fondness for the German liner *Kaiser Wilhelm der Grosse* of 1897, a brave pioneer of a liner that set the trend for luxury, speed and the four-stack profile: 'Each of those characteristics would inspire and affect the ocean liner industry for decades to come.'

Regarding the work that he produces, he says, 'Art is good if it triggers people's senses and steals some of their time to take a look. If I manage that, I've accomplished what I wanted.'

## LEVI ROURKE

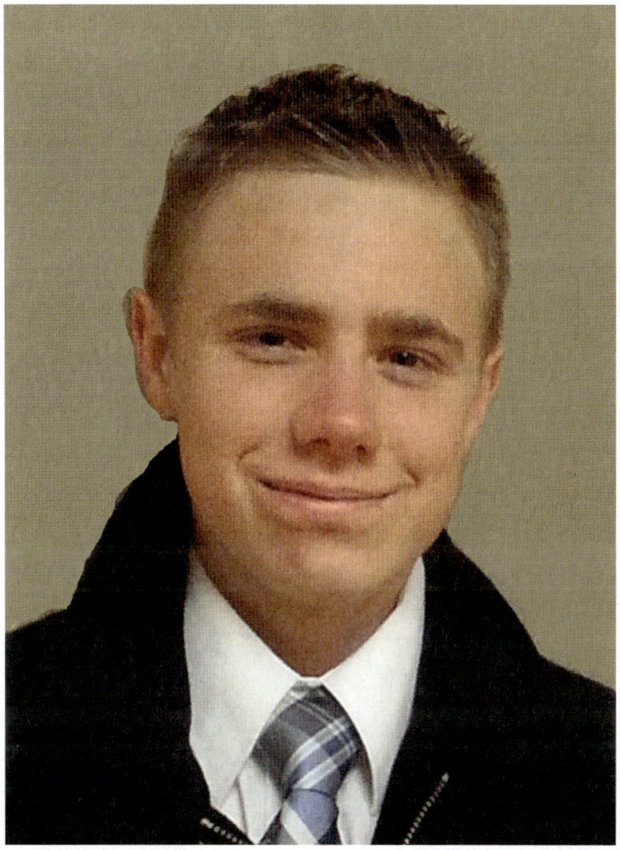

Levi Rourke

**Nationality:** Canadian

**Website or social media contact:**
**YouTube Channel:** Part-time Explorer, soon to be rebranded as 'HFX Studios' / *Titanic* Honor & Glory, for *Britannic*-specific videos.

Levi remembers, 'The 1997 [James Cameron] film without a question is what got the ball rolling for me on *Titanic*.' After that, even though he was still a child, his interest was fuelled by spending hours poring over various books and documentaries, and in particular studying Ken Marschall's masterful paintings of the ship.

Levi works on creating digital 3D models, specifically using programs such as Unreal Engine, Maya, SketchUp, Photoshop and Blender. He aims to create real-time animations and recreations of various ships, working with historians during every step of the process to create the most accurate depictions of the sinkings yet made. Levi is fascinated not only with *Titanic* herself, but particularly with the movements she made during the sinking and the way she broke up. He notes, 'Even after all the years that I and the historians have studied these disasters, as we work, we still make some new findings beyond what has been commonly portrayed.'

His favourite *Olympic*-class liner, however, is *Britannic*. This love grew from the more than 1,200 hours of work he did as the main programmer and animator for the game *Britannic: Patroness of the Mediterranean*, animating the sinking process as it had never been seen before.

He says that he hopes to 'help preserve the ships in a visual form', and also to 'leave behind reliable, factual information that is not altered for the sake of making it more interesting', as he works to keep the memories of those who perished in the sinking of these ships alive, adding, 'Since the actual ships are physically disintegrating, and less and less quality documentaries are coming out each year, I hope that these can teach a younger generation so these disasters are not forgotten.'

## LIAM SHARPE

**Nationality:** American

**Website or social media contact:**
**Facebook:** Liam Sharpe (liam.sharpe.3192)
**Instagram:** @Vintage_Liner_Archive

Liam uses computer graphics software to create digital models, with a particular focus on working with a program called Blender. The artwork he produces is a mixture of computer-generated models, animations and computer renders, and they are produced to help others with historical research and as a means of digitally preserving the vessel that he is recreating.

*Right:* Liam Sharpe

*Far right:* Derek Staines

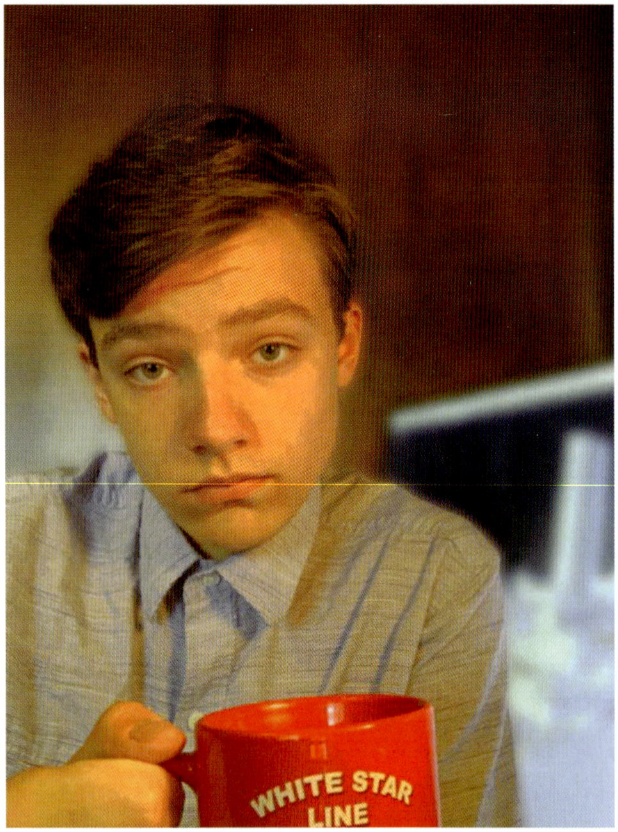


He was first drawn to *Titanic* at the age of 4, when he came across the book *Finding the Titanic*, by Robert Ballard, which was then popular among his classmates. He borrowed the book as soon as he had the chance to, interested in discovering the secret of its popularity. He could barely read at the time, but the photographs of *Titanic*, along with the artwork of Ken Marschall, which it featured, arrested his attention. He could not believe – and still has trouble imagining, at times – that a brand-new, beautiful ocean liner could sink on its maiden voyage in such a terrible way.

Later, in 2016, he saw a real-time *Titanic* sinking animation that was released online, and he was greatly inspired to resurrect ocean liners with the use of CGI software. Liam explains:

> The fact that the White Star Line ordered a class of three deluxe passenger vessels, where two of the newest ships in the class sank, is something that intrigues me greatly. Knowing that only one ship completed a commercial voyage out of the class of three ships is something that will always interest me.

In spite of his fascination with *Titanic* in particular, Liam's favourite ocean liner stands as the White Star liner *Atlantic* of 1871. 'Her gorgeous design is something that I will always admire greatly,' he says.

Liam hopes to continue helping to resurrect historical vessels through digital means so that everyone can access and virtually explore them. 'Personally,' he says, 'I would like to leave behind a legacy of inspiration. That is, for other ocean liner enthusiasts, to in their own way help keep the memory alive of the great passenger vessels that once sailed the seas.'

## DEREK STAINES

**Nationality:** English

**Website or social media contact:**
**Facebook:** Derek Staines (derek.staines.313)

While many of the artists who participated in this project worked in digital media or on canvas, Derek Staines undertook an incredible project that involved working with iron, steel, aluminum, brass and bronze to recreate a working scale version of *Titanic*'s port-side reciprocating engine.

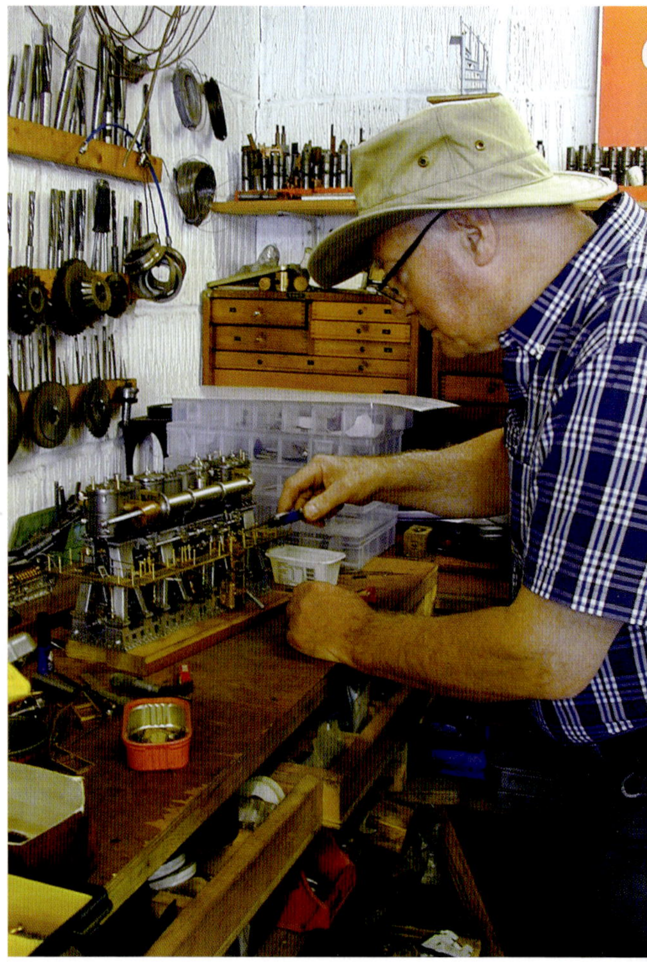

These sorts of projects are nothing new to Derek, whose workshop is full of amazing-looking tools of every description. 'Over the years I have built dozens of steam engines, and I wanted to do something a little bit different,' he says. 'But what??? One day, the idea for building *Titanic*'s port reciprocating engine just "popped" into my head. An annoying little voice in my head said, "Why not?" And that was that!'

However, the project was not easy. 'Detailed technical information is very difficult to find,' he reports. 'There seems to be far more material available about *Titanic* the ship than there is specifically about its propulsion system. Hopefully I can redress the balance a little.'

Derek says that he doesn't have a particular favourite among the *Olympic*-class, or even among Harland & Wolff-built ships. Yet as he did the research, he was very impressed with the quality of the castings, the design and the workmanship that went into each of *Titanic*'s massive engines.

'Hopefully when it's finished,' he says, 'the model will be a useful source of information.'

## CHRISTOPHER WALKER

**Nationality:** English

**Website or social media contact:**
**Facebook:** *Titanic* Design (rmstitanicdesign)
**Instagram:** @rmstitanic.design

Chris Walker discovered *Titanic* through books when he was 7 years old. Then, on 19 December 1997, when he was 9, the James Cameron film *Titanic* was released. This only fuelled his interest in the subject, and he was utterly fascinated by the ship, its design and style. 'The film sets really helped me visualise the spaces and how people interacted with them,' he recalls. 'This spurred me on to research more for myself, building models and drawing design details, and making me want to study interior architecture at university.'

At university, his eyes were opened to the world of 3D digital visuals and design, which aided his understanding of *Titanic*'s design and moved him to want to explore the relatively unknown areas of the ship. He now works in a variety of digital media to produce his beautiful renderings of *Titanic*, including SketchUp, Indigo Renderer and V-Ray.

Over the years, he has learned a great deal by following researchers in online forums and on social media:

> After finishing university, I was regularly drawing areas of the ship and starting to build up a portfolio, but I had nothing to do with it. I then decided to set up my Instagram page to showcase a few images, and it blew up from there. I've learnt so much more by opening up to the community. My Instagram is definitely a work in progress as I got better at design, detail and research.

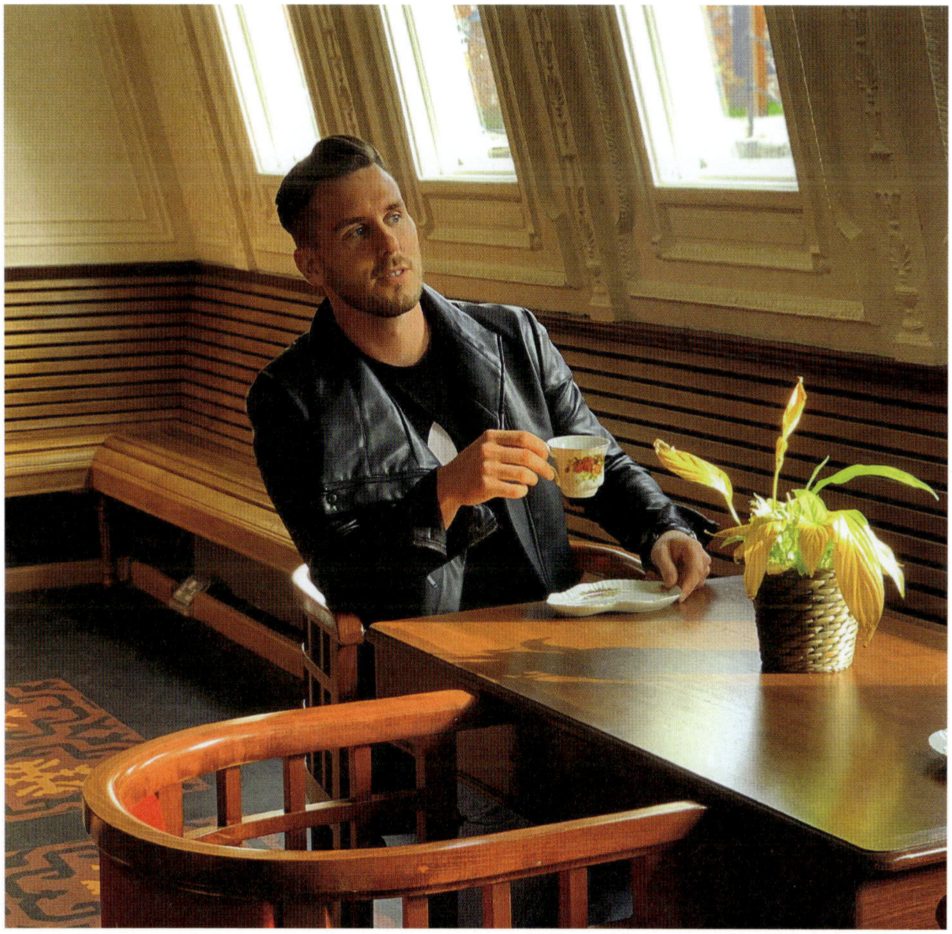

Christopher Walker aboard *Nomadic*.

Chris continues to refine his work.

His favourite *Olympic*-class ship is *Titanic*. 'I feel the design of *Titanic* is the refined version of *Olympic*, and *Britannic* was a reaction to the *Titanic* disaster,' he adds. In the future, he wants to release his own book filled with his images of *Titanic*, and he would also like to teach *Titanic* design, helping to inspire others to learn about her interiors and styles.

## TATIANA YAMSHANOVA

**Nationality:** Russian

**Website or social media contact:**
**Facebook:** *Titanic* Art Gallery by Tatiana Yamshanova (titanicartg)

Tatiana, or Tanya, became interested in *Titanic* through Walter Lord's classic book *A Night to Remember*. She had seen the James Cameron film long before reading the book, but as she recalls, 'Walter Lord definitely did the job!' After reading the book, her *Titanic* obsession began in earnest.

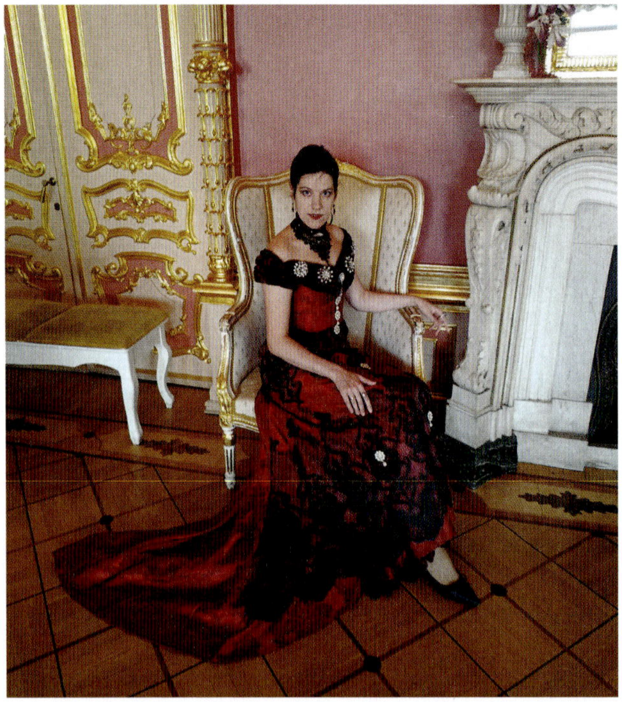

Tatiana Yamshanova

Tanya works as a character artist for game development in the fantasy and sci-fi genres, doing 2D digital paintings, working mainly in Clip Studio Paint. One day, an idea dawned on her as she realised that there were not many portraits of *Titanic*'s crew and the people connected with the ship. She wondered, 'What if I could paint them, using multiple photos of each as a reference? I'm a character designer after all! That's how the whole thing started.'

She adds, 'My creative licence is there and it's quite heavy. I try to be historically accurate, but in applying my skills to *Titanic*-related projects, of course my art can't be 100 per cent documentary.' She now also does maritime paintings, but not as often as the portraits of people connected with *Titanic*. For her, it's the stories of the people that really fascinates. Although she loves *Titanic* in particular, because of her unique and unprecedented story, the long and successful career of *Olympic* and all the people connected with her are also an area of interest.

'I'm trying to bring long-gone *Titanic* people back to life,' she says, 'even if it's just through the art. It's not only *Titanic* people, but also people of other ships connected to her, *Californian* in particular. That ship, with her Captain and officers, are my absolute favourites!'

# ACKNOWLEDGEMENTS

Tad Fitch would like to thank his father Jerry, Jackie, Jocelyn and the rest of his family, friends and readers for their support. You are why we put all of the research and work that we do into these projects.

J. Kent Layton would like to thank his wife, Tessa Anne, for her patience as *Titanic* consumed so much of his time and attention this year. Also, he appreciates the support and encouragement of his family and friends throughout his career as a maritime historian and author.

Bill Wormstedt would like to especially thank his wife Nancy for her support during this project.

We would like to extend a special thank-you to the incredible artists whose work is included in this book. You have all inspired us, and we hope that readers feel the same way after seeing your work.

Special thanks are also owed to Ken Marschall for his splendid Foreword, and for all of the inspiration he has given to us and to so many *Titanic* enthusiasts through the years. We also owe special thanks to our editor, Amy Rigg of The History Press, for believing in this book's merits, even before we could fully explain the concept in full.

The authors would also like to thank the following individuals, in alphabetical order, for all of their support, help and assistance during this project: Günter Bäbler, George Behe, Bruce Beveridge, Mark Chirnside, Steve Hall, Samuel Halpern, Will Kindler, Henning Pfeifer, Mike Poirier and Bill Sauder.

# BIBLIOGRAPHY

## BOOKS


Archbold, Rick, *Ken Marschall's Art of Titanic* (Hyperion, 1998).
Ballard, Robert, *The Discovery of the Titanic* (Warner/Madison Press, 1988).
Behe, George, *On Board RMS Titanic: Memories of the Maiden Voyage* (The History Press, 2018).
Behe, George, *Those Brave Fellows* (Lulu Books, 2020).
Behe, George, *Voices From the Carpathia: Rescuing RMS Titanic* (The History Press, 2015).
Beveridge, Bruce, Andrews, Scott, Hall, Steve, Klistorner, Daniel, & Braunschweiger, Art, *Titanic: The Ship Magnificent*, Vols 1 and 2 (The History Press, 2016).
Cameron, James, et al., *Exploring the Deep* (Insight Editions, 2013).
Chirnside, Mark, *The 'Big Four' of the White Star Fleet: Celtic, Cedric, Baltic & Adriatic* (The History Press, 2018).
Chirnside, Mark, *Oceanic: White Star's 'Ship of the Century'* (The History Press, 2019).
Chirnside, Mark, *Olympic, Titanic, Britannic: An Illustrated History of the Olympic-Class Ships* (The History Press, 2014).
Chirnside, Mark, *The Olympic-Class Ships: Olympic, Titanic, Britannic* (The History Press, 2011).
Chirnside, Mark, *RMS Olympic: Titanic's Sister* (The History Press, 2015).
Chirnside, Mark, Fitch, Tad, Georgiou, Ioannis, Hall, Steve, Halpern, Samuel, Layton, J. Kent, & Wormstedt, Bill, *Titanic: Solving the Mysteries* (Blurb Books, 2019).
Fitch, Tad, Layton, J. Kent, & Wormstedt, Bill, *On a Sea of Glass: The Life & Loss of the RMS Titanic* (Amberley Books, Third Edition, 2015).
Halpern, Samuel, *Strangers on the Horizon: Titanic and Californian – A Forensic Approach* (Independently Published, 2019).
Halpern, Samuel, et al., *Report into the Loss of the SS Titanic* (The History Press, Second Edition, 2017).
Klistorner, Daniel, Hall, Steve, Beveridge, Bruce, & Braunschweiger, Art, *Titanic in Photographs* (The History Press, 2014).
Lord, Walter, *A Night to Remember* (Griffin, 2005).
Lord, Walter, *The Night Lives On* (William Morrow and Company, 1986).
Lynch, Don, & Marschall, Ken, *Ghosts of the Abyss: A Journey Into the Heart of the Titanic* (Da Capo Press, 2003).
Lynch, Don, & Marschall, Ken, *Titanic: An Illustrated History* (Firefly Books, 2021).


## ARTICLES


Chirnside, Mark, et al., 'Abandoning the *Titanic*, Abandoning Reality' www.wormstedt.com/titanic/abandoning_the_titanic.pdf
Wormstedt, Bill, Fitch, Tad, & Behe, George, '*Titanic*: The Lifeboat Launching Sequence Re-Examined', www.wormstedt.com/Titanic/lifeboats/lifeboats.htm


## WEBSITES


Behe, George, www.wormstedt.com/GeorgeBehe
Halpern, Samuel, www.titanicology.com
Layton, J. Kent, www.atlanticliners.com
Wormstedt, Bill, www.wormstedt.com/Titanic